JET FUEL TOXICOLOGY

JET FUEL TOXICOLOGY

Edited by
Mark L. Witten
Errol Zeiger
Glenn D. Ritchie

CRC Press
Taylor & Francis Group
Boca Raton London New York

CRC Press is an imprint of the
Taylor & Francis Group, an **informa** business

CRC Press
Taylor & Francis Group
6000 Broken Sound Parkway NW, Suite 300
Boca Raton, FL 33487-2742

Printed in the United States of America on acid-free paper
10 9 8 7 6 5 4 3 2 1

International Standard Book Number: 978-1-4200-8020-9 (Hardback)

Library of Congress Cataloging-in-Publication Data

Jet fuel toxicology / editors, Mark L. Witten, Errol Zeiger, Glenn David Ritchie.
p. cm.
Includes bibliographical references and index.
ISBN 978-1-4200-8020-9 (hardcover : alk. paper)
1. Jet planes--Fuel--Toxicology. I. Witten, Mark L. (Mark Lee), 1953- II. Zeiger, Errol. III. Ritchie, Glenn David. IV. Title.

RA1242.H87J48 2011
615.9'5--dc22 2010021868

Visit the Taylor & Francis Web site at
http://www.taylorandfrancis.com

and the CRC Press Web site at
http://www.crcpress.com

Dedication

Dr. Mark Witten would like to dedicate this book to his wife, Christine, and son, Brandon. Thank you for being supportive as I have lived the life of a research scientist for the past 30 years! My fondest wish will soon be fulfilled when my son pursues a medical career as well.

Dr. Glenn Ritchie would like to dedicate this book to three very bright scientists, without whom my knowledge of the scientific method and my concern for the welfare of others would be far less:

To Dr. Donald R. Meyer, the former Director of the Physiological Psychology Department, Department of Psychology, The Ohio State University (Columbus, Ohio)

To his wife and Laboratory Supervisor, Dr. Patricia Meyer

To Commander John Rossi III, Ph.D., former Director of the Neurobehavioral Toxicology Laboratory at the U.S. Navy Detachment-Toxicology at Wright-Patterson Air Force Base (Dayton, Ohio).

These three individuals taught me to believe nothing until valid laboratory data confirmed it to be so.

Dr. Errol Zeiger would like to dedicate this book to his wife, Mardi, for her constant support.

Contents

Preface

Jet fuel exposure is the largest toxicant exposure for the U.S. Armed Forces. The Air Force Office of Scientific Research as well as the Naval Toxicology Detachment have sponsored jet fuel toxicology research over the past almost 20 years. This book is a summation of this research effort with major emphasis on the body systems of the lungs and skin, and the immune system. We believe that this book will enable all of the hydrocarbon industry to make better choices for alternative fuels over the next decade, accounting for their toxicological effect(s) as well as the natural choices for these fuels in terms of power generation, engine performance, and pollution by-products.

More importantly, this book demonstrates that the finest organization on planet Earth, the U.S. Armed Forces, has learned the lessons of Agent Orange exposure from the Vietnam War era. Their most important weapons are their volunteers who fill their ranks. For example, the Air Force now removes pregnant airwomen from any exposure to jet fuel during their pregnancies. Additionally, the U.S. Armed Forces have reduced their exposure levels to jet fuel due to the research presented in this book. We hope this book will serve as a template for any future toxicological issues for the U.S. Armed Forces.

About the Editors

Mark L. Witten, Ph.D., is a Research Professor and Director of the Joan B. and Donald R. Diamond Lung Injury Laboratory at the University of Arizona College of Medicine, Tucson, Arizona. Dr. Witten has been at the University of Arizona for 25 years and previously held a faculty position at Harvard Medical School in Boston, Massachusetts. He has over 300 manuscripts/abstracts in his research career focused on environmental toxicant(s) effects on lung function, substance P, microgravity research and fluid balance, and the environmental origins of childhood leukemia clusters throughout the United States. Dr. Witten earned his Ph.D. degree in Human Performance/Physiology from Indiana University in 1983 and a Bachelor of Science degree in the Physical Sciences at Emporia State University in Emporia, Kansas, in 1975.

Errol Zeiger, Ph.D., has an M.S. (1969) and Ph.D. (1973) in Microbiology from George Washington University and a J.D. (1991) from North Carolina Central University. He was with the U.S. FDA from 1969 to 1976 and then at the National Institute of Environmental Health Sciences, in Research Triangle Park, North Carolina, for 24 years, until retiring in December 2000. During this time, he served a year at the Organisation for Economic Co-operation and Development in Paris, France, where he wrote and edited health effects test guidelines and guidance documents, and helped manage their in vivo endocrine disruptor assay validation program. Since January 2001, he has served as a consultant to organizations in the United States, Canada, and Europe, providing services that include evaluation of data, report preparation, litigation support, test protocol design, and preparation and review of project and grant proposals.

Dr. Zeiger's scientific career has concentrated on the design and direction of laboratory validation studies of the effectiveness of short-term genetic toxicity tests and on developing standardized test protocols; the evaluation, interpretation, and integration of toxicological test data; the use of short-term genetic toxicity tests to predict chronic effects; and the study of mechanisms of chemical mutagenesis and carcinogenesis. Much of this work was performed under the auspices of the National Toxicology Program where he was responsible for developing and managing its genetic toxicology testing program and evaluating the test data. He is a member of a number of scientific societies, has authored more than 200 scientific publications, serves or has served on a number of editorial boards, was Editor-in-Chief of *Environmental and Molecular Mutagenesis*, and was co-editor and contributor to the 1997 *Handbook of Carcinogenic Potency and Genotoxicity Databases.*

Glenn D. Ritchie, Ph.D., served for over 10 years as Assistant Director of the Neurobehavioral Toxicology Laboratory at the U.S. Navy Detachment—Toxicology at Wright-Patterson Air Force Base (Dayton, Ohio). As Neurobehavioral Toxicologist, he collaborated with the U.S. Air Force Research Laboratory's 711th Human

Performance Wing, Human Effectiveness Laboratory (Tri-Service Toxicology Laboratory), to publish many studies of the effects of jet fuels, refrigerants, and other military-related chemical toxicants on military troops and laboratory animals. Dr. Ritchie was involved in investigations of various military-related illnesses, including Persian Gulf War Illnesses, Multiple Chemical Sensitivity, and the Childhood Leukemia Cluster in Fallon, Nevada. As a jet fuel specialist for the U.S. Navy, he was involved in extensive medical/epidemiological studies of career jet fuel workers assigned to seven domestic Air Force bases. Dr. Ritchie published two extensive reviews of hydrocarbon toxicity and jet fuel/additive neurobehavioral toxicity.

More recently, Dr. Ritchie serves as Group Leader of the Central Nervous System Safety Pharmacology Department at Battelle Memorial Institute in Columbus, Ohio, evaluating neurobehavioral toxicity during drug discovery testing. He also serves as Vice President of SciScout.com, an internet provider of toxicology literature surveys and report writing for pharmaceutical clients throughout the world. During his career, Dr. Ritchie has been integrally involved in developing and validating laboratory animal-based in vivo and in vitro test batteries that can be used to predict human risk. He earned his Masters degree and Ph.D. in Physiological Psychology at The Ohio State University in Columbus, Ohio, and has over 300 publications and major presentations to his credit.

Contributors

Aria Attia
Department of Biochemistry and Molecular Biology
Georgetown University School of Medicine
Washington, D.C.

R. Jayachandra Babu
Department of Pharmacal Sciences
Harrison School of Pharmacy
Auburn University
Auburn, Alabama

Eric M. Brandon
Department of Biochemistry and Molecular Biology
Georgetown University School of Medicine
Washington, D.C.

Tim Edwards
Air Force Research Laboratory
Wright-Patterson Air Force Base
Ohio

Luis A. Espinoza
Department of Biochemistry and Molecular Biology
Georgetown University School of Medicine
Washington, D.C.

Jeffrey W. Fisher
Interdisciplinary Toxicology Program
University of Georgia
Athens, Georgia

David T. Harris
Department of Immunobiology
University of Arizona
Tucson, Arizona

Alfred O. Inman
Center for Chemical Toxicology Research and Pharmacokinetics
North Carolina State University
Raleigh, North Carolina

Clement Kleinstreuer
Department of Mechanical and Aerospace Engineering
North Carolina State University
Raleigh, North Carolina

Walter J. Kozumbo
Air Force Office of Scientific Research
Arlington, Virginia

Sheppard A. Martin
Interdisciplinary Toxicology Program
University of Georgia
Athens, Georgia

James N. McDougal
Department of Pharmacology and Toxicology
Boonshoft School of Medicine
Wright State University
Dayton, Ohio

Nancy Monteiro-Riviere
Center for Chemical Toxicology Research and Pharmacokinetics
North Carolina State University
Raleigh, North Carolina

Ram Patlolla
College of Pharmacy and
Pharmaceutical Sciences
Florida A&M University
Tallahassee, Florida

Gerardo Ramos
The Department of Immunology and
the Center for Cancer Immunology
Research
MD Anderson Cancer Center
The University of Texas
Houston, Texas
and
Toxicology Program, The Graduate
School of Biomedical Sciences
The University of Texas Health
Sciences Center
Houston, Texas

Terence H. Risby
Department of Environmental Health
Sciences
Bloomberg School of Public Health
Johns Hopkins University
Baltimore, Maryland

Glenn D. Ritchie
Safety Pharmacology Department
Battelle Memorial Institute
Columbus, Ohio

James E. Riviere
Center for Chemical Toxicology
Research and Pharmacokinetics
North Carolina State University
Raleigh, North Carolina

James V. Rogers
Battelle Biomedical Research Center
Columbus, Ohio

Mandip Singh
College of Pharmacy and
Pharmaceutical Sciences
Florida A&M University
Tallahassee, Florida

Mark E. Smulson
Department of Biochemistry and
Molecular Biology
Georgetown University School of
Medicine
Washington, D.C.

Raphaël T. Tremblay
Interdisciplinary Toxicology Program
University of Georgia
Athens, Georgia

Raymond H. Tu
Department of Environmental Health
Sciences
Bloomberg School of Public Health
Johns Hopkins University
Baltimore, Maryland

Stephen E. Ullrich
The Department of Immunology and
the Center for Cancer Immunology
Research
MD Anderson Cancer Center
The University of Texas
Houston, Texas
and
Toxicology Program, The Graduate
School of Biomedical Sciences
The University of Texas Health
Sciences Center
Houston, Texas

Vijayalaxmi
Department of Radiation Oncology
University of Texas Health Science
Center
San Antonio, Texas

Mark L. Witten
Odyssey Research Institute
Tucson, Arizona

Frank A. Witzmann
Department of Cellular and Integrative Physiology
Indiana University School of Medicine
Indianapolis, Indiana

Simon S. Wong
Lung Injury Laboratory, Pediatrics
University of Arizona Health Sciences Center
Tucson, Arizona

Zhe Zhang
Department of Mechanical and Aerospace Engineering
North Carolina State University
Raleigh, North Carolina

1 Air Force-Related Jet Fuel Toxicology Research (1991–2010)

Walter J. Kozumbo

CONTENTS

INTRODUCTION

Since 1991 the Air Force Office of Scientific Research (AFOSR) has sponsored basic research to investigate the potential toxic effects resulting from exposure to the kerosene-based jet fuel, JP-8 (Jet Propellant 8, MIL-DTL-83133F). This book presents studies from many of the scientists who have been supported by AFOSR over the years. Each study (chapter) highlights one specific research area from the many topical areas comprising jet fuel toxicology, emphasizing recent findings and background information that are topic-specific. By contrast, this introductory chapter aims to provide a general overview, emphasizing a chronological review of those events and studies (funded by AFOSR or other organizations) that proved particularly important in motivating and directing research activities related to jet fuel toxicology. To provide an integrative perspective, this overview discusses research findings and activities as a product of organized efforts by the U.S. Air Force and other institutions to elucidate the potential health threats due to jet fuel exposures.

JET FUEL USE BY THE MILITARY

JP-8 jet fuel is essentially the same as commercial jet fuel (Jet A and A-1, ASTM D1655 and DEFSTAN 91-91, respectively) except for the inclusion of several additives in JP-8 that inhibit icing, reduce corrosion, and dissipate static electricity [1]. As a multipurpose fuel, JP-8 was developed by the U.S. military for use as a single "universal" fuel in ground vehicles, tanks, generators, tent heaters, cooking stoves, and other equipment, as well as in aircraft of the entire U.S. military and the forces of the North Atlantic Treaty Organization (NATO). JP-8 is the primary fuel used by the U.S. Department of Defense (DOD), although the Navy procures significant quantities of high-flash-point JP-5 (MIL-DTL-5624U) and naval distillate (marine diesel) fuel F-76 (MIL-DTL-16884L). The DOD also uses significant amounts of commercial jet fuel. A good source for data and information on military fuels are the annual reports of the Petroleum Quality Information System located at the Defense Energy Support Center, which procures all fuels for the DOD [2]. The Air Force started transitioning to JP-8 from JP-4 (Jet Propulsion Fuel 4) in the 1980s. JP-4 was a "wide-boiling" fuel with a much lower boiling point than current "kerosene" jet fuels [1]. (See Chapter 2 for a fuller discussion.)

JP-8 EXPOSURES AND HEALTH CONCERNS

The worldwide, multipurpose use of JP-8 jet fuel provides opportunities for many varied types of exposures to the unburned fuel, such as during cold engine starts, fueling and de-fueling, engine and fuel cell maintenance, fuel transportation, and accidental spills. JP-8 was developed as a replacement fuel for JP-4 and, with safety in mind, was designed to reduce the inherent risks of fire and toxicity associated with JP-4. In contrast to the gasoline-like nature of JP-4, JP-8 is more kerosene-like. That is, relative to JP-4, JP-8 is formulated to contain more long-chain aliphatic hydrocarbons and fewer low-molecular-weight constituents that tend to raise the flash point and lower the vapor pressure of the fuel. This formulation renders JP-8 less volatile than JP-4 and thus less flammable and less likely to catch fire or explode upon collision. Moreover, most of the carcinogenic (benzene) and neurotoxic (n-hexane) constituents found in JP-4 are eliminated from JP-8 when the low-molecular-weight fractions are removed during its formulation. For example, benzene has been reduced to a level less than 100 μg/ml [3]. The replacement of JP-4 with JP-8 has succeeded in yielding fewer fuel-related fires and lower exposures to benzene and n-hexane, ostensibly saving lives and reducing the potential for adverse health effects. The same properties of JP-8 (low vapor pressure and high flash point) that are responsible for saving lives and protecting health also slow its rate of evaporation, allowing JP-8 to remain in its liquid phase much longer than JP-4. This means that if JP-8 is unintentionally released into the environment, then fuel exposure times could increase along with health risks. For example, starting cold engines with JP-8 often can produce sizable exhaust plumes of unburned aerosolized particles that could be either inhaled or deposited on surfaces (including skin and clothing) for a time long enough to be easily absorbed—a problem not previously encountered with JP-4. On cold days, this is especially troublesome for military flight-line personnel whose necessary proximity to the aircraft during engine start-up virtually ensures some skin and clothing exposure to JP-8 even when proper protective gear is donned. Other exposure scenarios, such as leaks and spills of JP-8 during transportation or engine maintenance, have also produced situations where clothing, skin, and surfaces are contaminated with fuel that evaporates slowly, enhancing exposure time and therefore the likelihood of both the inhalation and dermal absorption of JP-8 constituents.

As the Air Force was nearing the completion of its 20-year conversion program from JP-4 to JP-8 in the early to mid-1990s, a series of individual complaints and concerns arose about possible health effects related to using and working with and around JP-8. The main complaints consisted of objectionable odors, dizziness, and skin irritations on parts of the body experiencing repeated contact with JP-8. Tasting or burping JP-8 long after exposure was another common complaint [4]. As a strong organic solvent, JP-8 can dissolve many of the "rubberized" gloves used by mechanics for protection as they work on engines leaking residual JP-8, thereby irritating and inflaming the skin on hands and arms to varying degrees of severity. JP-8's potential to irritate and inflame skin is thought to be linked to its solvent-like capacity for extracting fatty lipids from tissue.

1996 REPORT BY THE NATIONAL RESEARCH COUNCIL (NRC): RECOMMENDATIONS FOR JP-8 TOXICOLOGY RESEARCH

In 1996, the NRC completed a Navy-sponsored report entitled "Permissible Exposure Levels for Selected Military Fuel Vapors" [5], which included JP-4, JP-8, and JP-5—the Navy's kerosene-derived fuel that is similar to JP-8. At the time, the Navy was the only federal agency to propose standards of exposure for vapors of military jet fuels. These standards consisted of an interim 8-hr time-weighted-average permissible exposure limit (PEL) of 350 mg/m^3 and a 15-minute, short-term exposure level (STEL) of 1,800 mg/m^3. Toxicity data available on military fuels at the time was reported as sparse. Because no reliable data could be found to support changing the Navy's proposed PEL standard of 350 mg/m^3, it was recommended that this concentration be accepted as an interim PEL standard until further research was completed. For the STEL, however, it was recommended that the Navy consider lowering its proposed standard of 1,800 mg/m^3 to 1,000 mg/m^3. This was based on a study of Swedish jet-motor workers who displayed acute central nervous system effects, including dizziness, headache, nausea, and fatigue [6]. Although the 1996 NRC panel did not address the issue of toxicity resulting from fuel exposures to respirable aerosols, it emphatically warned that aerosols are "much more toxic than vapors" and recommended that future studies, consisting of breath analyses and neurological assessments, be conducted on military personnel to determine aerosol-vapor exposures and any resulting central nervous system effects. The fact that Air Force personnel were already being exposed to aerosols of JP-8 only served to underscore both NRC concerns about fuel-induced toxicity and the lack of toxicity data from exposures to aerosols of military fuels, including JP-8.

AFOSR JP-8 TOXICOLOGY PROGRAM: RESEARCH THRUSTS AND APPROACHES

As a result of the development and use of JP-8 by the military, the possibility of health and exposure risks to JP-8 were realized, strongly suggesting that additional research was needed to address the NRC-cited data gaps in the JP-8 toxicology literature. The AFOSR responded and initiated funding for one project in 1991, ramped up to ten in 1998, peaked at thirteen projects in 2003, and will end in 2010 when the final project is scheduled to expire. Although the program originated with only a single research project investigating the effects of JP-8 on the lungs, it soon grew to include investigations of jet fuel effects on other tissues, such as the brain, skin, and immune system. The AFOSR JP-8 Toxicology program essentially consisted of the following research thrusts:

- *Effects:* characterizing JP-8-induced alterations (via aerosol, vapor, and liquid exposures) at the biochemical, molecular, cellular, and histological levels on lung, immune system, and brain of rodents
- *Dose-response relationships:* identifying threshold exposure levels
- *Mechanisms:* biodynamic understanding of effects at the biochemical, molecular, and cellular levels with both in vitro and in vivo studies
- *Models:* whole animal biokinetic models of JP-8 exposures via dermal and inhalation routes, and lung deposition models of aerosolized JP-8
- *Biomarkers:* biochemical and molecular markers in lung, brain, and immune system, and human breath analysis following JP-8 exposures
- *Protective strategies:* toxicity inhibitory agents and their mechanisms of action

The program also placed emphasis on the development of collaborations, especially among the AFOSR-sponsored researchers, and held annual program reviews to evaluate the research as well as promote further interactions with other scientists and managers from both inside and outside the program. Outside interactions and collaborations included the American Petroleum Institute, Navy, Army, Veterans Affairs, National Institutes of Environmental Health Sciences, Environmental Protection Agency (EPA), and the NRC of the National Academy of Sciences. Because of general budget limitations at the AFOSR, the intra- and extramural collaborations proved an effective mechanism for leveraging research activities and sharing costs and facilities across both universities and federal agencies involved with issues related to jet fuel toxicology. The remainder of this chapter attempts to summarize highlights, provide context, and integrate some important findings from many (but not all) of the researchers who have been directly or indirectly supported by this program over the years.

INHALATION OF JP-8 AEROSOLS, LUNG DAMAGE, AND IMMUNE SYSTEM SUPPRESSION

In 1991, the first AFOSR JP-8 toxicology grant was awarded to Mark Witten (University of Arizona) to investigate a toxicological data gap that involved the unknown actions of ambient levels of aerosolized particles of JP-8 on the lung. Experiments were originally conducted on both mice and rats inhaling aerosolized JP-8 particles at various occupationally relevant concentrations and durations in a nose-only exposure chamber. To facilitate the experimental characterization of multiple toxicity endpoints in both the immune and respiratory systems, the decision was made to use mostly one species—C57BL/6 mice—at mostly one exposure concentration—1,000 mg/m^3 of JP-8 aerosol for 1 hour/day for 7 days.

As it turns out, the lung damage produced by the aerosolized fuels was quite remarkable, especially when compared to the absence of pulmonary damage that had been previously reported for animals exposed to fuel vapors. The key lung injuries consisted of increases in Type II cell damage, airway resistance, lung leakage, lavage fluid proteins, LDH (lactate dehyrdrogenase) levels, and neutral endopeptidases; and decreases in airway compliance and levels of Substance P (SP)—a neuropeptide whose main purpose in the lung is uncertain [7]. The lung function indicators of airway resistance and compliance were found to be sensitive to concentrations of JP-8 as low as 50 mg/m^3 and thus were more easily affected by JP-8 exposures than were the structural (biochemical and histological) entities of the lung [8]. Curious about the JP-8-induced depletion of SP in the lungs, Witten and co-workers decided to re-supplement (via aerosolization) the SP-depleted lungs with a biologically active SP-analog following JP-8 exposure. To their amazement, they observed a complete inhibition of nearly all JP-8-induced toxicity. In fact, inhibition by SP re-supplementation proved an effective strategy even when implemented for up to 6 hours post JP-8 exposure [9, 10]. Unfortunately, the mechanism by which SP inhibits JP-8-induced lung toxicity remains unknown.

Another significant data gap involved the absence of rigorous animal studies to determine the impact of inhaled aerosolized JP-8 on the systemic immune system, in terms of both function and cellularity. In 1994, David Harris (University of Arizona) was funded to conduct a battery of immunological assessments on the mice that Witten exposed to JP-8 aerosols. The collaborative studies of Harris and Witten produced a number of remarkable immunotoxicological findings. Essentially, these included significant decreases in immune organ weights (thymus and spleen), in immune cell numbers (70% for thymus, spleen, lymph nodes, blood, and bone marrow), in mitogenic function (95%), and in natural killer cells and cytotoxic T lymphocytes (both nondetectable). All effects were concentration dependent, reversible, and affected only T cell- (and not B cell-) mediated immune reactions [11]. Times to onset of the toxic effects were determined to be 1 to 2 days for reduced mitogenic function, 2 to 3 days for cell numbers, and 4 to 6 days for organ weights. Times to recovery were found to be 1 day for organ weights, 4 days for cell numbers, and 4 to 6 days for cell function—just the reverse of the onset times [12]. The lowest JP-8 exposure concentration to mediate an effect was at 100 mg/m^3 in the thymus [13]. When commercial jet fuel (Jet A) and a modified military JP-8 fuel called JP-8+100 were tested, both were found equally as toxic as JP-8, thus indicating that fuel additives were not playing a significant role in producing systemic suppression of the immune system in mice [14]. As with the lung, SP was a powerful and complete inhibitor of nearly all the JP-8-induced immune effects [15]. A later study by Witten and co-workers [16] showed that JP-8 aerosol exposures potentiated the infectivity of mice by Hong Kong virus and that SP could also effectively inhibit the potentiated response. Immunotoxicological studies in rats have been undertaken by others to examine species specificity of response to JP-8.

DERMAL APPLICATION OF JP-8 AND IMMUNE SYSTEM SUPPRESSION

As the immune effects produced by inhaling aerosolized JP-8 were being characterized and reported by Harris and co-workers, an interest developed in knowing whether dermal contact with JP-8 in its liquid phase could elicit immunotoxicological responses similar to those elicited by respirable JP-8 aerosols in the lung. At the time, the immunosuppressive profile observed by Harris and co-workers appeared to resemble the immunosuppressive profile observed by Stephen Ullrich and co-workers (MD Andersen Cancer Research Center) in mice exposed to ultraviolet (UV) light. In 1998, Ullrich was funded by the AFOSR to explore the possibility that dermal application of JP-8 to mice could produce systemic immune responses similar to those produced by ultraviolet (UV) light on skin or by inhaled JP-8 aerosols. Ullrich's initial studies showed that JP-8 applied to skin as multiple small doses (50 μl each) or as one large dose (300 μl) would preferentially suppress T-cell-mediated immune reactions, such as delayed and contact hypersensitivity as well as T-cell proliferation, while antibody formation by B cells remained unaffected [17, 18]. Also, dermal absorption of JP-8 was found to suppress the secondary immune reactions and immunological memory cells that are typically vaccine induced and essential in mounting a lethal and rapid host response to infectious agents [19]. This observation appears to have special relevance for those in the military who are often vaccinated against deadly infectious agents prior to deployment. The concern is that exposure to JP-8 during deployment may delay or prevent a vaccine-induced immune response to deadly microbes. Ullrich's findings clearly demonstrate that skin absorption of JP-8 induces a systemic T-cell suppression, and this suppression closely resembles the lung/aerosol-induced suppression previously described by Harris.

MECHANISMS AND INHIBITORS OF JP-8-INDUCED IMMUNE SYSTEM SUPPRESSION

Other important studies were conducted by Ullrich that helped explain the mechanisms and identify the molecular entities that may be involved in down-regulating many of the T-cell-mediated events in mice. The following encapsulates key findings by Ullrich and co-workers that yielded mechanistic insight. First, JP-8 application to the skin was shown to trigger a release of the biological response modifiers prostaglandin E_2 (PGE_2) and interleukin (IL)-10, both known for their suppressive effects on immune T-cell reactions [17, 18]. Second, blocking the biological activity of IL-10 with monoclonal antibodies or recombinant IL-12 reversed the jet-fuel-induced immune suppression, as did blocking PGE_2 formation with cyclooxygenase inhibitors such as ibuprofen or aspirin [17, 18]. Thus, the release and blocking of these biological response modifiers correspondingly suppressed and protected the immune system, indicating a role for PGE_2 and IL-10 in mediating JP-8-induced immune suppression. Third, because the release of PGE_2 and IL-10 had been shown to be controlled by platelet-activating factor (PAF) [20], Ullrich and co-workers were able to use PAF antagonists to abrogate JP-8-induced immune suppression [21]. This indicated that the roles of PGE_2 and IL-10 were actually secondary to PAF. Finally, because radiation and chemicals (i.e., JP-8) are known to produce oxidative stresses that activate PAF, antioxidants such as vitamin C and E were injected into mice prior to a suppressive dose of JP-8. As one might guess, the antioxidants completely inhibited the immune suppression response [21]. Taken together, these findings suggest a mechanism of toxicity consisting of the following series of events: JP-8 exposure ® oxidative stress ® PAF ® PGE_2 ® IL-10 ® T-cell suppression. Interestingly, this mechanistic sequence closely parallels and mimics the systemic immunosuppressive action induced by UV light on skin, as also elucidated by Ullrich and co-workers [22]. Finally, it should be noted that safe and low-cost inhibitors/drugs are currently available for human use to provide possible protection against JP-8-mediated immune suppression. These include the cyclooxygenase inhibitors (ibuprofen, aspirin, and Celebrex), PAF antagonists, and over-the-counter antioxidants (vitamins C and E).

MECHANISMS AND INHIBITORS OF JP-8-INDUCED CELLULAR TOXICITY IN LUNG, IMMUNE, AND SKIN CELLS

The interactions of JP-8 at the cellular and molecular levels are important in helping explain the mechanisms by which JP-8 disrupts the subcellular biochemical processes leading to the injury or death of cells and eventually the pathological and dysfunctional alterations observed at the organ/system level, such as increased airway resistance, enhanced lung permeability, or systemic immune suppression. Mark Smulson and co-workers (Georgetown University) were funded to employ molecular and biochemical techniques on cultured cells treated with JP-8 and to explore the mechanisms underlying JP-8-induced cytotoxicity. Various cell lines and cultured primary cells were used to represent the target cells of JP-8-induced toxicity in the lung, immune system, and skin. Results indicated that alveolar type II cells, T lymphocytes, and monocytes in culture underwent apoptotic cell death when treated with a cytotoxic concentration of JP-8. Signs of apoptosis occurred as early as 2 hr for caspase-3 activation and 4 hr for DNA ladder formation. But the manifestation of even earlier increases in levels of cytochrome c protein suggest that mitochondrial damage probably occurred first and served to initiate the apoptotic response. The fact that caspase inhibitors increased the survival of JP-8-treated cells to the level of untreated controls reinforced the idea that both epithelial type II lung cells and immune cells utilize an apoptotic mechanism to self-destruct [23]. In contrast to immune and lung cells, immortalized human skin cells failed to display any signs of apoptosis following treatment with a range of cytotoxic concentrations of JP-8, and apoptotic inhibitors had no effect on survival of skin cells [24]. From these observations, Smulson and co-workers concluded that skin kerotinocytes, unlike epithelial lung cells and lymphocytes, utilized a necrotic rather than apoptotic mechanism of death.

Another study by Smulson and colleagues explored the underlying events responsible for damaging mitochondria and thus inducing apoptosis in lung epithelial cells. They discovered that inducers of oxidative stress, such as glutathione depletion and hydrogen peroxide supplementation, significantly enhanced JP-8-induced death. Conversely, supplementation with the antioxidants glutathione or N-acetyl-cysteine protected against cell death, inhibited caspase-3 activation, and prevented the loss in mitochondrial membrane potential. These findings supported the notion that JP-8-induced apoptosis is mediated by oxidative stress caused by the generation of reactive oxygen species proximate to mitochondrial membranes [25].

Indications that JP-8 induces severe oxidative stress in cells and that it concentrates at relatively high levels in bone marrow cells of rats [26] provided ample justification to investigate the genotoxic potential of JP-8 in animals. To address this issue, Vijayalaxmi et al. [27] determined the incidence of micronuclei formation in bone marrow and peripheral blood cells of mice treated topically with JP-8 at levels sufficiently high to induce immune suppression. Results from this 2006 study showed that JP-8 produced no statistically significant elevations in micronuclei formation in these cells [27]. These negative genotoxicity data are important because of the implication of this cytogenetic endpoint to cancer. The detailed report is contained herein.

AIR FORCE (AF) CONCERN AND RESPONSE TO POTENTIAL HEALTH EFFECTS OF JP-8: THE AF JP-8 ACUTE HUMAN STUDY

By 1995 the early research of Witten and Harris was beginning to elicit concerns about the potential health effects of JP-8. In response to these concerns, the AFOSR decided to hold its first of 13 Annual Jet Fuel Toxicology Reviews at the University of Arizona. The purpose of these reviews was to bring together AFOSR-funded researchers, collaborators, and representatives from interested outside parties to share and review JP-8 toxicology data, encourage collaborations, and develop research thrusts for the following year. The first meeting consisted of only four researchers, but in a few years the annual meetings included 25 to 30 researchers and continued through 2007 when the program began phasing down.

By 1996 and partly in response to the findings of Witten and Harris, the AF Surgeon General's Office created a special AF team (Integrated Process Team or IPT) responsible for coordinating and initiating AF research activities related to the environmental, health, and safety issues of JP-8. The military scientists and officers Sandy Zelnick, Les Smith, and Roger Gibson were succeeding IPT leaders who actively interacted with the Air Force Research Laboratory (AFRL); the Air Force Institute for Environment, Safety and Occupational Health Risk Analysis (AFIERA); and AFOSR to coordinate, propose, and fund research activities. They also promoted and established collaborations with the EPA, the National Institute of Occupational Safety and Health (NIOSH), and the National Institute of Environmental Health Sciences (NIEHS) through a special representative. Together, the IPT and AFIERA conducted JP-8 occupational exposure studies that identified the worst-case exposure scenarios within the AF [4]. In 1997, the Centers for Disease Control (CDC) released a report on the toxicology profiles of JP-8 and JP-5 that identified data gaps and called for further research into the possible health effects of JP-8. Subsequently, the EPA was able to use this report to identify additional data gaps and issue its own set of research recommendations [4, 28]. These reports plus results from the IPT/AFIERA studies, the 1996 NAC study, and the AFOSR animal toxicology studies merged to justify and validate the need for conducting other JP-8 health-related research activities. One such activity involved co-sponsoring and organizing, together with AFOSR, an *International Conference on the Environmental Health and Safety of Jet Fuel* in 1998 in San Antonio, Texas [29]. In 1999 and with information previously obtained from AF studies, the IPT took the lead in organizing researchers from the EPA, Navy, Army, CDC, NIOSH and NIEHS, as well as the AF and six universities, to conduct a multi-million-dollar acute human epidemiological study of JP-8 [30]. The Strategic Environmental Research and Development Program (SERDP) funded the study through an award granted to Texas Tech University, one of the university collaborators. Around the same time, the IPT requested and funded the NRC to conduct a follow-up to the 1996 JP-8 health effects report by the NRC (see above). The formation of the AF IPT and its proactive involvement in promoting JP-8 research activities can be traced in part to results obtained from the early AFOSR studies of Witten and Harris.

The JP-8 acute human-exposure study supported by the SERDP deserves special mention because it represented the most comprehensive jet fuel study on exposure and human health effects ever conducted in a working population. The purpose of the study was to assess the influence of acute jet fuel exposure on the health, safety, and operational capability of the AF population and evaluate the potential risks. Data were collected from military personnel who worked with fuel (a total of 250 males and 90 females) across six AF bases in the continental United States from the spring through the fall of 2000. Blood, urine, skin, and breath samples were taken to assess exposures on an individualized basis. Neurological tests (cognitive, reaction response, eye blink, and postural sway), immune studies, and hormone level tests were performed along with assessments of self-reported symptoms and medical visits. Preliminary results from this study were reported in August 2001 at a JP-8 conference sponsored by the AF and held in San Antonio, Texas. The preliminary findings indicated that there were acute effects on cognitive and motor processes associated with jet fuel exposure and significant chronic carry-over effects on conditioned response, balance, cognitive, and motor responses associated with fuel maintenance work in the AF [30, 31]. Most regrettably, funding ended in September 2001 and additional funding could not be acquired for completing the analyses of many samples from over a dozen researchers and collecting, processing, and cross-referencing the data on a project-wide basis.

Preliminary findings from the uncompleted AF JP-8 acute human study indicated that effects were occurring in the high exposure group, that is, the fuel workers. To generate preliminary findings expeditiously, observed effects were compared to general exposure categories based on job types rather than to the actual exposure levels measured in individuals. Interestingly, the control group of "unexposed" workers, who did not work directly with fuel on a daily basis, also displayed detectable but lower levels of JP-8, indicating that exposure to JP-8 on AF bases was difficult to avoid and widespread. In fact, one of the main conclusions was that exposure groups could not be accurately represented by lumping according to job categories and that exposed workers from any category should not be considered as a homogeneous group [4, 30, 31]. Thus, looking for effects by comparing across job classifications (as was done in developing the preliminary findings) would tend to diminish or underrepresent the magnitude of effects. Ironically, when the preliminary findings were released, many of the study samples and much of the data on the exposures and effects of each individual (no matter what job classification) remained stored, unanalyzed, or unprocessed. Had the study been completed, a correlation of exposure and effect on a person-by-person individualized basis would have undoubtedly emerged. Such a study would have provided an eloquent and powerful human dose-response relationship of potentially great value in assessing risk to JP-8 exposures. In 2004, these points and others were discussed in a subsequent *Jet Fuel Toxicology Workshop* sponsored by the AFOSR and AFRL for the benefit of updating and sharing information with the military logistics communities and industry.

2003 REPORT BY THE NATIONAL RESEARCH COUNCIL: RECOMMENDATION FOR FURTHER RESEARCH ON JP-8 TOXICOLOGY

Only 4 years after publication of the Navy-funded NRC study of 1996 ("Permissible Exposure Levels for Selected Military Fuel Vapors"), the Air Force requested and funded the NRC to revisit the jet fuel topic in view of recent studies suggesting the possibility of adverse health effects from exposures to JP-8. During the next couple of years, the new NRC panel evaluated many endpoints and concluded that JP-8 jet fuel was potentially toxic to the immune system, respiratory tract, and nervous system. Evidence stemming from the animal toxicology studies of Harris, Ullrich, Witten, Ritchie [32] and others, as well as from the AF acute human study, was used to justify the NRC conclusions and formulate constructive recommendations for future studies. The 2003 NRC study (Toxicologic Assessment of Jet-Propulsion Fuel 8) [4] offered the following recommendations:

1. Lower the PEL from 350 to 200 mg/m^3 for inhalation exposures to JP-8.
2. Accurately recalibrate vapor/aerosol concentrations in animal exposure chambers using advanced chemical and physical techniques.
3. Test for and correlate indicators of immunotoxicity in the blood of humans and animals exposed to JP-8.
4. Correlate JP-8 constituents to toxic endpoints.
5. Validate two preliminary positive neurological findings from the Air Force acute human study with standard neurological tests.
6. Conduct health effects assessments and blood analyses of JP-8 in humans.
7. Conduct toxicokinetic animal models to improve interpretation of existing human studies on JP-8.
8. Correlate adverse effects and JP-8 exposures in Air Force personnel for validation of toxicokinetic models.

9. Conduct longer-term inhalation studies on vapor/aerosol exposures.
10. Conduct 2-year inhalation carcinogenicity bioassays on two animal species.

RESPONSE OF AFOSR-FUNDED JP-8 TOXICOLOGY RESEARCH TO THE 2003 NRC RECOMMENDATIONS (ITEMS 1 TO 8*)

To varying degrees, many of the above recommendations (NRC Items 1 to 8) were currently or subsequently being acted upon by the AFOSR and the Air Force. The AF immediately adopted the first NRC recommendation (Item 1) in 2003, which strengthened the safety standard for JP-8 aerosol/vapor exposures by lowering the PEL from a concentration of 350 to 200 mg/m^3. To address Item 2, an "in-line", real-time, total hydrocarbon analysis system and aerosolizer were purchased by the AFOSR to calibrate accurate concentrations of vapors/aerosols in Witten's exposure chamber. Validation studies were conducted in the chamber and indicated that only 5% to 15% of the JP-8 aerosol levels previously reported by Witten were actually composed of aerosol particles, confirming the NRC panel's suspicion of a tenfold overestimation of JP-8 aerosol concentrations in Witten et al.'s earlier studies [33]. This overestimation did not affect the accuracy of the PEL recommendation however, because the NRC panel had already factored it into the panel's PEL calculations.

During and following the NRC study, Ullrich and, to some extent, Harris had shown that IL-10, PGE_2, and PAF were important factors in mediating JP-8-induced immune suppression and could theoretically serve as blood biomarkers of effect, especially IL-10 [17–19, 34]. However, the opportunity to correlate levels of these factors in the blood of animals and humans exposed to JP-8 (Item 3) was lost along with the unanalyzed human blood samples from the uncompleted AF acute human study. Another study by Ullrich et al. [35] showed that, unlike JP-8, Fischer-Tropsch (FT) synthetic jet fuel failed to mediate immune suppression in mice when topically applied. Because FT fuel contains aliphatic hydrocarbons like JP-8, but none of the aromatic fractions of JP-8, it was concluded that the aromatic component of JP-8 was necessary to mediate systemic immune suppression via dermal administration (Item 4). On the other hand, when mice inhaled FT synthetic fuel in the form of mixed aerosol/vapor fractions, Witten, Harris, et al. [36–38] still observed injury to the lung and suppression of the immune system, albeit with slightly different toxicity profiles. Shortly before the AFOSR decided to wind down the JP-8 toxicology program in 2007, other approaches were being considered to identify more narrowly the specific classes of toxic JP-8 constituents. Due to limitations on funds and time constraints however, the development and implementation of these approaches were never completely realized.

The NRC recommendation to validate preliminary neurological findings from the AF acute human study with standard neurological tests (Item 5) was addressed in part by Terence Risby (Johns Hopkins) in his AFOSR-funded human study. Much smaller in scope than the AF acute human study, the AFOSR study was conducted on

* Because standardized toxicology assays are not considered basic research, Items 9 and 10 would not be supported by the AFOSR, the basic research organization of the USAF.

one Air National Guard (ANG) base rather than on several AF bases. The study goal was simple and direct: to correlate breath analysis measurements of JP-8 exposure to neurocognitive function as determined by a standardized test battery (CogScreen) in chronically exposed fuel workers. Chronic exposure to JP-8 levels on an ANG base was found to be substantially lower than on AF bases and below current JP-8 exposure standards. Even at these lower exposure levels however, the ANG group performed significantly poorer on 20 of the 47 CogScreen tests as compared to controls, and performances on the tests correlated with individual JP-8 exposure levels. JP-8 exposure significantly reduced performance on complicated tasks, response time, and accuracy, with suggested decreases in attention, tracking, memory, and concentration. Furthermore, the indication that vapor inhalation and dermal absorption—and probably not aerosol inhalation—of JP-8 were the dominant means of exposure in the ANG study suggests that the aromatics in JP-8, which are more volatile and have greater skin penetration than the aliphatics, were significant in causing the cognitive effects. Routine tests of liver function and urine chemistry were performed on blood and urine samples with no observed effects [39]. In hindsight, it may have been useful to have tested the blood of ANG workers for factors related to JP-8 exposures, such as IL-10, PAF, or PGE_2, and to have compared the ANG study findings to those from the unfinished AF acute human study (Item 6). Nevertheless, Risby et al.'s study showed that long-term exposure to JP-8, at levels lower than the current standard and many times lower than that found on AF bases, correlated with diminished neurocognitive functions, especially those associated with complicated tasks, response time, and accuracy.

Another important neurological effect of JP-8 was demonstrated in an animal study jointly supported by the AFOSR, Veterans Affairs, and the American Petroleum Institute, and conducted by Laurence Fechter (Veterans Affairs) and coworkers. Fechter's team found that inhalation by rats of JP-8 aerosol/vapor (1,000 mg/m^3 for 4 hours/day for 5 days) produced inconsistent and variable impairment of outer hair cell auditory function that only partially recovered after 4 weeks. In other experiments, the potentiating effect of JP-8 exposure on noise-induced damage was investigated. Compared to the damage caused by noise alone (at 97 to 105 dB), combined treatment of JP-8 plus noise produced greater impairment of auditory function and a significant increase in outer hair cell death [40]. Findings from this animal study support the notion from the ANG human study that JP-8 may adversely affect neurological functions and, furthermore, may potentiate neuronal damage caused by other neurotoxicants, such as loud jet engine noises experienced by AF flight-line workers (Item 5). The importance of compliance and enforcement in utilizing safety protection gear is underscored by these studies.

Since about 2002, the AFOSR has supported research to develop a whole-animal biokinetic model (Item 7) to predict the uptake, distribution, metabolism, and elimination of complex mixtures, such as JP-8. With skin and lung serving as the portals of entry, the model would enable the quantitative assessment of JP-8 body burden, the correlation of internal dose of individual hydrocarbons to external dose and toxic endpoints, and the generation of viable hypotheses concerning the mode of action for JP-8-induced toxicity in animals and humans. Scientists from five universities and the AFRL have worked individually and collaboratively to develop novel techniques for the biokinetic modeling of complex mixtures. To model the inhalation and deposition of JP-8 in the lung, Clement Kleinstreuer (North Carolina State University) has conducted accurate multi-scale computational simulations of the transport and deposition of JP-8 aerosol/vapor in representative human respiratory systems under realistic ambient inhalation conditions [41, 42]. Other scientists, such as Jeffrey Fisher (University of Georgia) and David Mattie (AFRL) have conducted theoretical mathematical simulations and experimental inhalation studies on rats exposed to JP-8 and its components. They aimed to develop a whole-animal, physiologically based toxicokinetic model that may eventually utilize components of the lung-deposition model being developed separately by Kleinstreuer and co-workers.

Acting as both a target organ of toxicity and a major route of exposure to JP-8, skin can be irritated and inflamed as it absorbs, retains, and allows JP-8 and its constituents to enter the body. Modeling the biokinetic and biodynamic actions of JP-8 in skin will help not only in developing a whole-animal biokinetic model, but also in understanding the irritation response, as it may occur in the skin or other tissues such as the lung (Item 7). Given these considerations, several scientists were funded to model the dermal biokinetics of JP-8 and its constituents and the molecular biodynamics responsible for triggering the irritation response in skin. To determine JP-8 interactions with and absorption by the skin, James Riviere and colleagues (North Carolina State University) have been investigating the percutaneous absorption rates of JP-8 and its components as well as modeling the effect of individual components of JP-8 on dermal absorption and kerotinocyte toxicity [43–45]. In one particularly important study by Riviere and co-workers [46], the long-chain aliphatic hydrocarbons tridecane and tetradecane were identified as the key constituents in JP-8 responsible for dermal irritation, suggesting that these same constituents may also be responsible for causing the sensory irritation response in upper airways of mice exposed to JP-5 aerosols, as observed by John Hinz (AF Institute of Operational Health) [47]. Other studies by Mandip Singh (Forida A&M University) have supported the notion that long-chain aliphatic hydrocarbons may be critical in causing irritation and eventually inflammation. Singh reported that the skin irritation effects (cytokine release and transepidermal water loss) were strongest following unocclusive dermal exposures of rats to tetradecane as compared to dodecane or nonane [48]. These findings indicate that the longer chain aliphatics C-13 and C-14, which are found at much higher concentrations in JP-8 than JP-4, may be important in causing the irritation responses in skin as well as in lungs exposed to JP-8 aerosols (Item 4). These findings also justify the studies by James McDougal (Wright State University), who has been characterizing the molecular events responsible for triggering skin irritation and developing a predictive, biologically based, molecular model of the irritant cascade [49–51]. Together, these modeling activities have resulted in numerous publications (and chapters in this book) that only begin to advance the fundamental understanding needed to develop and validate the biokinetic and biodynamic models for use in both animals and humans. Once developed, these models will help in predicting and assessing JP-8-induced human health risks (Item 8) as well as in developing possible protective strategies.

COMMENTARY ON JET FUEL TOXICITY AND COMPLIANCE WITH PEL STANDARDS AND EXPOSURE SAFEGUARDS

Animal studies funded by the AFOSR and other organizations have clearly shown that JP-8 and its constituents can not only injure skin and lung upon entering the body, but also damage the immune and nervous systems before being excreted or expired. And furthermore, AF studies have confirmed the potential of JP-8 to adversely affect skin and the nervous system of humans. Many of these studies were taken into account by the 2003 NRC panel and thus influenced their recommendation to strengthen the PEL standard, which the AF quickly adopted. However, one important study, the ANG human study on JP-8 exposure and cognition, was not completed until after the NRC report had been published. In this ANG study, cognitive deficiencies were found in safety-compliant fuel workers who had been chronically exposed to JP-8 at levels lower than the NRC PEL and approximating the levels to which non-fuel workers were exposed on AF bases [39]. This implies that some non-fuel workers on AF bases may risk cognitive effects from long-term exposure to JP-8. Hence, the usefulness of the NRC-adopted PEL in safeguarding against neurological effects produced by long-term, low-level exposures to JP-8 is questioned by this study. Moreover, unlike the several inhibitors that can protect lung and immune cells from JP-8-induced toxicity, there are no inhibitors known for blocking the subtle neurotoxic actions of JP-8. Mechanistic studies would help in understanding how JP-8 affects cognition and in providing clues for developing protective agents.

It is also worth noting that the lower levels of JP-8 exposure in ANG versus AF fuel workers were attributed to the ANG workers' compliance with safety guidelines and with standard operating procedures [39]. Given the low levels of exposure achieved by ANG fuel workers, it is reasonable to expect that complying with safety measures as well as adhering to the new stronger PEL standard, together, would protect the skin, lungs, and immune systems of AF fuel workers from the adverse effects of JP-8. At the same time, however, it is disturbing that safety compliance was ineffective in providing long-term protection against significant loss of mental processing in ANG fuel workers. The cognitive effects resulting from long-term, low-level exposures to JP-8 necessitate further research, especially human studies, on the neurological and cognitive effects of short- and long-term exposures to JP-8 and its constituents.

With regard to the immune system, JP-8 has been shown to be suppressive in only one animal species thus far, mice. Therefore, it would be helpful to determine the uniqueness of the immune-suppressive effect by exposing other species, such as rats, to JP-8 via inhalation and dermal application. Furthermore, immune-suppressive doses of topically applied JP-8 failed to induce cytogenetic damage in peripheral blood and bone marrow cells of mice [27]. This is an important finding because JP-8 has been suspected of producing a cluster of acute lymphocytic leukemias in children from the town of Fallon, Nevada—the home of a naval air base where JP-8 is used and stored in proximity to town [4]. This finding would tend to support the notion that JP-8 is neither genotoxic to immune cells nor leukemogenic.

TOXICOLOGY AND FUTURE JET FUELS

The future of jet fuel is going to be determined, in part, by AF interests in national security, sustainability, affordability, and the environment. To help break its reliance on foreign oil, the AF began a number of years ago to develop its own synthetic fuel from coal and natural gas using the FT process. More recently, however, the potential consequences of climate change on future national security issues were realized and have prompted the AF to explore the utilization of biomass—instead of coal and natural gas—as a possible carbon-neutral feedstock for the production of prospective aviation biofuels. The AF had previously shown that a 50/50 blend consisting of JP-8 and either biomass-derived or fossil-fuel-derived fuel can successfully power aircraft without causing any significant engineering difficulties [52]. Currently, the overall AF goal for alternative fuels is to "prepare to cost competitively acquire 50% of the AF's domestic aviation fuel requirement via an alternative fuel blend in which the alternative component is derived from domestic sources produced in a manner that is greener than fuels produced from conventional petroleum" (Air Force Policy Directive 90-17, 16 July 2009). A supporting goal is to have all aircraft certified for use of the 50/50 bio-blend by as early as 2012. Although the compositions of fossil fuel-derived and biomass-derived fuels are very similar (mostly of n- and iso-paraffins) [52, 53], the composition of either of these 50/50 blends is very different from that of JP-8 or Jet A. Essentially, the blended fuels will exhibit a 50% volume increase in n- and iso-paraffins and a proportional percent volume decrease in aromatics and cycloparaffins. This would imply that the FT synthetic- and bio-blended jet fuels are likely to produce toxicological profiles that are similar to each other but quite different from that of JP-8, as some reports in this book indicate. David Mattie and co-workers at AFRL/RH have designed and are now conducting a battery of preliminary studies to explore the potential toxicity of the FT synthetic fuel. Knowledge gained from the many AF-funded toxicology investigations over the past 19 years has influenced the design of this initial test battery and will undoubtedly help in interpreting its results [54].

AFOSR JET FUEL TOXICOLOGY PROGRAM: SUMMATION AND CONCLUSION

During the 1990s, health complaints by jet-fuel-exposed personnel in the Air Force were not well understood because of existing gaps in the knowledge base of jet fuel toxicology. This lack of understanding served to justify and motivate the development of a fundamental research program at AFOSR on the toxicology of jet fuels. The goal of such a program was to enrich the knowledge base in jet fuel toxicology by filling in some of those critical gaps. To accomplish this goal, research was funded that would further characterize toxic effects due to jet fuel exposures, explain key molecular/cellular events causing the toxicity, and determine and model the exposure levels necessary for inducing the toxicity. Essentially, the program funded studies to identify toxic *Effects*, understand toxic *Mechanisms*, and develop computational *Models* for predicting biomolecular responses to and internal doses of jet fuel and its components. Because enough annual AFOSR funding was available to support only a few of the many studies recommended by two NRC toxicology panels on jet fuels, opportunities were sought to use AFOSR research funds more efficiently. Some opportunities included limiting the number of target tissues studied to four (lung, skin, immune, and nervous tissues), reducing animal and aerosol exposure costs by sharing tissue samples among collaborators, conducting studies sequentially instead of in parallel, and creating collaborations and research partnerships across AF organizations, universities, other federal agencies, and industry. Collaborations proved especially productive by stimulating discussions and new research ideas, serving to cross-check and validate research findings, and sharing in the costs of conducting research. For example, studies funded and conducted by the Navy (see Chapter 4 by Ritchie and co-workers), determined effects of jet fuel on neurological and cognitive responses—areas complementary to but not emphasized by the AFOSR—that helped the 2003 NRC panel formulate its research recommendations. Nearly all the partnerships proved extremely productive and rewarding with perhaps one unfortunate, but important, exception. Not being able to obtain extra funding to complete the AF acute human study meant that a rare and valuable opportunity of acquiring human information to better explain the human health risks associated with jet fuel exposures was lost.

As the program progressed from 1991 through 2010, balance among the research areas of *Effects, Mechanisms*, and *Models* varied over time. During the initial phase of the program, most of the research focused on identifying effects and characterizing dose-responses. Several years into the program, a second research phase was introduced that sought to understand mechanisms of action and to modulate/inhibit observed effects with bioactive agents. In fact, most of the findings highlighted in this introductory chapter represent results from the earlier research phases involving effects and mechanisms rather than from the more recent phase involving model development. Not until 2002, a little more than a decade into the program, was modeling research beginning to be emphasized in such areas as physiologically based toxicokinetics, aerosol deposition in lungs, skin penetration of JP-8 components, and biomolecular modeling of the jet fuel-induced irritation response in skin. The modeling effort evolved into a major focus area of the program only a few years ago, just before the AFOSR decided to phase out toxicology-related research. However, despite the limited time and funding that was available for modeling research, a number of significant findings were contributed by several AFOSR-funded researchers, including Jeffery Fisher, Clement Kleinstreuer, James Riviere and James McDougal. (They, along with their co-workers, have authored chapters in this book.) Nevertheless, the work of developing highly accurate predictive models for jet fuel toxicology remains unfinished and will require many more years of sustained research activity.

Acquiring the requisite information on *Effects, Mechanisms,* and *Models* is not only challenging, but also important. It enables the development of various protective strategies involved in both mitigating the occurrence and augmenting the treatment of jet fuel-induced toxicity. Knowing the tissue sites of toxicity and threshold levels of exposure translates into better predictive biokinetic models for use in health risk assessments and for establishing more scientifically based and accurate exposure standards. Likewise, identifying specific biomolecular events or entities involved in mediating the tissue-damaging or -repairing processes can serve as biomarkers for use in developing personal monitoring devices for early warnings to toxic exposures and effects. Furthermore, identifying inhibitors of damage mechanisms or promoters of repair processes can lead to discovery opportunities for developing new prophylactics or treatments. In fact, as a result of this research, some potential strategies to safeguard AF personnel exposed to jet fuels have already been identified, including, for example, a new AF PEL safety standard, antioxidant protectors of cell death, molecular biomarkers and inhibitors of immune system suppression, and Substance P-induced inhibition of lung injury. Moreover, in the case of the last example, Substance P has even demonstrated potent anti-radiation, -viral, and -cancer activities in unpublished studies used to influence the start of a new biotech company [55]. Understanding the protective action of Substance P against jet fuel-induced toxicity may reveal mechanisms that could also advance technology in the field of drug development, thus benefiting both civilian and military interests alike. Such potential beneficial applications are the direct result of AFOSR-funded basic research in jet fuel toxicology and should serve to justify and motivate future research.

As previously discussed, AFOSR-funded basic research in jet fuel toxicology comes to a close in 2010. Nevertheless, as this overview has documented, scientific accomplishments from this program have been quite substantial—the result of interactions among many AFOSR-funded researchers and their collaborators across federal agencies, universities, and industry. Many significant findings have spawned from countless publications to enhance the knowledge base in toxicology that can now be used to inform future studies on kerosene-based jet fuels as well as on the new bio-based and FT synthetic fuels being developed by the AF. It is hoped that support by federal agencies, industry, and foundations will be available in the future to carry on and extend at least some of the interesting basic research activities initiated by AFOSR since 1991 in the area of jet fuel toxicology.

REFERENCES

[1] Martel, C.R., Military Jet Fuels, 1944–1987, Air Force Wright Aeronautical Laboratory, Report AFWAL-TR-87-2062, November 1987.

[2] http://www.desc.dla.mil/DCM/DCMPage.asp?pageid=99.

[3] Shafer, L.M., Striebich, R.C., Gomach, J., and Edwards, T., Chemical Class Composition of Commercial Jet Fuels and Other Specialty Kerosene Fuels, AIAA 2006-7972.

[4] NRC (National Research Council), *Toxicologic Assessment of Jet-Propulsion Fuel 8*. Washington, D.C.: National Academies Press, 2003.

[5] NRC (National Research Council), *Permissible Exposure Levels for Selected Military Fuel Vapors*. Washington, D.C.: National Academies Press, 1996.

[6] Knave, B., Olson, B.A., Elofsson, S., Gamberale, F., Isaksson, A., Mindus, P., Persson, H.E., Struwe, G., Wennberg, A., and Westerholm, P., Long-term exposure to jet fuel. II. A cross-sectional epidemiologic investigation on occupationally exposed industrial workers with special reference to the nervous system. *Scandinavian J. Work Environ. Health*, 1978, 4(1): 19–45.

[7] Pfaff, J.K., Parton, K., Lantz, R.C., Chen, H., Hays, A.M., and Witten, M.L., Inhalation exposure to JP-8 jet fuel alters pulmonary function and Substance P levels in Fischer 344 rats. *J. Appl. Toxicol.*, 1995, 15: 249–256.

[8] Robledo, R.F., and Witten, M.L., Acute pulmonary response to inhaled JP-8 jet fuel aerosol in mice. *Inhalat. Toxicol.,* 1998, 10: 531–553.

[9] Robledo, R.F., Young, R.S., Lantz, R.C., and Witten, M.L., Short-term pulmonary response to inhaled JP-8 jet fuel aerosol in mice. *Toxicol. Pathol.,* 2000, 28: 656–663.

[10] Robledo, R.F., and Witten, M.L., NK1 receptor activation prevents hydrocarbon-induced lung injury in mice. *Am. J. Physiol.: Lung Cell. Mol. Physiol.,* 1999, 276: L229–L238.

[11] Harris, D.T., Sakiestewa, D., Robledo, R., and Witten, M., Immunotoxicological effects of exposure to JP-8 jet fuel. *Toxicol. Indus. Health,* 1997a, 13(1): 43–56.

[12] Harris, D.T., Sakiestewa, D., Robledo, R., and Witten, M., Short-term exposure to JP-8 jet fuel results in long-term immunotoxicity. *Toxicol. Indus. Health,* 1997b, 13(5): 559–570.

[13] Harris, D.T., Sakiestewa, D., Robledo, R.F., Young, R.S., and Witten, M., Effects of short term JP-8 jet fuel exposure on cell-mediated immunity. *Toxicol. Indus. Health,* 2000, 16: 78–84.

[14] Harris, D.T., Sakiestewa, D., Titone, D., Robledo, R.F., Young, R.S., and Witten, M., Jet fuel induced immunotoxicity. *Toxicol. Indus. Health,* 2000, 16: 261–265.

[15] Harris, D.T., Sakiestewa, D., Titone, D., Robledo, R.F., Young, R.S., and Witten, M., Substance P as prophylaxis for JP-8 jet fuel-induced immunotoxicity. *Toxicol. Indus. Health,* 2000, 16: 253–259.

[16] Fastje, C.D., Hyde, J.D., Goot, B., Meigs, E., Sun, N.N., Wong, S.S., and Witten, M.L., Substance P agonist attenuates pulmonary injury induced through exposure to A/Hong Kong/8/68 influenza virus and JP-8 jet fuel. *FASEB J.,* 2004, A461.

[17] Ullrich, S.E., Dermal application of JP-8 jet fuel induces immune suppression. *Toxicol. Sci.,* 1999, 52: 61–67.

[18] Ullrich, S.E., and Lyons, H.J., Mechanisms involved in the immunotoxicity induced by dermal application of JP-8 jet fuel. *Toxicol. Sci.,* 2000, 58: 290–298.

[19] Ramos, G., Nghiem, D.X., Walterscheid, J.P., and Ullrich, S.E., Dermal application of jet fuel suppresses secondary immune reactions. *Toxicol. Appl. Pharmacol.,* 2002, 180: 136–144.

[20] Walterscheid, J.W., Ullrich, S.E., and Nghiem, D.X., Platelet activating factor, a molecular sensor for cellular damage, activates systemic immune suppression. *J. of Exp. Med.,* 2002,195: 171–179.

[21] Ramos, G., Nghiem, D.X., Walterscheid, J.W., Ullrich, S.E., Platelet activating factor receptor binding plays a critical role in jet fuel-induced immune suppression. *Toxicol. Appl. Pharmacol.,* 2004, 195: 331–338.

[22] Ullrich, S.E., Mechanisms underlying UV-induced immune suppression. *Mutat. Res.,* 2005, 571: 185–205.

[23] Stoica, B.A., Boulares, A.H., Rosenthal, D.S., Iyer, S., Hamilton, I.D.G., and Smulson, M.E., Mechanism of JP-8 jet fuel toxicity. I. Induction of apoptosis in rat lung epithelial cells. *Toxicol. Appl. Pharmacol.,* 2001, 171: 94–106.

[24] Rosenthal, D.S., Simbulan-Rosenthal, C.M., Liu, W.F., Stoica, B.A., and Smulson, M.E., Mechanism of JP-8 jet fuel cell toxicity. II. Induction of necrosis in skin fibroblasts and keratinocytes and modulation of levels of Bcl-2 family members. *Toxicol. Appl. Pharmacol.,* 2001, 171: 107–116.

[25] Boulares, A.H., Contreras, F.J., Espinoza, L.A., and Smulson, M.E., Roles of oxidative stress and glutathione depletion in JP-8 jet fuel-induced apoptosis in rat lung epithelial cells. *Toxicol. Appl. Pharmacol.,* 2002, 180: 92–99.

[26] Personal communication, Jeffery Fisher, University of Georgia.

[27] Vijayalaxmi, Kligerman, A.D., Prihoda, T.J., and Ullrich, S.E., Micronucleus studies in the peripheral blood and bone marrow of mice treated with jet fuels, JP-8 and Jet-A. *Mutat. Res.,* 2006, 606: 82–87.

[28] ATSDR, Toxicological Profiles for Jet Fuels (JP-5 and JP-8). 1998. U.S. Department of Health and Human Services, Public Health Service, Agency for Toxic Substances and Disease Registry, Atlanta, GA.
[29] Zeiger, E., and Smith, L., The First International Conference on the Environmental Health and Safety of Jet Fuel. *Environ. Health Perspect.,* 1998, 106(11): 763–764.
[30] TIEHH (The Institute of Environmental and Human Health), JP-8 Final Risk Assessment. The Institute of Environmental and Human Health, Lubbock, TX. August 2001.
[31] Anger, W.K., Oregon Health and Science University, at the *Air Force Logistics Jet Fuel Toxicol. Workshop,* National Composite Center's Education and Conference Center in Kettering, OH, 16–17 November 2004.
[32] Ritchie, G.D., Rossi, J. III, Nordholm, A.F., Still, K.R., Carpenter, R.L., Wenger, G.R., and Wright, D.W., Effects of repeated exposures to JP-8 jet fuel vapor on learning of simple and difficult operant tasks by rats. *J. Toxicol. Environ. Health,* 2001b, Part A, 64(5): 385–415.
[33] Herrin, B.R., Haley, J.E., Lantz, R.C., and Witten, M.L., A reevaluation of the threshold exposure level of inhaled JP-8 in mice. *Toxicol. Sci.,* 2006, 31(3): 219–228.
[34] Harris, D.T., Sakiestewa, S., Titone, D., and Witten, M., JP-8 jet fuel exposure induces high levels of IL-10 and PGE2 secretion and is correlated with loss of immune function. *Toxicol. Indus. Health,* 2007, 23: 223–230.
[35] Ramos, G., Limón-Flores, A.Y., and Ullrich, S.E., Dermal exposure to jet fuel suppresses delayed type hypersensitivity: A critical role for aromatic hydrocarbons. *Toxicol. Sci.,* 2007; 100: 415–422.
[36] Wong, S.S., Vargas, J., Thomas, A., Heys, J., McLaughlin, M., Camponovo, R., Lantz, R.C., and Witten, M.L., In vivo comparison of epithelial responses for S-8 versus JP-8 jet fuels below permissible exposure limit. *Toxicology,* 2008, 254: 106–111.
[37] Wong, S.S., Desmaris, T., Lantz, R.C., Thomas, A., Witten, M.L., Pulmonary evaluation of permissible exposure limit of S-8 synthetic jet fuel in mice. *Toxicol. Sci.,* 2009, 109: 312–320.
[38] Harris, D.T., The effects of aerosolized JP-8 jet fuel exposure on the immune system: A review. In *JP-8 Jet Fuel,* M. Witten, Ed., New York: Taylor & Francis, 2010.
[39] Tu, R.H., Mitchell, C.S., Kay, G.G., and Risby, T.H., Human exposure to jet fuel, JP-8. *Aviat., Space, Environ. Med.,* 2004, 75(1): 49–59.
[40] Fechter, L.D., Gearhart, C., Fulton, S., Campbell, J., Fisher, J., Na, K., Cocker, D., Nelson-Miller, A., Moon, P., and Pouyatoz, B., JP-8 can promote auditory impairment resulting from subsequent noise exposure in rats. *Toxicol. Sci.,* 2007, 98(2): 510–525.
[41] Zhang, Z., Kleinstreuer, C., Kim, C.S., and Cheng, Y.S., 2004, Vaporizing micro-droplet inhalation, transport and deposition in a human upper airway model, *Aerosol Sci. Technol.,* 2004, 38: 36–49.
[42] Li, Z., Kleinstreuer, C., and Zhang, Z., Simulation of airflow fields and microparticle deposition in realistic human lung airway models. I. Airflow patterns. *Eur. J. Mechanics B/Fluid,* 2007, 26: 632–649.
[43] Chou, C.C., Riviere, J.E., and Monteiro-Riviere, N.A., Differential relationship between carbon chain-length of jet fuel aliphatic hydrocarbons and their ability to induce cytotoxicity versus interleukin-8 release in human epidermal keratinocytes. *Toxicol. Sci.,* 2002, 69: 226–233.
[44] Xia, X.R., Baynes, R.E., Monteiro-Riviere, N.A., Leidy, R.B., Shea, D., and Riviere, J.E., A novel in vitro technique for studying percutaneous permeation with a membrane coated fiber and gas chromatography/mass spectrometry. I. Performance of the technique and determination of the permeation rates and partition coefficients of chemical mixtures. *Pharmacolog. Res.,* 2003, 20: 275–282.

[45] Muhammad, F., Baynes, R.E., Monteiro-Riviere, N.A., Xia, X.R., and Riviere, J.E., Dose related absorption of JP-8 jet fuel hydrocarbons through porcine skin with QSPR analysis. *Toxicol. Mechan. Methods,* 2004, 14: 159–166.

[46] Muhammad, F., Monteiro-Riviere, N.A., and Riviere, J.E., Comparative *in vivo* toxicity of topical JP-8 jet fuel and its individual hydrocarbon components: identification of tridecane and tetradecane as key constituents responsible for dermal irritation. *Toxicol. Pathol.,* 2005, 33: 258–266.

[47] Whitman, F.T., and Hinz, J.P., Sensory Irritation Study in Mice, JP-5, JP-TS, JP-7, DFM, JP-10. Air Force Institute of Operational Health, Report No. IOH-RS-BR-SR-2004-0001.

[48] Babu, R.J., Chatterjee, A., Ahaghotu, E., and Singh, M., Percutaneous absorption and skin irritation upon low-level prolonged dermal exposure to nonane, dodecane and tetradecane in hairless rats. *Toxicol. Indus. Health,* 2004, 20: 108–118.

[49] Kabbur, M.B., Rogers, J.V., Gunasekar, P.G., Garrett, C.M., Geiss, K.T., Brinkley, W.W., and McDougal, J.N., Effect of JP-8 jet fuel on molecular and biological parameters related to acute irritation. *Toxicol. Appl. Pharmacol.,* 2001, 175: 83–88.

[50] McDougal, J.N., Garrett, C.M., Amato, C.M., and Berberich, S.J., Effects of brief cutaneous JP-8 fuel exposures on time course of gene expression in the epidermis. *Toxicol. Sci.,* 2007, 95(2): 495–510.

[51] McDougal, J.N., and Garrett, C.M., Gene expression and target tissue dose in the rat epidermis after brief JP-8 and JP-8 aromatic and aliphatic component exposures. *Toxicol. Sci.,* 2007, 97(2): 569–581.

[52] Moses, C., Comparative Evaluation of Semi-Synthetic Jet Fuels, Coordinating Research Council final report on CRC Project No. AV-2-04a, September 2008.

[53] Rahmes, T.F., Kinder, J.D., Henry, T.M., Crenfeldt, G., LeDuc, G.F., Zombanakis, G.P., Abe, Y., Lambert, D.M., Lewis, C., Juenger, J.A., Andac, M.G., Reilly, K.R., Holmgren, J.R., McCall, M.J., and Bozzano, A.G., Sustainable Bio-Derived Synthetic Paraffinic Kerosene (Bio-SPK) Jet Fuel Flights and Engine Tests Program Results, AIAA Paper 2009-7002, September 2009.

[54] Personal communication, David Mattie, Wright-Patterson Air Force Base, Ohio.

[55] Personal communication, Mark Witten, University of Arizona.

2 Jet Fuel Composition

Tim Edwards

CONTENTS

JET FUEL COMPOSITION: JP-8, JET A, AND JET A-1

The composition of jet, gasoline, and diesel fuels is not controlled explicitly by their specifications, except for a few specific components. Rather, their specifications are used to control quality and to confine the fuel's physical properties to acceptable ranges. For example, most jet fuels are allowed to have a specific gravity anywhere between 0.775 and 0.84. Often, minimum or maximum properties are specified, such as a mass heat of combustion no less than 43.2 MJ/kg, a flash point not less than 38 °C, and a freezing point not greater than −47°C. These specifications have implicitly limited jet fuel hydrocarbons to carbon numbers between about 8 and 16, but the fuels still consist of a mixture of hundreds of hydrocarbons. For example, a list of all the compounds in a sample of Jet A with GC (gas chromatography) area fractions of greater than 0.1% was recently published (Table 2.1) [Bruno et al., 2006]. This list would be different for other jet fuels—it is a function of the crude oil itself and the processing conditions. Jet fuel specifications do have one major compositional limit—aromatics are limited to less than 25% by volume. As shown in Figure 2.1, JP-8 typically averages about 18% of volume aromatics [PQIS, 2008]. To better understand the average composition of jet fuels, ASTM D2425 Hydrocarbon Composition was employed on the World Survey of Jet Fuels sample set (55 fuels) [Hadaller and Johnson, 2006; Shafer et al,, 2006]. As shown in Table 2.2, D2425 separates the fuel into paraffins (n- + iso-), cycloparaffins (1- and 2-ring), and several types of aromatics. A separate analysis was used to determine n-paraffins directly, so the iso-paraffins could be determined by difference measurement. The chemical classes in jet fuels generally had fairly unimodal distributions, as shown in Figure 2.2.

ALTERNATIVE (NON-PETROLEUM) JET FUELS

Two types of synthetic (non-petroleum) fuels have been flight tested in the 2006 to 2009 time frame. Both Fischer-Tropsch (FT) "synthetic paraffinic kerosene" from coal and natural gas and "hydrotreated renewable jet" from fats/oils consist primarily of n- and iso-paraffins. The lack of cycloparaffins and aromatics creates issues with density and fuel system seals, so these fuels will be initially implemented as

TABLE 2.1
Analysis of Jet-A by Gas Chromatography/Mass Spectrometry

Retention Time (min)	Correlation Coefficient	Name	CAS No.	Area Percentage
1.726	72.9	n-Heptane	142-82-5	0.125
1.878	76.9	Methylcyclohexane	108-87-2	0.198
2.084	71.6	2-Methylheptane	592-27-8	0.202
2.144	29.2	Toluene	108-88-3	0.320
2.223	41.9	*cis*-1,3-Dimethylcyclohexane	638-04-0	0.161
2.351	44.0	n-Octane	111-65-9	0.386
2.945	31.1	1,2,4-Trimethylcyclohexane	2234-75-5	0.189
3.036	12.4	4-Methyloctane	2216-34-4	0.318
3.169	37.6	1,2-Dimethylbenzene	95-47-6	0.575
3.527	33.9	n-Nonane	111-84-2	1.030
3.921	NA	?		0.321
4.066	NA	x-Methylnonane	NA	0.597
4.576	7.97	4-Methylnonane	17301-94-9	0.754
4.655	35.8	1-Ethyl-3-methylbenzene	620-14-4	1.296
4.764	10.7	2,6-Dimethyloctane	2051-30-1	0.749
4.836	5.27	1-Methyl-3-(2-methylpropyl) cyclopentane	29053-04-1	0.285
5.012	27.8	1-Ethyl-4-methylbenzene	622-96-8	0.359
5.049	13.7	1-Methyl-2-propylcyclohexane	4291-79-6	0.370
5.291	26.3	1,2,4-Trimethylbenzene	95-63-6	1.115
5.325	37.7	n-decane	124-18-5	1.67
5.637	36	1-Methyl-2-propylbenzene	1074-17-5	0.367
5.825	36	4-Methyl decane	2847-72-5	0.657
5.910	26.9	1,3,5-Trimethylbenzene	108-67-8	0.949
6.073	NA	x-Methyldecane	NA	0.613
6.176	5.01	2,3-Dimethyldecane	17312-44-6	0.681
6.364	25.7	1-Ethyl-2,2,6-trimethylcyclohexane	71186-27-1	0.364
6.516	35.6	1-Methyl-3-propylbenzene	1074-43-7	0.569
6.662	NA	aromatic	NA	0.625
6.589	20.4	5-Methyldecane	13151-35-4	0.795
6.728	22.9	2-Methyldecane	6975-98-0	0.686
6.862	23.2	3-Methyldecane	13151-34-3	0.969
7.110	NA	Aromatic	NA	0.540
7.159	NA	Aromatic	NA	0.599
7.310	17.9	1-Methyl-(4-methylethyl)benzene	99-87-6	0.650
7.626	22.0	n-Undecane	1120-21-4	2.560
7.971	NA	x-Methylundecane	NA	1.086
8.875	22.3	1-Ethyl-2,3-dimethylbenzene	933-98-2	1.694
9.948	19.6	n-Dodecane	112-40-3	3.336
10.324	19.0	2,6-Dimethylundecane	17301-23-4	1.257

TABLE 2.1 (*Continued*)
Analysis of Jet-A by Gas Chromatography/Mass Spectrometry

Retention Time (min)	Correlation Coefficient	Name	CAS No.	Area Percentage
12.377	10.8	n-Tridecane	629-50-5	3.998
12.901	24.1	1,2,3,4-Tetrahydro-2,7-dimethylnaphthalene	13065-07-1	0.850
13.707	3.5	2,3-Dimethyl dodecane	6117-98-2	0.657
14.138	14.5	2,6,10-Trimethyldodecane	3891-98-3	0.821
13.834	NA	x-Methyltridecane	NA	0.919
13.998	NA	x-Methyl tridecane	NA	0.756
14.663	29.8	n-Tetradecane	629-59-4	1.905
16.86	24.7	n-Pentadecane	629-62-9	1.345

Source: From Bruno, T.J., Laesecke, A., Outcalt, S.L., Seelig, H.-D., and Smith, B., Properties of a 50/50 Mixutre of Jet A and S-8, NIST-IR-6647, NIST, Boulder, CO, 2006.

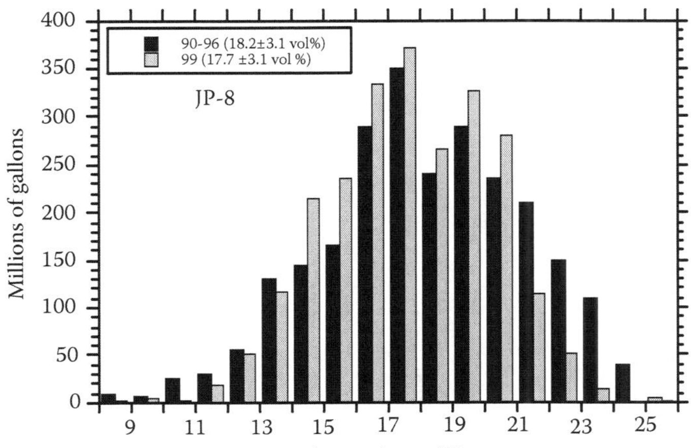

FIGURE 2.1 Aromatics distribution in JP-8 jet fuel. (*Source:* From Petroleum Quality Information System, annual reports, Defense Energy Support Center, available on-line at http://www.desc.dla.mil/DCM/DCMPage.asp?pageid=99.)

TABLE 2.2
Mean Hydrocarbon Composition from ASTM D2425 in World Survey of Jet Fuels

Paraffins (n- + iso-)	58.78
Monocycloparaffins	10.89
Dicycloparaffins	9.25
Tricycloparaffins	1.08
Alkyl benzenes	13.36
Indans/tetralins	4.9
Naphthalene	0.13
Substituted naphthalenes	1.55

Source: From Hadaller, O.J., and Johnson, J.M., World Fuel Sampling Program, Coordinating Research Council Report No. 647, June 2006; Shafer, L.M., Striebich, R.C., Gomach, J., and Edwards, T. Chemical Class Composition of Commercial Jet Fuels and Other Specialty Kerosene Fuels, AIAA 2006-7972.

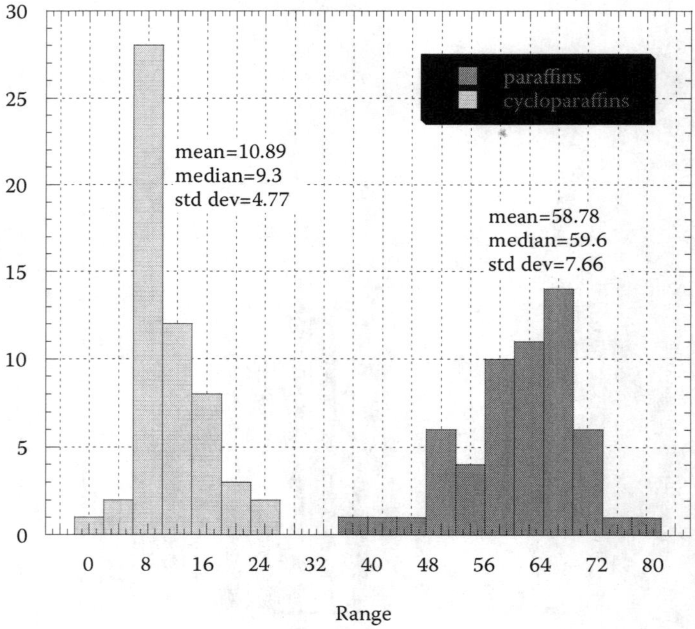

FIGURE 2.2 Distribution of mono-cycloparaffins and paraffins (n- plus iso-) from World Survey of Jet Fuels. (*Source:* From Hadaller, O.J., and Johnson, J.M., World Fuel Sampling Program, Coordinating Research Council Report No. 647, June 2006; Shafer, L.M., Striebich, R.C., Gomach, J., and Edwards, T. Chemical Class Composition of Commercial Jet Fuels and Other Specialty Kerosene Fuels, AIAA 2006-7972.)

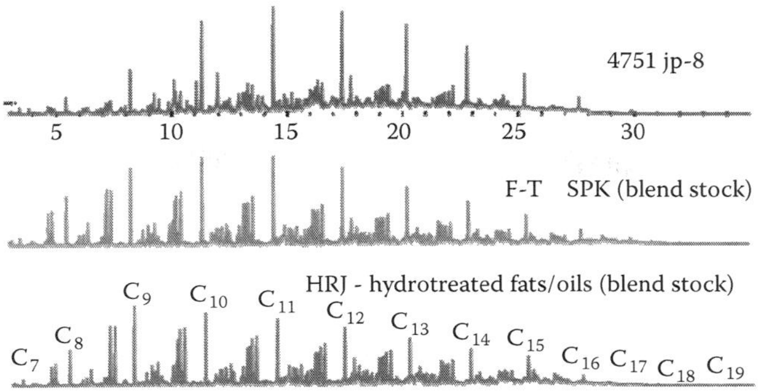

FIGURE 2.3 GC traces of one petroleum-derived jet fuel, one Fischer-Tropsch synthetic paraffinic kerosene, and one hydrotreated renewable jet.

50/50 blends. The hydrocarbon distribution of the alternative fuels is similar to that of current fuels, as shown in Figure 2.3 via GC traces of examples of these three fuels. Jet fuels in the R&D stage also include kerosenes derived from ligno-cellulosic feedstocks through biomass pyrolysis or fermentation. These fuels potentially may have compositions and boiling ranges much different from those shown in Figure 2.3. Ethanol (and higher alcohols) and FAME (fatty acid methyl ester, biodiesel) type fuels are not acceptable jet fuels, per extensive discussions with the military and commercial aviation industry. The military and commercial (ASTM D7566) specifications for alternative fuels limit non-hydrocarbon components to trace levels. These alternative fuels are expected to be approved as classes, differentiated by feedstock and processing. Commercial approval of the alternative fuels is usually preceded by the publication of a Research Report, which includes composition, property, and performance data obtained from commercial and military testing. The Research Report for Fischer-Tropsch "synthetic paraffinic kerosenes" has been published [Moses, 2008, 2009]; the Research Report for HRJ fuels is in preparation at the time of this writing (October 2009). However, some results of testing and flight demonstrations have been published [Rahmes et al., 2009].

REFERENCES

Bruno, T.J., Laesecke, A., Outcalt, S.L., Seelig, H.-D., and Smith, B., Properties of a 50/50 Mixture of Jet A and S-8, NIST-IR-6647, NIST, Boulder, CO, 2006.

Hadaller, O.J., and Johnson, J.M., World Fuel Sampling Program, Coordinating Research Council Report No. 647, June 2006.

Martel, C.R., Military Jet Fuels, 1944-1987, Air Force Wright Aeronautical Lab., Rept. AFWAL-TR-87-2062, November 1987.

Moses, C., Comparative Evaluation of Semi-Synthetic Jet Fuels, Coordinating Research Council final report on CRC Project No. AV-2-04a, September 2008.

Moses, C., Comparative Evaluation of Semi-Synthetic Jet Fuels. Addendum: Further Analysis of Hydrocarbons and Trace Materials to Support DXXXX, Coordinating Research Council report on CRC Project No. AV-2-04a, April 2009.

Petroleum Quality Information System, annual reports, Defense Energy Support Center, available on-line at http://www.desc.dla.mil/DCM/DCMPage.asp?pageid=99.

Shafer, L.M., Striebich, R.C., Gomach, J., and Edwards, T. Chemical Class Composition of Commercial Jet Fuels and Other Specialty Kerosene Fuels, AIAA 2006-7972.

Rahmes, T.F., Kinder, J.D., Henry, T.M., Crenfeldt, G., LeDuc, G.F., Zombanakis, G.P., Abe, Y., Lambert, D.M., Lewis, C. , Juenger, J.A., Andac, M.G., Reilly, K.R., Holmgren, J.R., McCall, M.J., and Bozzano, A.G., Sustainable Bio-Derived Synthetic Paraffinic Kerosene (Bio-SPK) Jet Fuel Flights and Engine Tests Program Results, AIAA Paper 2009-7002, September 2009.

3 The Toxicity and Underlying Mechanism of Jet Propulsion Fuel-8 on the Respiratory System

Simon S. Wong and Mark L. Witten

CONTENTS

INTRODUCTION

Jet Propulsion Fuel-8 (JP-8) is a kerosene-based fuel and contains approximately 228 long-/short-chain aliphatic/aromatic hydrocarbons (C_6–C_{14}), including benzene; toluene; ethylbenzene; xylene; 1,2,4-trimethlybenzene; cyclohexylbenzene; and dimethylnaphthalene. Its low volatility characteristics allow it to be a potential toxic and irritant on the respiratory system in aerosol, vapor, or liquid forms. Because of the produced volumes and the multipurpose nature of the fuel, occupational exposures to JP-8 occur during fuel transport, aircraft fueling and defueling, aircraft maintenance, cold aircraft engine starts, maintenance of equipment and machinery, use of tent heaters, and cleaning or degreasing with fuel. It is estimated that over 2 million people worldwide are exposed to 60 billion gallons of JP-8 annually. The U.S. Navy Occupational Safety and Health Standards Board (NAVOSH) has proposed interim exposure limits of 350 mg/m^3 and 1800 mg/m^3 as the 8-hour permissible exposure limit (PEL) and the 15-minute short-term exposure limit (STEL), respectively. These limits are based on the U.S. National Institute of Occupational Safety and Health (NIOSH) regulatory levels for more volatile petroleum distillates.

Most of the body of literature on respiratory toxicities of JP-8 has substantially been reviewed by the newly formulated toxicological assessment of JP-8 by the National Research Council. Because jet fuel vapors and aerosols are mainly an inhalational hazard, the major route of jet fuel exposure to flight and ground crew personnel is via the respiratory tract. Studies on JP-8 pulmotoxicology have found significant physiological [1–3], cellular [4, 5], and proteomic/genomic changes [4, 6] resulting from jet fuel exposure. Changes in airways were characterized by loss of epithelial barrier integrity and alterations of ventilatory function in bronchial and bronchiolar airways [1, 3, 7, 8]. These parameters at the high jet fuel levels may be important indicators for lung injury at the low level as well. For example, acute 1-hr inhalation exposures to aerosolized JP-8 have been shown to induce cellular and morphological indications of pulmonary toxicity that were associated with increased respiratory permeability to 99mTc-DTPA [1, 3, 7, 9, 10]. Most recently, morphological examination and morphometric analysis of distal lung tissue demonstrated that alveolar type II epithelial cells showed a notable increase in the volume density of lamellar bodies (vacuoles), which is indicative of increased surfactant production at 45 mg/m^3 [11]. The morphometric analysis techniques appear to provide an increased sensitivity for detecting the deleterious effects of JP-8 as compared to the evidence offered by physiological and biochemical tests. With rapidly increased evidence of jet fuel toxicities for JP-8 and new biofuels, this chapter summarizes our recent findings from *in vivo* and *in vitro* studies.

MORPHOLOGICAL ALTERATIONS

In a simulated military flight-line exposure protocol, histopathological changes of the lung following acute and chronic exposure in rats and mice have been conducted in several studies. Exposures of Fischer 344 rats to JP-8 (500 mg/m^3/hr, and 813 to –1094 mg/m^3/hr) for 1 hour/day for 7, 28, and 56 days, respectively, were performed [1]. Electron micrographs showed that all groups had interstitial edema resulting from endothelial damage. There was apparent thickening of the alveolar septa, and alveolar macrophages were activated in all groups. Lung permeability data, as determined by 99^{m}TcDTPA alveolar clearance, indicated that lung injuries peaked at 28 days of jet fuel exposure, and this finding corresponded with the histology data. A similar 28-day study in rats showed a jet fuel-induced effect on terminal bronchiolar airways that was accompanied by sub-endothelial edema [12]. The observed effects were manifested by JP-8 exposures ranging from 500 to 1000 mg/m^3, which are above safety limits. An additional investigation of the acute 1-hour exposure to NAVOSH permissible JP-8 aerosols also showed mild to moderate pulmonary toxicity [2]. Morphological alterations were characterized by the targeting of bronchiolar epithelium and consisted of perivascular edema, Clara cell vacuolization, and necrosis. Alveolar changes included sporadic pulmonary edema, intra-alveolar hemorrhage, and alterations in type II epithelial cells. These alterations have been confirmed consistently by several recent studies [8, 11]. Taken together, these results indicated that repeated inhalation of aerosolized JP-8 induces morphological lung injury. These studies also provide solid evidence for the reevaluation of the 350 mg/m^3 PEL for more volatile petroleum distillates with regard to respirable aerosol, vapor, or liquid.

We also examined if exposure to JP-8 concentrations below the current permissible exposure level (PEL) of 350 mg/m^3 still have adverse effects in lungs. C57BL/6 mice were nose-only exposed to JP-8 at average concentrations of 45, 267, and 406 mg JP-8/m^3 for 1 hour/day for 7 days. The study also validates a new "in-line, real-time" total hydrocarbon analysis system capable of measuring both JP-8 vapor and aerosol concentrations. The only significant effect observed at 406 mg/m^3 was a decrease in inspiratory dynamic lung compliance. Morphological examination and morphometric analysis of distal lung tissue demonstrated that alveolar type II epithelial cells showed limited cellular damage with the notable exception of a significant increase in the volume density of lamellar bodies (vacuoles), which is indicative of increased surfactant production, at 45 and 406 mg/m^3. The terminal bronchial epithelium showed initial signs of cellular damage, but the morphometric analysis did not quantify these changes as significant. The morphometric analysis techniques appear to provide an increased sensitivity for detecting the deleterious effects of JP-8 as compared to the physiological evidence offered by pulmonary function or respiratory permeability tests. These observations suggest that the current 350-mg/m^3 PEL for both JP-8 jet fuel and for other more volatile petroleum distillates should be reevaluated and that a lower, more accurate PEL should be established with regard human occupational exposure limits.

To examine the hypothesis that JP-8 inhalation potentiates influenza virus-induced inflammatory responses, we randomly divided female C57BL/6 mice (4 weeks old, weighing approximately 24.6 g) into the following groups: air control, JP-8 alone (1023 mg/m^3 of JP-8 for 1 hour/day for 7 days); A/Hong Kong/8/68 influenza virus (HKV) alone (a 10-μl aliquot of 2000 viral titer in the nasal passages); and a combination of JP-8 with HKV (JP-8 + HKV) [3]. HKV infection induced a substantial (18.3-fold) increase in total BALF cell number compared to controls, whereas JP-8 alone (76.7 $\pm$ 27.1 $\times$ 10^4 cells/ml) induced no change. There was an additional threefold increase in the total cell numbers in JP-8 + HKV (Mean $\pm$ SEM, 237 $\pm$ 65.5 $\times$ 10^4 cells/ml) compared to HKV alone (76.7 $\pm$ 27.1 $\times$ 10^4 cells/ml). Obviously, prior JP-8 exposure exacerbated HKV-induced inflammatory cell infiltration. The profile in cell differentials indicated that both granulocytes and lymphocytes may be the most important activated inflammatory cells in response to HKV infection.

In addition, we observed the ectasia of respiratory bronchioles and alveoli due to exposure of the mice to JP-8 [3]. Mice exposed to HKV consistently demonstrated the respiratory tract infection with obvious peribronchiolitis, sporadic airsacculitis, and interstitial edema. Mice exposed to JP-8 + HKV had significantly more severe inflammatory responses, characterized by airway epithelium sloughing, alveolar edema, and inflammatory cell infiltration. Furthermore, the mice exhibited increased lung injury, from interstitial edema observed in HKV alone to substantial alveolar edema in JP-8 + HKV exposure. This study provides initial evidence that JP-8 inhalation may partially synergize influenza virus-induced inflammatory responses.

Neuro-Immunological Responses

In response to JP-8 inhalation for 7, 28, or 56 days at two JP-8 concentrations (low dose = 469–520 mg/m^3/hr, high dose = 814–1263 mg/m^3/hr), exposed Fischer 344

TABLE 3.1
Genes Whose Expression Was Shown to be Up-Regulated by JP-8 in Jurkat Cells by Microarray Analysis

Gene Bank[a]	Gene Name	JP-8/V[b]	*t*-Test[c]
	Cell Cycle-Related Proteins		
L13698	Growth arrest specific protein 1 (GAS1)	2.36	0.018
L35255	p38 mitogen-activated protein kinase (p38 MAPK)	3.27	0.017
	Apoptosia-Related Proteins		
U13737	Caspase-3	2.92	0.003
U56390	Caspase 9	3.67	0.016
X17620	Nucleoside diphosphate kinase A (NDKA)	2.46	0.002
X66362	Serine-threonine protein kinase PCTAIRE 2 (PCTK3)	2.0	0.026
	Transcriptional Activators or Repressers		
L23959	E2F dimerization partner-1 (DP-1)	2.11	0.017
X56134	E2F-1	2.32	0.015
U15642	E2F-5	2.66	0.012
	Stress Response-Related Proteins		
U90313	Glutathione transferase omega (GSTO1)	2.51	0.012
Y00371	Heat shock 70-kDa protein (HSC70)	2.56	0.013
M34664	Heat shock 60-kDa protein (HSP60)	2.39	0.0029
M16660	Heat shock 90-kDa protein beta (HSP90)	6.01	0.016
	Metabolic Enzymes and DNA Repair Proteins		
U35835	DNA-dependent protein kinase (DNA-PK)	2.14	0.0016
K00065	Cytosolic superoxide dismutase 1 (SOD1)	3.47	0.013
U51166	G/T mismatch-specific thymine DNA glycosylase (TDG)	2.30	0.0025

Source: From Espinoza, L.A., and Smulson, M.E. *Toxicology,* 189: 181–190, 2003.
Note: Data are means of values from four independent experiments.

rats exhibited a dose-dependent as well as duration-determined reduction in broncho-alveolar lavage fluid (BALF) Substance P (SP) concentration [12, 13]. SP has been demonstrated to play an important role in inflammatory response and wound healing, mediated through neurokinin-1 receptor (NK-1R) on effector cells, modulated by neutral endopeptidase (NEP) through degradative cleavage of SP. Subsequent to its release from afferent nerve endings or immuno-inflammatory cells (such as macrophages, lymphocytes, and leukocytes), SP increases substantial responses, such as an increase in microvascular permeability, promotion of plasma extravasation, and priming of other inflammatory mediators. Therefore, decreased SP in lungs could be attributed to the irritant effects of afferent nerve or down-regulation of NEP following JP-8 exposure.

To investigate the role of SP on JP-8-induced pulmotoxicity, B6.A.D. (Ahr d/Nats) mice received subchronic exposures to JP-8 at 50 mg/m^3 [2]. Lung injury was assessed by the analysis of pulmonary physiology, bronchoalveolar lavage fluid, and morphology. Exposure to JP-8 concentrations of 50 mg/m^3, with saline treatment, was characterized by enhanced respiratory permeability to 99mTc-labeled diethylenetriaminepentaacetic acid, alveolar macrophage toxicity, and bronchiolar epithelial damage. Mice administered [Sar^9,Met(O$_2$)^{11}] SP (SP's agonist) after each JP-8 exposure had the appearance of normal pulmonary values and tissue morphology. In contrast, endogenous SP's receptor antagonism by CP-96345 administration exacerbated JP-8-enhanced permeability, alveolar macrophage toxicity, and bronchiolar epithelial injury. This study indicated that SP may have a protective role in preventing the development of JP-8-induced pulmotoxicity, possibly through the modulation of bronchiolar epithelial function.

To examine other JP-8-induced inflammatory mechanisms, we utilized C57BL/6 mice (young, 3.5 months; adult, 12 months) that were nose-only exposed to either room air or atmospheres of 1000 mg/m^3 JP-8 for 1 hour/day for 7 days (Wang, 2001, 40 /id). This study showed that exposure of these mice to JP-8 resulted in multiple inflammatory mediator releases, such as cytokines and arachidonic acid, which may be associated with JP-8-induced physiological, cellular, and morphological alterations. Data showed that bronchoalveolar lavage fluid PGE$_2$ and 8-iso-PGF$_2\alpha$ levels were decreased in both the young and adult JP-8 jet fuel-exposed groups compared with control values, suggesting that the metabolites of arachidonic acid may be important mediators in both models of JP-8 jet fuel exposure. These mediators are very potent regulators of cellular signaling and inflammation and can be further metabolized by lipoxygenase or cyclooxygenase enzymes to yield the family of leukotrienes or prostaglandins and thromboxanes.

CELLULAR TOXICITY

In vitro studies further revealed that exposure to JP-8 induces increases in paracellular permeability in an SV-40 transformed human bronchial epithelial cell line (BEAS-2B) [7]. Incubation of confluent BEAS-2B cells with concentrations of JP-8 or n-tetradecane (a primary constituent of JP-8) induced dose-dependent increases in paracellular flux. Following exposures of 0.17, 0.33, 0.50, or 0.67 mg/ml of JP-8, mannitol flux increased above vehicle controls by 10%, 14%, 29%, and 52%, respectively, during a 2-hour incubation period. N-tetradecane also caused higher mannitol flux increases of 37%, 42%, 63%, and 78%, respectively, following identical incubation concentrations. The transepithelial mannitol flux reached a maximum at 12 hours and spontaneously reversed to control values over a 48-hour recovery period, for both JP-8 and n-tetradecane exposure. These data indicate that exposures to JP-8 exert a noxious effect on bronchial epithelial barrier function that may preclude pathological lung injury. N-tetradecane exposure could partially initiate JP-8-induced lung injury through a disruption in the airway epithelial barrier function.

A follow-up study [5] examined the effect(s) of JP-8 on cytokine secretion in a transformed rat alveolar type II epithelial cell (AIIE cell) line (RLE-6TN) and primary alveolar macrophages (AM, from Fischer 344 rats). The cell co-culture study indicated that the balance of cytokine release could be regulated possibly by cross-communication of AIIE and alveolar macrophages (AM), in close proximity to each other. The concentration ranges from 0 to 0.8 μg/ml JP-8, which may actually be encountered in the alveolar spaces of lungs exposed in vivo, were placed in cell culture for 24 hours. Cultured AIIE alone secreted spontaneously interleukin (IL)-1β and IL-6 (below detectable limits for IL-10 and tumor necrosis factor-alpha [TNF-α]), whereas cultured AM alone secreted IL-1β, IL-10, and TNF-α, in a concentration-dependent manner. These data suggest that the release of cytokines, not only from AM but also from AIIE cells, may both contribute to the JP-8-induced inflammatory response in the lungs. However, the co-cultures of AIIE and AM showed no significant changes in IL-1β, IL-6, and TNF-α at any JP-8 concentration when compared to controls. These cytokine levels in co-cultures of AIIE and AM were inversely related to those of cultured AIIE or AM alone. Interestingly, IL-10 levels in the co-culture system were concentration-dependently increased up to 1058% at JP-8 concentrations of 0.8 μg/ml, although under detectable limits in cultured AIIE alone and no significant concentration change in cultured PAM alone. It is likely, we speculate, that AM may possibly act via paracrine or autocrine pathways to signal AIIE cells to regulate cytokine release. Further studies are required to identify the potential signaling mechanism(s) of cytokine expression and release between these two lung cells.

The effects of combined JP-8 and SP on cell growth and survival, and cytokine and chemokine production, were also studied using co-cultures of rat alveolar macrophages and type II alveolar epithelial cells. These cell types were selected because the macrophages appear to communicate with the type II cells, which are the first to be affected by inhalation exposure to JP-8. A rat transformed type II cell line was allowed to form a monolayer and primary pulmonary alveolar macrophages were layered on top and allowed to settle on the epithelial monolayer. JP-8 was applied 1 hour later, and the effects of [$Sar^9met(O_2)^{11}$] SP on cell numbers and on responses of cellular factors were examined. Control cultures consisted of each individual cell type.

SP, alone, significantly depressed alveolar cell numbers when applied in the range of 10^{-4} to 10^{-9}M; however, the depression was not concentration related. Macrophage and co-culture cell numbers were not affected by any SP concentration. All studies of cytokine or chemokine induction used 10^{-4} to 10^{-6}M SP. A single concentration of 16 μg JP-8/ml was selected based on a preliminary study showing a non-significant increase in macrophage numbers, a significantly decreased survival of type II cells, and the absence of an effect on the cell combination at this concentration. It was noted in the discussion that although most of the effects were observed at SP concentrations of 10^{-5} or 10^{-6}M, these levels are much higher than the 40 to 50 fmol/ml (40–50 × 10^{-15}M) of endogenous SP observed in untreated rat lungs. This value decreases to less than 5 fmol (below the level of detection) following JP-8 inhalation.

SP significantly increased the JP-8-induced release of IL-1α, IL-1β, and IL-18 by macrophages, alone, but there was no effect in the co-culture. SP significantly decreased IL-6 production by the co-culture, but not in the individual cell cultures. The results for the other cytokines and chemokines were less clear, and tended to be more variable. IL-10, MCP-1, and GRO-KC releases were significantly depressed in the co-culture, but the responses were not clearly dose dependent. TNF-α release in the co-culture was increased, but only at the high SP concentration. The results were interpreted to show the presence of cell-to-cell communication between the alveolar epithelial cells and macrophages, and that this type of co-culture system could be used to measure the effects of SP on JP-8-induced lung injury. It was suggested that IL-12 also be measured because it is produced by macrophages under stress.

An effort has been made to affirm validation of tissue slice models [14]. As is known, interactions between the heterogeneous cell types comprising the lung parenchyma are maintained, thus providing a controlled system for the study of pulmonary toxicology in vitro. Previous in vivo studies have demonstrated that Clara cells and alveolar type II cells are the targets following inhalation of JP-8 jet fuel. We have utilized the lung slice model to determine if cellular targets are similar following in vitro exposure to JP-8. Agar-filled adult rat lung explants were cored and precision cut, using the Brende/Vitron tissue slicer. Slices were cultured on titanium screens located as half-cylinders in cylindrical Teflon cradles that were loaded into standard scintillation vials and incubated at 37°C. Slices were exposed to JP-8 jet fuel (0.5, 1.0, and 1.5 mg/ml in medium) for up to 24 hours. We determined ATP content using a luciferin-luciferase bioluminescent assay. No significant difference was found between the JP-8 jet fuel doses or time points, when compared to controls. Results were correlated with structural alterations following aerosol inhalation of JP-8. As a general observation, ultrastructural evaluation of alveolar type cells revealed an apparent increase in the number and size of surfactant secreting lamellar bodies that was JP-8 jet fuel-dose dependent. These results are similar to those observed following previous aerosol inhalation exposure [1, 2, 8–10]. Thus, the lung tissue slice model appears to mimic the in vivo effects of JP-8 and therefore is a useful model system for studying the mechanisms of pulmotoxicity following JP-8 exposure.

Considering that alveolar epithelial type II cells are critical in the regulation of airway inflammation in response to JP-8 exposure, we examined if the expression of several proinflammatory cytokines were associated with activation of NF-κB [15]. The findings show that a low threshold dose (8 μg/ml) of the jet fuel JP-8 induces in a rat alveolar epithelial cell line (RLE-6TN) a prolonged activation of NF-κB as well as the increased expression of the proinflammatory cytokines TNF-α and IL-8, which are regulated by NF-κB. The up-regulation of IL-6 mRNA in cells exposed to JP-8 appears to be a reaction of RLE-6TN cells to reduce the enhancement of proinflammatory mediators in response to the fuel. Moreover, lung tissues from rats exposed to occupational levels of JP-8 by nasal aerosol also showed dysregulated expression of TNF-α, IL-8, and IL-6, thus confirming the in vitro data. The poly(ADP-ribosyl)ation of PARP-1, a coactivator of NF-κB, was coincident with the prolonged activation of NF-κB during JP-8 treatment. These results demonstrated that expression of JP-8-induced inflammatory cytokine genes in AEII cells was mediated by NF-κB and PARP-1.

PROTEOMIC AND GENOMIC CHANGES

Proteomic analysis of lungs in male Swiss-Webster mice exposed for 1 hour/day for 7 days to JP-8 at concentrations of 1000 and 2500 mg/m^3 showed significant quantitative and qualitative changes in tissue cytosol proteins [6, 16]. We analyzed protein expression in the cytosolic fraction prepared from whole lung tissue (Table 3.3). Lung cytosol samples were solubilized and separated via large-scale, high-resolution, two-dimensional electrophoresis (2-DE) and gel patterns scanned, digitized, and processed for statistical analysis. Significant quantitative and qualitative changes in tissue cytosol proteins resulted from jet fuel exposure. Several of the altered proteins were identified by peptide mass fingerprinting, confirmed by sequence tag analysis, and related to impaired protein synthetic machinery, toxic/metabolic stress and detoxification systems, ultrastructural damage, and functional responses to CO2 handling, acid-base homeostasis, and fluid secretion. These results demonstrate a significant but comparatively moderate JP-8 effect on protein expression and corroborate previous morphological and biochemical evidence. Further molecular marker development and mechanistic inferences from these observations await proteomic analysis of whole-tissue homogenates and other cell compartments, that is, mitochondria, microsomes, and nuclei of lung and other targets.

Recent data have also shown that exposure of several human and murine cell lines, including in a rat lung alveolar type II epithelial cell (RLE-6TN), to JP-8 in vitro induces biochemical and morphological markers of apoptotic cell death such as activation of caspase-3, cleavage of poly(ADP-ribose) polymerase, chromatin condensation, membrane blebbing, release of cytochrome c from mitochondria, and cleavage of genomic DNA. Generation of reactive oxygen species (ROS) and depletion of intracellular reduced glutathione (GSH) were also shown to play important roles in the induction of programmed cell death by JP-8. At 250 mg/m^3 JP-8 concentration, 31 proteins exhibited increased expression, while 10 showed decreased expression. At 1,000 mg/m^3 exposure levels, 21 lung proteins exhibited increased expression and 99 demonstrated decreased expression. At 2,500 mg/m^3, 30 exhibited increased expression, while 135 showed decreased expression. Several of the proteins were identified by peptide mass fingerprinting, and were found to relate to cell structure, cell proliferation, protein repair, and apoptosis. These data demonstrate the significant stress that JP-8 jet fuel puts on lung epithelium. Furthermore, there was a decrease in alpha1-anti-trypsin expression, suggesting that JP-8 jet fuel exposure may have implications for the development of pulmonary disorders.

Microarray analysis [17] has been utilized to characterize changes in the gene expression profile of lung tissue induced by exposure of rats to JP-8 at a concentration of 171 or 352 mg/m^3 for 1 hour/day for 7 days, with the higher dose estimated to mimic the level of occupational exposure in humans. The expression of most (48/56, 86%) of the genes affected by JP-8 at 171 mg/m^3 was down-regulated, whereas that of 42% (28/66) of the genes affected by JP-8 at 352 mg/m^3 was up-regulated. The affected genes were classified according to their biological function, being divided into eight or nine groups for the low and high doses of JP-8, respectively (Table 3.1 and Table 3.2).

The lungs of rats exposed to 352 mg/m^3 JP-8 manifested a 5.5-fold increase in expression of the gene for γ-synuclein, a centrosomal protein that plays an important role in the regulation of cell growth. The expression of genes whose products contribute to the cellular response to oxidative stress or to toxicants, including glutathione S-transferases (Gsta1, GstaYc2) and cytochromes P450 (CYP2C13, Cyp2e1),

TABLE 3.2
Genes Whose Expression Was Shown to be Down-Regulated by JP-8 in Jurkat Cells by Microarray Analysis

Gene Bank[a]	Gene Name	JP-8/V[b]	*t*-Test[c]
	Cell Cycle-Related Proteins		
X05360	Cell division control protein 2 homologue (CDC2)	0.49	0.0037
	Apoptosis-Related Proteins		
X76104	Death-associated protein kinase 1 (DAPK1)	0.43	0.0026
X79389	Glutathione S-transferase theta 1 (GSTT1)	0.38	0.004
	Stress Response-Related Proteins		
M64673	Heat shock transcription factor 1 (HSTF1)	0.47	0.0048
U07550	Mitochondrial heat shock 10-kDa protein (HSP10)	0.35	0.005
M86752	Stress-induced phosphoprotein 1 (STIP1)	0.48	0.015
D43950	T-complex protein 1 epsilon subunit (TCP1 epsilon)	0.38	0.018
U83843	T-complex protein 1 eta subunit (TCP 1-eta)	0.40	0.016
	Metabolic Enzymes and DNA Repair Proteins		
D49490	Protein disulfide isomerase-related protein 1 (PDIR)	0.45	0.014
	Others		
X56134	Vimentin	0.39	0.005

Source: From Espinoza, L.A., and Smulson, M.E. *Toxicology,* 189: 181–190, 2003.
Note: Data are means of values from four independent experiments.

was also prominently increased in lung tissue from rats exposed to JP-8 at the higher dose. The expression of these genes has been shown to be induced by Nrf2, a transcription factor that protects against oxidative or chemically induced cellular damage in the lungs. In contrast, the expression of none of the apoptosis-related genes represented on the microarray was affected by JP-8 at either dose, which is consistent with the results of the histologic analysis showing minimal cell damage in the lungs. The abundance of mRNAs for various structural proteins, including myosin heavy chain 7 (Myh7), α-actin, and β-actin, was increased by exposure to 352 mg/m^3 JP-8, whereas that of the mRNA for high-molecular-weight microtubule-associated protein 2 (HMW-MAP2) was decreased. Expression of the gene for the inositol 1,4,5-trisphosphate receptor (InsP3R1), a ligand-gated Ca^{2+} channel that mediates the release of Ca^{2+} from intracellular stores, was reduced by a factor of 10 in rat lungs exposed to 352 mg/m^3 JP-8. In this regard, studies have previously shown that JP-8-induced fragmentation of DNA during apoptosis in RLE-6TN cells is Ca^{2+} dependent. Expression of the genes for aquaporin 1 (Aqpl) and aquaporin 4 (Aqp4), proteins that mediate water transport in various tissues including the lungs, was increased and decreased, respectively, in the rat lungs exposed to 352 mg/m^3 JP-8. The microarray data were confirmed by quantitative RT-PCR analysis. Data may provide important insight into

TABLE 3.3
250 mg/m³ JP-8 Jet Fuel Aerosol-Mediated Protein Expression Alterations in Identified Proteins

		Protein Abundance		
SSP	Identity	Control	JP-8	Prob.
3314	Actin	1,038.3	1,546.7	0.001
2414	Actin, alpha	5,348.4	5,987.3	0.5
4404	Actin-related protein	400.6	4,91.5	0.4
6516	Aldehyde dehydrogenase	4,277.9	4,558.7	0.4
8311	Aldose reductase	1,532.6	1,708.8	0.6
1602	AAT 1–4	2,869.6	2,296.2	0.02
4304	Annexin III	2,057.7	2,414.2	0.06
1307	Annexin V	4,054.4	5,011.8	0.4
6107	Antioxidant protein 2	7,325.2	7,231.8	0.9
3102	Apolipoprotein A-l	5,235.0	4,492.5	0.2
2501	ATP synthase, beta chain	2,926.3	3,275.9	0.2
5820	Brefeldin A-inhibited guanine nucleotide-exchange protein	140.1	161.9	0.5
712	Calreticulin	1,710.9	1,835.8	0.6
8409	Creatine kinase	2,203.6	1,889.5	0.6
4517	CKA	1,007.9	1,365.2	0.03
3501	Desmin fragment	1,02.3	1,37.4	0.03
6612	Dihydropyrimidinase related protein-2	1,278.0	1,681.3	0.04
7502	Enolase, alpha	3,325.3	3,117.5	0.8
6511	Enolase, alpha (charge variant)	1,274.6	1,387.0	0.4
5604	ER60	1,534.8	1,754.9	0.1
5720	Ezrin	991.7	1,149.5	0.2
2523	FAF1	349.9	503.6	0.05
2701	grp78	2,755.6	3,009.0	0.1
1802	grp94 (eudoplasmin)	1,931.8	2,513.2	0.3
3708	Hexokinase, type 1	1,131.6	714.0	0.05
3706	hsc70	3,634.9	3,919.1	0.4
3607	HSP60	1,295.2	1,753.1	0.007
2506	Keratin, type 1 cytoskeletal 9	190.4	266.7	0.02
7707	Moesin	1,233.5	1,265.6	0.9
7715	Moesin	873.8	964.0	0.6
2105	Myosin light chain 1	385.6	108.7	0.1
3609	P21 activated kinase 1B	198.5	278.5	0.0004
4217	PA28, alpha subunit	950.8	1,173.3	0.06
5513	Selenium binding protein 1	2,335.5	2,698.8	0.05
5503	Selenium binding protein 2	1,346.0	1,528.2	0.3
6002	Superoxide dismutase	3,944.0	4,635.8	0.1
5204	Thioether S-methyltransferase	5,806.5	6,883.5	0.08
3901	Ubiquitin carboxyl-terminal hydrolase 2	3,048.0	2,818.7	0.6

TABLE 3.3 (*Continued*)
250 mg/m³ JP-8 Jet Fuel Aerosol-Mediated Protein Expression Alterations in Identified Proteins

		Protein Abundance		
SSP	Identity	Control	JP-8	Prob.
2611	Vimentin	1,444.2	1,217.4	0.3
5817	Vinculin	3,413.0	3,185.2	0.5
5811	Vinculin (charge variant)	571.6	632.3	0.4

Source: From Drake, M.G., Witzmann, F.A., Hyde, J., and Witten, M.L., *Toxicology,* 191(2-3): 199–210, 2003.

the toxicological basis of the pulmonary response to occupational exposure to JP-8 at the molecular level.

SYNTHETIC JET FUEL S-8

With increasing oil prices and decreasing oil reserves, the U.S. military is searching for fuel alternatives to conserve energy and reduce dependence on foreign oil supplies. Although JP-8 is currently the primary jet fuel for both the U.S. Armed Forces and NATO ground-based operations, the U.S. Air Force is hopeful that a synthetic jet fuel will soon replace it. A synthetic jet fuel known as S-8 has been created by Syntroleum. This synthetic jet fuel is produced from the Fisher-Tropsch (FT) process, a process that relies on pressure and heat to convert natural gas or coal into a clean-burning liquid fuel. In an effort to learn about the synthetic fuel's effects, the Air Force Advanced Power Technology Office has conducted studies at both Selfridge Air National Guard Base in Michigan and Edwards Air Force Base (AFB) in California. When a TF33 engine testbed utilized both a 50/50 blend of synthetic and JP-8 and 100% JP-8, no differences were observed in performance. These results prompted Air Force personnel to test the synthetic jet fuel's effects in a B-52H Stratofortress in flight. During the first test on 19 September 2006, the B-52 flew from Edwards AFB with two of its eight engines powered by a 50/50 blend of both S-8 and JP-8. Flights on 27 September 2006 and 29 September 2006 completed the initial test cycle, and all performed similarly to other engines that had utilized only JP-8. Ground tests revealed that the synthetic and JP-8 blend burn at the same rate as petroleum-based fuels, yet the blend shows a 20% to 40% reduction in particulate emission.

To gain an understanding about the threshold concentration in which lung injury is observed, C57BL/6 male mice were nose-only exposed to S-8 for 1 hour/day for 7 days at average concentrations of 0 (control), 93, 352, and 616 mg/m³ [20]. Evaluation of pulmonary function, airway epithelial barrier integrity, and pathohistology was performed 24 hours after the final exposures. Significant decreases were detected in expiratory lung resistance and total lung compliance of the 352-mg/m³ group, for which no clear concentration-dependent alterations could be determined. No significant changes in respiratory permeability were exhibited, indicating that there was no loss of epithelial barrier integrity following S-8 exposure. However, morphological examination and morphometric analysis of distal lung tissue, using transmission electron microscopy (TEM), revealed cellular damage in alveolar type II epithelial cells, with significant increases in volume density of lamellar bodies/vacuoles at 352 and 616 S-8 mg/m^3. In particular, terminal bronchiolar Clara injury, as evidenced by apical membrane blebs, was observed at relatively low concentrations, suggesting that if this synthetic jet fuel is utilized, the current permissible exposure limit (PEL) of 350 mg/m^3 for hydrocarbon fuels should cautiously be applied.

We also compared the cytotoxic or injury effects of the distal lungs after nose-only exposure of C57BL/6 mice to S-8 and JP-8 at 53 mg/m^3 (80 ppm THC) 1 hour/day for 7 days, respectively [20]. The results showed that there was a significant increase in expiratory lung resistance in the S-8 group. In the JP-8 mice, there were significant increases in both inspiratory and expiratory lung resistance compared to control values. In ultrastructure examination, differences were observed in cellular injury or organelle alterations of the Clara cells and alveolar type II cells for the two groups. JP-8 inhalation resulted in structural damage to the extended alveolar area involving cytoplasmic vacuolization and apical membrane blebs. These alterations, especially bleb formation, represented an early sign of cell injury that may precede changes in respiratory permeability after JP-8 exposure. This finding is similar to a previous study that revealed that bleb formation preceded increases in membrane permeability after acute naphthalene injury to Clara cells in vivo. This finding suggests that JP-8-induced Clara cell blebbing may be a uniform response observed in other P-450 bioactivated toxicants through one or several possible mechanisms, such as transformation of preexisting microvilli, changes in the cortical cytoskeleton, and disturbances in both thiol and calcium homeostasis within the injured cell. However, these alveolar alterations were not observed in the S-8 group, which exhibited mild ultrastructure alterations in bronchiolar Clara cells. Our findings suggest that there were site-selective and cell-specific epithelial responses after the two jet fuel exposures. The morphometric analysis provided further evidence for differences in targeted cellular effects of the two fuels: bronchioles in S-8 and alveoli/terminal bronchioles in JP-8. These data suggest that surfactant synthesis and (or) secretory processes were affected by JP-8 exposure, as indicated by an increase in volume density of surfactant-producing lamellar bodies. Consequently, the type II epithelial cells may compromise their normal capabilities to control the volume and composition of the epithelial lining fluid and to maintain the integrity of the alveolar wall. However, S-8 exposure significantly decreases the volume density of mitochondria, as indicative of the onset of cell swelling, when compared to the JP-8 group.

Note that the two jet fuels also exhibited a similar bronchiolar epithelial response, as indicated by the decreases in vacuole volume density of secretory granules in Clara cells. This phenomenon was similar to previous data, in that Clara cells' secretory granules disappeared 1 hr after propranolol plus isproterenol administration, and the granules reappeared at 4 hours. As we know, Clara cells represent the predominant secretory cell containing characteristic secretory granules and constitute up to 80% of the epithelial cell populations of the distal airways. Clara cell secretory protein deficiency after jet fuel exposure could lead to alterations in the composition of the epithelial lining fluid and to enhanced susceptibility to environmental agents. As for the relevance of human toxicity, however, the relevance of functional alteration of Clara cells remains to be addressed because there are species-dependent differences histologically and physiologically, such as cell density and metabolizing capacity.

The differences for two jet fuels may be attributed to the chemical differences between the two fuels *per se* or their dynamics in airways such as diffusion and deposition, consequently resulting in differences of ventilatory function of the small respiratory tract between S-8 and JP-8 exposure. JP-8 consists of a hydrocarbon-rich kerosene base commercial jet fuel (Jet A) plus additives [corrosion inhibitor DC1-4A, antistatic compound Stadis 450, and an icing inhibitor diethylene glycol monomethyl either (DIGME)] that make up less than 2% of the formulation. S-8, an aliphatic hydrocarbon (HC) fuel, is synthesized using the Fischer-Tropsch (FT) synthetic fuel process. Synthetic gas produced from natural gas, coal, or biomass is converted, using heat and pressure, into a clean-burning liquid fuel made of aliphatic HCs (hydrocarbons).

The two studies demonstrated that exposure of mice to both S-8 and JP-8 caused the morphological/morphometric and functional alterations around the permissible exposure limit. Data showed that the relative differences in epithelial deposition and response between the two jet fuels leads to different adverse effects in the distal lungs, possibly through their deposition patterns and chemical compositions. However, additional experiments with the dose-effect design are needed to confirm the effects generated by these two fuels over a wider range of jet fuel concentrations. Moreover, it is important to correlate human studies with actual field exposure in order to develop the best formulation of S-8 for both Air Force personnel and aircraft.

SUMMARY

This chapter provided an overview of the recent findings in our lung injury laboratory or from collaborations with us. Emphasis was given to the respiratory effects of JP-8 as well as an emerging synthetic-8 fuel (S-8). It appears that additional adverse effects of JP-8 on the respiratory system have been found in vivo and in vitro. Moreover, the molecular mechanisms responsible for inflammatory response have been explored with new molecular tools, such as proteomics and microarray. In addition, pulmonary evaluation of permissible exposure limit of S-8 has been initiated in mice, as measured by pulmonary function, airway epithelial barrier integrity, and pathohistology. The pulmonary effects in distal lung from a low-level exposure to JP-8 and S-8 have been compared. However, the action mechanisms responsible for these effects remain to be studied, especially below the current PEL and STEL. Studies of exposure, absorption, and metabolism of JP-8 need to be explored despite the complex nature of the studies. In exposure analysis, ambient exposure and breathing-zone concentrations of JP-8 and its constituents (such as naphthalene and toluene) should be determined in relation to lung burden through assays of biological samples for the concentrations and constituents of the jet fuel. Therefore, the need exists for additional studies with appropriate cellular and molecular strategies that should generate enough data as required for human risk assessment.

REFERENCES

[1] Hays, A.M., Parliman, G., Pfaff, J.K., Lantz, R.C., Tinajero, J., Tollinger, B., et al. Changes in lung permeability correlate with lung histology in a chronic exposure model. *Toxicol. Indus. Health,* 1995; 11(3): 325–336.

[2] Robledo, R.F., Young, R.S., Lantz, R.C., and Witten, M.L. Short-term pulmonary response to inhaled JP-8 jet fuel aerosol in mice. *Toxicol. Pathol.*, 2000; 28(5): 656–663.

[3] Wong, S.S., Hyde, J., Sun, N.N., Lantz, R.C., and Witten. M.L. Inflammatory responses in mice sequentially exposed to JP-8 jet fuel and influenza virus. *Toxicology,* 2004; 197(2): 139–147.

[4] Espinoza, L.A., Tenzin, F., Cecchi, A.O., Chen, Z., Witten, M.L., and Smulson, M.E. Expression of JP-8-induced inflammatory genes in AEII cells is mediated by NF-kappaB and PARP-1. *Am. J. Respir. Cell Mol. Biol.,* 2006; 35(4): 479–487.

[5] Wang, S., Young, R.S., Sun, N.N., and Witten, M.L. In vitro cytokine release from rat type II pneumocytes and alveolar macrophages following exposure to JP-8 jet fuel in co-culture. *Toxicology,* 2002; 173(3): 211–219.

[6] Witzmann, F.A., Bauer, M.D., Fieno, A.M., Grant, R.A., Keough, T.W., Kornguth, S.E., et al., Proteomic analysis of simulated occupational jet fuel exposure in the lung. *Electrophoresis,* 1999; 20(18): 3659–3669.

[7] Robledo, R.F., Barber D.S., and Witten M.L. Modulation of bronchial epithelial cell barrier function by in vitro jet propulsion fuel 8 exposure. *Toxicol. Sci.*, 1999; 51(1): 119–125.

[8] Wong, S.S., Vargas, J., Thomas, A., Fastje, C., McLaughlin, M., Camponovo, R., et al. In vivo comparison of epithelial responses for S-8 versus JP-8 jet fuels below permissible exposure limit. *Toxicology,* 2008; 254(1-2): 106–111.

[9] Robledo, R.F., and Witten, M.L. NK1-receptor activation prevents hydrocarbon-induced lung injury in mice. *Am. J. Physiol.,* 1999; 276(2 Pt. 1): L229–L238.

[10] Wang, S., Young, R.S., and Witten, M.L. Age-related differences in pulmonary inflammatory responses to JP-8 jet fuel aerosol inhalation. *Toxicol. Indus. Health,* 2001; 17(1): 23–29.

[11] Herrin, B.R., Haley, J.E., Lantz, R.C., and Witten, M.L. A reevaluation of the threshold exposure level of inhaled JP-8 in mice. *J. Toxicol. Sci.,* 2006; 31(3): 219–228.

[12] Pfaff, J.K., Tollinger, B.J., Lantz, R.C., Chen, H., Hays, A.M., and Witten, M.L. Neutral endopeptidase (NEP) and its role in pathological pulmonary change with inhalation exposure to JP-8 jet fuel. *Toxicol. Indus. Health,* 1996; 12(1): 93–103.

[13] Pfaff, J., Parton, K., Lantz, R.C., Chen, H., Hays, A.M., and Witten, M.L. Inhalation exposure to JP-8 jet fuel alters pulmonary function and substance P levels in Fischer 344 rats. *J. Appl. Toxicol.,* 1995; 15(4): 249–256.

[14] Hays, A.M., Lantz, R.C., and Witten, M.L. Correlation between in vivo and in vitro pulmonary responses to Jet Propulsion Fuel-8 using precision-cut lung slices and a dynamic organ culture system. *Toxicol. Pathol.,* 2003; 31(2): 200–207.

[15] Fernandez, A., Wendt, J.O., Wolski, N., Hein, K.R., Wang, S., and Witten, M.L. Inhalation health effects of fine particles from the co-combustion of coal and refuse derived fuel. *Chemosphere,* 2003; 51(10): 1129–1137.

[16] Drake, M.G., Witzmann, F.A., Hyde, J., and Witten, M.L. JP-8 jet fuel exposure alters protein expression in the lung. *Toxicology,* 2003; 191(2-3): 199–210.

[17] Espinoza, L.A., Valikhani, M., Cossio, M.J., Carr, T., Jung, M., Hyde, J., et al. Altered expression of gamma-synuclein and detoxification-related genes in lungs of rats exposed to JP-8. *Am. J. Respir. Cell. Mol. Biol.,* 2005; 32(3): 192–200.

[18] Stoica, B.A., Boulares, A.H., Rosenthal, D.S., Iyer, S., Hamilton, I.D., and Smulson, M.E. Mechanisms of JP-8 jet fuel toxicity. I. Induction of apoptosis in rat lung epithelial cells. *Toxicol. Appl. Pharmacol.,* 2001; 171: 94–106.

[19] Boulares, A.H., Contreras, F.J., Espinoza, L.A., and Smulson, M.E. Roles of oxidative stress and glutathione depletion in JP-8 jet fuel-induced apoptosis in rat lung epithelial cells. *Toxicol. Appl. Pharmacol.,* 2002; 180: 92–99.

[20] Wong, S.S., Thomas, A., Lantz, R.C., and Witten, M.L. Pulmonary evaluation of permissible exposure limit of Syntroleum S-8 synthetic jet fuel in mice. *Toxicologic. Sci.,* 2009; 109(2): 312–320.

4 Neurotoxicological and Neurobehavioral Effects from Exposure to Jet Fuels

Glenn D. Ritchie

CONTENTS

INTRODUCTION

Well over 2.5 million military and civilian personnel per year (over 1 million in the United States) are occupationally exposed to the military fuels Jet Propulsion Fuel-8 (JP-8), JP-8+100, or Jet Propulsion Fuel-5 (JP-5), or to the civil aviation fuel equivalents Jet A, Jet A-1, or Aviation Gas (AVGAS). Research is presently ongoing to introduce S-8 synthetic jet fuels (S-8 Jet Fuel, Synthetic Jet A, and Synthetic Jet A-1), 50/50 blends of synthetic jet fuel (C_{7+}), and kerosene. Over 60 billion gallons of kerosene-based jet fuels are annually consumed worldwide (over 26 billion gallons in the United States), including more than 5 billion gallons of JP-8 by the militaries of the United States and other North Atlantic Treaty Organization (NATO) countries. JP-8, for example, may be the largest single chemical exposure in the

U.S. military (2.53 billion gallons in 2000), while Jet A and Jet A-1 are among the most common sources of non-military occupational chemical exposure (Ritchie et al., 2001b, 2003).

Kerosene-based hydrocarbon fuels are complex mixtures of 228+ aliphatic and aromatic hydrocarbon compounds (C_6 to C_{30+}; possibly 2,000+ isomeric forms), including varying concentrations of potential toxicants such as benzene, n-hexane, toluene, xylenes, ethylbenzene, trimethylpentane, methoxyethanol, naphthalenes (including polycyclic aromatic hydrocarbons [PAHs]), and certain other C_9 to C_{12} fractions (n-propylbenzene, trimethylbenzene isomers, etc.). The vast majority of these constituent chemicals have been minimally tested for human or animal neurotoxicity, and virtually no research has been conducted on the effects from concurrent exposure to more than one of the constituents. Different jet fuels, differing primarily in carbon fraction constituents, also may contain up to 0.5% (v/v) additive packages (e.g., static dissipaters, corrosion inhibitor/lubricity improvers, icing inhibitors, metal deactivators, antioxidants, detergents/dispersants, biocides, octane enhancers, ignition controllers, etc.) that differ to meet fuel performance requirements, further providing unique exposure scenarios. In some cases, these additive formulations may contain additional benzene or toluene, or may contain proprietary chemicals with undetermined toxicity potential. JP-8 formulations, for example, can vary substantially in content based on manufacturer, manufacturing season and standards, and the additive package requested. Thus, most workers are exposed dermally and by inhalation to a number of different jet fuel formulations during a career, as well as to a wide diversity of other hydrocarbon and non-hydrocarbon industrial chemicals (solvents, paints, cleaners, etc.). Typical inhalation exposures emphasize only the lighter fractions of the jet fuel ($<C_{10}$), while significant dermal exposures occur typically to the nonvolatile, heavier fractions of the fuel that do not easily volatilize from the skin. Hence, any conclusion about the human neurotoxicity of a particular hydrocarbon fuel may be confounded by effects of numerous controllable and uncontrollable exposure factors. Even animal research, where conditions can be more closely controlled, have been found to depend on such factors as whether the inhalation exposure occurs to vapor, aerosol, or vapor-aerosol phase (Ritchie et al., 2001b, 2003). There have also been a few published studies suggesting toxicological additivity, synergism, or antagonism when hydrocarbon fuel formulations contain or omit specific carbon fractions (Andrews et al., 1977; Olsen et al., 1985; Yin et al., 1987). Considering the extensive acute and repeated human exposure to jet fuels, there has been minimal research conducted to evaluate possible neurotoxicological, neurohistopathological, and neurobehavioral consequences. There has been far more intensive human and animal research conducted to evaluate the neurotoxicological effects of single and combined exposures to a number of the common chemical constituents of hydrocarbon fuels, including xylenes, toluene, benzene, naphthalene, and n-hexane. Much of this research was reviewed by Ritchie et al. (2001b, 2003) and allows guarded conclusions about the neurotoxicity expected when humans are exposed to hydrocarbon fuels containing the typically lowered concentrations of these well-known neurotoxicants. Ritchie et al. (2001b, 2003) concluded that, in many cases, deficits attributed to neat jet fuel exposures may parallel those observed when exposure occurred to only one or two chemical constituents.

Hydrocarbon fuel exposures may occur once or repeatedly to raw (neat) fuel, vapor, aerosol or vapor-aerosol phases, and fuel combustion products by dermal, pulmonary, or (infrequently) oral ingestion routes. While there is little epidemiological evidence for fuel-induced death, cancer, developmental deficits, or other serious organic disease in fuel-exposed workers, large numbers of self-reported health complaints in this cohort appear to justify study of more subtle health consequences (Pleil and Smith, 2000; Ritchie et al., 2001b, 2003). A number of recently published studies have reported acute or persisting health effects from acute, subchronic, or chronic exposure of humans or animals to kerosene-based hydrocarbon fuels or to fuel combustion products. This review provides an in-depth summary of human, animal, and *in vitro* studies of neurotoxicological, neurohistopathological, or neurobehavioral effects from exposure to jet fuels.

JET FUEL EXPOSURE STUDIES IN HUMANS AND ANIMALS

There is an extensive literature reporting the acute and long-term neurobehavioral effects of (high-level) occupational or laboratory exposure to the major known neurotoxic constituents of jet fuels (e.g., benzene, toluene, xylenes, n-hexane, etc.) (Hanninen et al., 1976; Andrews et al., 1977; Gamberale et al., 1978; Ikeda and Miyake, 1978; Andersson et al., 1981, Glowa, 1981; 1983; Formazzari et al., 1983; Lazar et al., 1983; Miyake et al., 1983; Savolainen et al., 1984; Rosenberg et al., 1988; Foo et al., 1990; Gralewicz et al., 1995; Koschier, 1999; Wada, 1999; Gralewicz and Wiaderna, 2001). In two separate review articles, Ritchie et al. (2001b, 2003) provided substantial evidence that many of the neurobehavioral and neurohistopathological effects identified in humans for acute or chronic exposure to jet fuels can be related to studies of exposure to one or more of the hydrocarbon constituents (generally C_6 to C_{10}, and especially the aromatic hydrocarbons) of the fuel mixtures.

There is a very limited scientific literature evaluating human or animal neurobehavioral toxicity from exposure to neat JP-8 or JP-5, and virtually no data for such fuels as Jet A, Jet –A-1, AVGAS, and the new S-8 synthetic jet fuels. The majority of the human exposure data comes from accidental acute exposure of a few individuals to high fuel concentrations. The vast majority of the chronic exposure data derives from epidemiologic studies of military or civil aviation mechanics or manufacturing personnel who are occupationally exposed to various jet fuels and exhaust byproducts, as well as to a wealth of other industrial/military chemicals. While neurobehavioral animal studies are generally limited to inhalation (vapor, aerosol, or aerosol-vapor), dermal or oral gavage exposures or oral gavage treatments, human occupational exposure studies generally include simultaneous inhalation and dermal routes. As will be shown, the route of exposure (inhalation versus dermal) as well as the predominant exposure phase (aerosol versus vapor, versus aerosol-vapor) may be critically related to the neurobehavioral consequences. It is also important to note that toxicological studies of jet fuels, commonly utilizing the oral gavage route of administration, may tend to minimize CNS (central nervous system) effects that might be expected to occur with inhalation exposures (Ritchie et al., 2001b, 2003).

ANIMAL STUDIES

Cognitive and Neurobehavioral Studies in Animals

Most recently, Baldwin et al. (2007) exposed Fischer Brown Norway rats for 1 hour/day for 5, 10, 15, or 20 days to low concentrations of aerosolized-vapor JP-8 by nose-only inhalation. Treated rats and (free-moving or sham-restrained) controls were tested for neurobehavioral effects using the U.S. Environmental Protection Agency Functional Observational Battery (EPA FOB) (Moser and MacPhail, 2007). Exploratory ethological factor analysis identified only three salient factors: (1) CNS excitability, (2) autonomic 1 signs, and (3) autonomic 2 signs. Exposed rats, as compared to controls, exhibited more rearing and other hyperarousal behaviors, replicating prior JP-8 exposure findings. Exposed animals also showed increasing (but rapidly decelerating) stool output (autonomic 1 sign), and a significant increasing linear trend for urine output (autonomic 2 sign). Sub-groups of the exposed and control rats were assayed for CNS neurotransmitter levels from seven brain regions. Hippocampal dihydroxyphenylacetic acid (DOPAC) was significantly elevated after 4-week JP-8 aerosolized vapor exposures compared to the control groups, suggesting increased dopamine release and metabolism. Baldwin et al. (2007) reported that the observed neurobehavioral changes did not appear until weeks 3 and 4 of exposure, suggesting the need for longitudinal studies to determine if these behaviors occur due to cumulative exposure or result from behavioral sensitization related to repeated exposure to aerosolized vapor JP-8.

Previously, Baldwin et al. (2001) exposed male rats for 28 days to aerosolized vapor JP-8 or to (sham-confined) fresh room air. In addition to the EPA FOB, the Morris Water Maze (MWM) swim test was used to evaluate visual and spatial learning, as well as possible memory effects. The MWM test included memory for the original target location following 15 days of JP-8 exposure, as well as a 3-day "new target location" learning paradigm implemented the day (day 29) following the final day of exposure. JP-8-exposed animals had a significant mean weight loss by the second week of exposure compared with control rats. On the EPA FOB, rats exposed to JP-8 exhibited significantly greater rearing (hyperactivity) and less grooming behavior than did controls during open-field testing. On the MWM test, exposed rats swam significantly faster than controls during the new target location training and testing, possibly supporting the increased activity reported during EPA FOB testing. There were no significant differences between the exposed and control-group MWM performances during acquisition, retention, or learning in either the visual discrimination or spatial version of the MWM task. The data suggest that although visual discrimination, spatial learning, and memory were not disrupted by brief daily exposures to JP-8 in the aerosol/vapor phase, arousal indices and activity measures were significantly increased.

Ritchie et al. (2001a) exposed male Sprague-Dawley rats by whole-body inhalation to vapor-phase JP-8 at approximately the highest vapor concentration possible without formation of aerosol (1,000 ± 10% mg/m^3); to 50% of this concentration (500 ± 10% mg/m^3); or to treated room air for 6 hours/day, 5 days/week, for 6 weeks (180 hours total). Although two male rats died of apparent kidney complications during the study, no other change in the health status of exposed rats was observed, including rate of weight gain. Following a 65-day period of rest, rats were evaluated for their capacity to learn and perform a series of operant tasks. These tasks ranged in difficulty from learning a simple food-reinforced lever pressing response to learning a highly complex Incremental Response Acquisition (IRA) task in which subjects were required to emit up to four-response chains of pressing three different levers (e.g., the rat must press levers Center, Right, Left, then Center again for reinforcement). It was shown that repeated exposure to 1,000-mg/m^3 JP-8 vapor induced statistically significant deficits in acquisition or performance of moderately difficult or difficult tasks, but not simple learning tasks, as compared to those animals exposed to 500 mg/m^3. Learning/performance of complex tasks by the 500-mg/m^3 exposure group generally exceeded the performance of control animals, while learning by the 1,000-mg/m^3 group was nearly always inferior to controls, indicating possible neurobehavioral hormesis. Neurobehavioral hormesis, in this application, refers to the apparent capacity of JP-8 vapor to improve learning capacity (e.g., activate the CNS) in rats following lower concentration exposures, yet induce an opposite effect (learning deficits) with higher concentration exposures. These findings appear consistent with previously reported data for operant performance following acute exposure to certain hydrocarbon constituents of JP-8 (i.e., toluene, xylenes) (Ritchie et al., 2001b, 2003). It might be concluded that repeated lower-level exposures to JP-8 vapor (e.g., 500 mg/m^3) may "activate" the CNS, including the capacity to learn complex tasks, while repeated exposure to higher vapor concentrations (e.g., 1,000 mg/m^3) may result in significant learning deficits on complex tasks. Examination of regional brain tissues from vapor-exposed rats indicated significant changes in levels of dopamine (DA) in the cerebral cortex and 3,4-dihydroxyphenylacetic acid (DOPAC) in the brainstem, measured as long as 180 days post exposure, as compared to controls. Based on a comparison with the previously discussed MWM learning data provided by Baldwin et al. (2001), using aerosolized vapor delivery of JP-8, it would appear that exposure of rats to JP-8 in vapor phase may induce measurable learning enhancements (500-mg/m^3 group) or deficits (1,000-mg/m^3 group) on tasks of moderate and high difficulty. Similar to the results reported by Baldwin et al. (2007), JP-8-induced alterations in the dopamine neurotransmitter system were observed. The authors hypothesized that "lower" dose JP-8 exposures may have activated portions of the CNS (up to 200 days post exposure), an effect previously reported for operant responding following acute exposures to toluene (Glowa, 1981; Wada et al., 1999).

Rossi et al. (2001) compared the possible neurobehavioral toxicity of repeated exposure of male rats to JP-5 vapor versus JP-8 vapor. In this study, groups of 32 male Sprague-Dawley rats each were exposed for 6 hours/day, 5 days/week for 6 weeks (180 hours) to JP-8 jet fuel vapor (1,000 ± 10% mg/m^3), JP-5 vapor (1,200 ± 10% mg/m^3), or room air control conditions. Following a 65-day rest period, rats completed ten tests selected from the Neurobehavioral Toxicity Assessment Battery (NTAB) to evaluate changes in performance capacity. The NTAB is similar to the EPA FOB, but contains an expanded number of neurobehavioral tests that do rely upon subjective scoring (Rossi et al., 1996). Repeated exposure to JP-5 vapor resulted in significant effects on only one test, forelimb grip strength (FGS). The significantly increased FGS observation in rats exposed to 1,200 mg/m^3 JP-5 vapor might suggest general CNS activation, including increased neuromuscular integrity. Repeated exposure to JP-8 vapor resulted in a significant difference versus controls on appetitive reinforcer approach sensitization (ARAS). In this test, rats are placed for several minutes in a rectangular open field in which two opposite corners are baited with raw hamburger meat, while the other two corners are non-baited. The percentage of time a subject spent in close proximity to the baited corners was used to measure ARAS. The number of times each rat crosses a centerline divider in the apparatus was used to measure total locomotor activity. The increased ARAS scores in JP-8-exposed rats suggest general approach sensitization to novel stimuli (e.g., raw hamburger meat) and included increased general locomotor activity, consistent with previously described studies of JP-8. Rats were further evaluated for concentrations of major neurotransmitters and metabolites in five brain regions and in the blood serum. Examination of neurotransmitter levels in the brains and blood of JP-8-exposed rats indicated (1) a significantly increased level of DA in the cortex; (2) a significantly increased level of DOPAC, the major metabolite of DA, in the brainstem; and (3) a decreased level of the serotonin (5-HT) metabolite 5-HIAA (5-hydroxyindole acetic acid) in the serum as long as 200+ days post exposure. This result was interpreted as possibly indicating sensitization of the forebrain dopaminergic system. The authors hypothesized that this fuel-induced change in neurobehavioral capacity may reflect long-term modulation of brainstem inhibitory systems.

Nordholm et al. (1999) exposed male Sprague-Dawley rats to JP-4 vapor (2,000 mg/m^3) for 6 hours/day for 14 consecutive days and then measured changes in performance capacity using a battery of eight tests selected from the NTAB (Ritchie et al., 1995). The authors reported significantly increased approach sensitization to an appetitive stimulus (ARAS to raw hamburger meat), significantly reduced prepulse inhibition (PPI) of the acoustic startle response, significantly decreased mean time to tail flick nociceptive responses, and significantly decreased spontaneous, horizontal locomotor activity in JP-4-exposed rats, as compared to controls. These changes in performance capacity, measured 35 to 80 days post exposure, were consistent with persisting, significant increases in levels of 5-HT and the major metabolite 5-HIAA in several brain regions, as well as in the blood serum. Because JP-4 contains higher aromatic hydrocarbon fractions (e.g., benzene, toluene, xylenes, etc.) than JP-8, these results appear consistent with previously discussed neurobehavioral effects studies. It may be hypothesized that the higher xylene and toluene contents in JP-4 resulted in "activated" responses on the ARAS test (increased sensitization to a novel stimulus) and the tail flick nociception test (more rapid tail withdrawal), and reduced PPI to an auditory startle stimulus (a greater startle response). The reduced spontaneous locomotor activity observed in an open field test was perhaps unexpected based on the published research results for JP-8, and may have reflected a depressive effect during the 60-minute test from exposure to the higher levels of benzene, toluene, and xylenes present in JP-4.

Bogo et al. (1984) compared the relative neurobehavioral toxicity, for U.S. Navy applications, of petroleum-derived (P-JP-5) versus shale-derived (S-JP-5) JP-5 jet fuel. S-JP-5 contained higher percentages of alkanes, partially hydrogenated polynuclear aromatics, and other nitrogen-containing compounds than did P-JP-5. Adult Sprague-Dawley rats were either orally (3 to 8 ml/kg or 24 ml/kg) dosed one time, or were exposed by inhalation for 30 days (1,100 mg/m^3) to JP-5 vapor/aerosol. Among the orally gavaged groups, subjects lost up to 30% body weight within 4 days post exposure. This change, however, was reversed to normal levels by 7 days post exposure. Water volume intake, among both the S-JP-5 and P-JP-5 groups, was reduced by up to 30% during the first 2 days post exposure, but increased by day 4 to significantly above control levels. However, S-JP-5-exposed rats exhibited water volume intakes of nearly 200% of baseline on day 4 post exposure, and continued to show increased water consumption throughout the remainder of the 36-day study. Additionally, P-JP-5-exposed rats exhibited significantly increased home cage activity for 2.5 to 6 hours post exposure, but a subsequent significant reduction in home cage activity, relative to controls, during the remainder of the study. Additional studies of rotarod performance, somatosensory evoked potential (SEP), and shock-induced aggression produced no differences between fuel-exposed and control animals. From the inhalation studies, a significant polydipsia (increased water intake) was recorded, relative to controls, on exposure day 8 for P-JP-5-exposed rats, and on day 9 for S-JP-5-exposed animals. From days 13 to 36, both exposure groups consumed significantly more water than did controls. While the increased water volume may have reflected fuel-induced renal toxicity, it remains unknown if this observation may have reflected a behavioral sensitization of the CNS. This apparent polydipsea may be consistent with the fuel-induced changes in appetitive approach (ARAS) reported by Rossi et al. (2001).

No histopathological changes were noted in the nervous system of rats or dogs exposed to up to 100 mg/m^3 deodorized kerosene vapor for 6 hours/day, 5 days/week for 13 weeks (Carpenter et al., 1975, 1976). No significant clinical signs of toxicity were evident in mice exposed continuously to airborne JP-8 (500 or 1,000 mg/m^3) for 90 days, except for an increased incidence of fighting (Mattie et al., 1991). A number of animal studies of kerosene-based jet fuel neurotoxicity have been published. Mattie et al. (1995) reported no clinical signs of neurotoxicity, as assessed using an FOB, in female rats treated orally with 0 to 2,000 mg/kg/day JP-8 from gestational days 6 to 15. Also this study indicated no histopathological changes in the brains or sciatic nerves of male rats administered 750; 1,500; or 3,000 mg/kg/day JP-8 by oral gavage for 90 days.

Ocular and Dermal System Toxicity in Animals

Kinkead et al. (1992a) reported JP-8 ocular exposure (0.1 ml/eye) of female rabbits for 2 minutes to be non-irritating. Similarly, dermal exposure of female rabbits to 0.5 ml JP-8 for 24 hours was found to be only mildly irritating (mild erythma). Finally, dermal sensitization testing of male guinea pigs indicated a sensitization response in 10% of animals (weak sensitizing potential). The authors also reported minimal eye irritation (Draize scoring technique) in rabbits following corneal surface exposure to 0.1 ml neat Jet A for 20 to 30 seconds.

Wolfe et al. (1996) exposed male or female rats orally to 5 mg/kg JP-8 or JP-8+100 (Mobil or BetzDearborn additive packages), male rabbits dermally to 0.5 ml/kg JP-8 or JP-8+100, male guinea pigs to 0.1 ml JP-8 or JP-8+100 for 10 days, and male and female rats by whole-body inhalation to 3,700 mg/m^3 JP-8 or

JP-8+100 vapor for 4 hours. With oral exposures, there were no deaths or persisting signs of toxicity observed, although post-exposure lethargy, shallow breathing, and minor weight loss were commonly observed. In rabbits, 4-hour dermal exposure with JP-8 resulted in slight erythema, although animals exposed to JP-8+100 were generally normal. Rats exposed by whole-body inhalation for 4 hours to 3,700 mg/m^3 JP-8 or JP-8+100 (Mobil additive package) vapor exhibited eye and upper respiratory irritation. All exposed animals survived for 14 days post exposure with no obvious lesions.

McGuire et al. (2000) evaluated whether brief (1 hour/day for 7 days) inhalation exposure of mice to 1,000 or 2,500 mg/m^3 JP-8+100 (containing 2,6-di-*tert*-butyl-4-methyl-phenol as the antioxidant) would result in possible toxicity of the retina. It was shown that JP-8+100 exposures resulted in significant increases in reactivity to anti-glutathione-*S*-transferase *mu*-1 (GSTM1) antibodies in the radial glial Muller cell perikarya and fibers, as compared to controls. It was concluded that increased expression of GST detoxification isoenzymes in the retinas of JP-8+100-exposed mice indicated possible toxicity in this sensory organ, with the possibility of resultant increased passage of xenobiotics across the retinal-brain barrier.

Peripheral Auditory System Effects in Animals

There is a significant literature detailing auditory system deficits in animals exposed repeatedly to doses of toluene higher than those found typically in jet fuels, especially when the chemical exposures are accompanied by exposure to noise or other stressors (Crofton et al., 1994; Campo et al., 1999; Hougaard et al., 1999, 2005; Lataye et al., 1999; McWilliams et al., 2000).

There are two published animal studies suggesting that JP-8 exposure can result in auditory system deficits. Fechter et al. (1990) hypothesized that exposure to the combination of ethylbenzene and toluene, two constituents of jet fuels, may promote susceptibility to noise-induced hearing loss. Rats received ethylbenzene and toluene by inhalation at concentration levels expected from typical human occupational jet fuel exposures. Half of the exposed rats were subjected simultaneously to an octave band of noise (OBN) of 93 to 95 decibels (dB). Another group received only the noise exposure that was designed to produce a small but permanent auditory impairment, while the control group was unexposed. In two separate experiments, exposures occurred either repeatedly on five successive days for 1 week or for 5 days on 2 successive weeks to 4,000 mg/m^3 total hydrocarbons for 6 hours. The concentration of toluene was approximately 400 ppm, and the concentration of ethylbenzene was approximately 660 ppm. Impairments in auditory function were assessed using distortion product oto-acoustic emissions and compound action potential testing. Following completion of these tests, the organs of Corti were dissected to permit evaluation of hair cell loss. The uptake and elimination of the solvents was assessed by harvesting key organs at two time points following ethylbenzene and toluene exposure from additional rats not used for auditory testing. Similarly, glutathione (GSH) levels were measured to determine if oxidative stress from the solvent-noise exposures were possibly involved in the deficits observed. Ethylbenzene and toluene exposure by itself at 4,000 mg/m^3 for 6 hours did not impair cochlear function or yield a loss of hair cells. However, when combined with a 93-dB OBN exposure, solvent and noise exposure resulted in a loss of auditory function and a clear potentiation of outer hair cell death that exceeded the loss produced by noise alone. No evidence was found for a loss in total GSH in lung, liver, or brain as a consequence of ethylbenzene and toluene exposure, suggesting that the observed effects did not result from oxidative stress.

In a follow-up study, Fechter et al. (2007) exposed rats to JP-8 (1,000 mg/m^3) by nose-only inhalation for 4 hours. Fifty percent of the rats were immediately subjected to an OBN ranging between 97 and 105 dB in different experiments. Exposures occurred either on one day or repeatedly on five successive days. Impairments in auditory function were assessed using distortion product oto-acoustic emissions and compound action potential testing. A single JP-8 exposure at 1,000 mg/m^3 did not disrupt auditory function. However, exposure to JP-8 and noise produced a disruption in outer hair cell function. Repeated 5-day JP-8 exposures at 1,000 mg/m^3 for 4 hours produced impairment of outer hair cell function that was most evident at the first post-exposure assessment time. Partial, although not complete recovery was observed over a 4-week post-exposure period. Major constituents of the fuel were largely eliminated in all tissues by 1-hour post exposure, with the exception of fat, minimizing the possibility that systemic toxicant accumulation accounted for the observed effects. In this study, JP-8 exposure resulted in a significant depletion of total GSH that was observable in liver, with a nonsignificant trend toward depletion in the brain and lung, raising the possibility that the promotion of noise-induced hearing loss by JP-8 may have resulted at least partially from oxidative stress.

HUMAN STUDIES

Cognitive and Neurobehavioral Studies in Humans

In a very early study of six volunteers, slight olfactory fatigue was induced in three subjects, and one reported "tasting something," following a 15-minute exposure to 140 mg/m^3 deodorized kerosene vapor (Carpenter et al., 1976). Acute exposure to kerosene in humans, as a function of increasing dose, has been reported to result progressively in irritability, restlessness, ataxia, drowsiness, convulsions, semi-consciousness, unconsciousness, or death (Aldy et al., 1978; Majeed et al., 1981; Akamaguna and Odita, 1983; Mahdi, 1988; Dudin et al., 1991; Lucas, 1994). Similarly, long-term exposure to low kerosene vapor concentrations has been reported to produce nonspecific CNS symptoms such as nervousness, loss of appetite, and nausea that may not be related to possible hypoxia effects (WHO Working Group, 1985).

The human exposure literature for jet fuels generally began in Sweden in the late 1970s when a series of studies were conducted to determine the effects of chronic exposure of aircraft manufacturing personnel to European military fuels and other hydrocarbon solvents (Knave et al., 1976, 1978, 1979; Mindus et al., 1978; Struwe et al., 1983; Holm et al., 1987). These studies were complicated by a number of uncontrolled factors, and included chronic exposure to jet fuels containing high levels of aromatic fractions (e.g., benzene, toluene, xylenes, etc.), n-hexane, and unleaded gasoline, as well as by poorly documented exposures to various hydrocarbon solvents.

Thus, the dramatic health effects reported in these studies cannot necessarily be applied to modern jet fuel exposure scenarios. In these studies, the neurobehavioral effects of acute and chronic exposure to Swedish Armed Forces MC 75 and MC 77 ("JP-4-like") jet fuels were evaluated in Scandinavian jet engine manufacturing and installation plant workers. In this plant, jet engines were both manufactured and tested, such that the majority of workers (controls) seldom or never experienced jet fuel, while employees in the fuel system manufacturing, installation, and testing areas were exposed repeatedly by inhalation or dermal routes. In the first study (Knave et al., 1976), based on personal interviews and company records, 29 workers were determined to be "considerably" exposed to jet fuel for a minimum of 5 years (between 1960 and 1974) and to be without existence of chronic disease that might account for symptoms of neuroasthenia (fatigue, depressed mood, lack of initiative, dizziness, sleep disorders) or polyneuropathy. Based on spectral parameter analysis (SPA) of the EEG (electroencephalogram) from exposed workers, the authors concluded that repeated jet fuel exposure may modulate thalamic control of cortical activity, resulting in a decreased frequency stability and increased time variability of activity emanating from these units. Of these 29 workers, 13 were assigned to Group A, the "high" exposure group (exposed to high concentrations of jet fuel from a maximum of 1 to 2 hours/day, 5 days/week, to a minimum of 20 to 30 minutes, once every 3 weeks), and the remaining 16 were assigned to Group B, the "lower" exposure group (20- to 30-minute- exposure for a maximum of once per month). In the worst-case scenario, workers spent approximately 50% of the workday 20 to 40 cm from objects perfused in jet fuel, or from 1 to 2 hours/day with the head and neck within a semi-enclosed tank containing residual jet fuel. The estimates of approximate exposure concentrations were not provided, and no control group was evaluated, but it was known that workers never used respiratory protection devices. For acute and chronic exposure to jet fuel, the symptoms listed in Table 4.1 were either self-reported or determined by medical examination.

The results of this study would appear to indicate substantial incidences of self-reported and examination-based symptomology, especially in Group A. Due to the study design, there were no statistical comparisons between Groups A and B, or among exposure and appropriate control groups. To establish the validity of the previously discussed study, Knave et al. (1978) conducted an additional study providing approximate jet fuel exposure concentrations and a comparison of results from 30 jet-fuel-exposed workers, as compared to data from 60 nonexposed controls working in the same manufacturing plants. The 30 exposed workers were selected as being the "most heavily exposed," and consisted of 15 fuel system testers, 7 engine testers, and 8 mechanics (19 male workers and 11 male foremen) with an average of 17.1 years of occupational exposure (mean age = 46.4 years). Initially, 30 workers from the same plant were matched for general health status, age, and duration of employment. A second control group was subsequently selected and matched for age, education, duration of employment, and trade union involvement. Two quantitative air sampling studies indicated that fuel-exposed personnel were routinely exposed to an "average jet fuel concentration" of 128 to 423 mg/m^3 or 85 to 974 mg/m^3 with a range of 41 to 3226 mg/m^3, for a mean time of 10 to 97 minutes per operation (Knave et al., 1978; Struwe et al., 1983). The fuels used for testing in this plant were jet fuels

TABLE 4.1
Percentages of Workers Exposed to MC 75 and MC 77 with Self-Reported or Diagnosed Disorders

Symptom[a]	Group A (%)	Group B (%)	Groups A&B (%)
Acute Symptoms	**100**	**55**	**69**
Dizziness	77	31	52
Respiratory symptoms	46	19	31
Headache	23	31	28
Nausea	25	29	26
Palpation/thoracic oppression	23	13	17
Chronic Symptoms	**92**	**56**	**72**
Neuropathy symptoms	85	50	59
Distal paresthenia	77	25	48
Dizziness	77	19	45
Thoracic oppression	62	25	41
Depression/anxiety	62	25	41
Restless legs	62	19	38
Headache	31	31	31
Sleep disturbances	38	19	28
Respiratory symptoms	46	6	24
Irritability	23	19	21
Extremity pain	31	13	21
Memory impairment	15	13	14
Examination signs of neuropathy	85	50	65
Temp. sensibility (foot and knee cap)	69	44	55
Pain sensibility (foot and leg needle prick)	31	6	17
Discrimination sensibility	15	6	10
Tendon reflexes	0	19	10
Motor paresis	8	6	7
Touch sensibility	8	0	3
Joint kinesthesia	0	0	0

Source: Adapted from Knave, B., Persson, H.E., Goldberg, J.M., et al. *Scand. J. Work Environ. Health*, 2: 152–64, 1976; Ritchie, G.D., Still, K.R., Alexander, W.K., et al. *J. Toxicol. Environ. Health. B Crit. Rev.*, 4: 223–312, 2001.

[a] Disorders normally associated with the nervous system are boldfaced.

MC-75 and MC-77, combinations of kerosene and unleaded gasoline containing high concentrations of toluene, benzene, n-hexane, and xylenes at approximately 1% (weight/weight) each. When compared to the matched controls, fuel-exposed workers reported or exhibited, from personal interview or medical records, significantly greater chronic incidence of symptoms or medical diagnoses as listed in Table 4.2.

In the United States, the jet fuels used predominantly before 1992 were JP-4 (USAF and U.S. Army) and JP-5 (U.S. Navy). Since 1992 through 1996, JP-8 has been used predominately by the USAF and U.S. Army, while the U.S. Navy has maintained use of JP-5. There is very limited human clinical or experimental data available for JP-4 (40–50% kerosene: 50–60% unleaded gasoline) or JP-5 (kerosene based with minimal C_3 to C_6 fractions) exposures. Nuttall (1958) provided a 2-year report of the neurobehavioral consequences of documented USAF jet fuel exposures. Symptoms self-reported from such exposures included irritation of the eyes and nose, nausea, headache, visual disturbance, and "disturbance of consciousness." Davies (1964) reported the neurobehavioral effects of the accidental exposure of a USAF pilot to JP-4 (estimated 3,000 to 7,000 ppm) jet fuel vapor/aerosol during a routine flight. An experienced 32-year-old male pilot began breathing cabin air mixed with oxygen (21% to 100%, as a function of altitude) through a facemask. Due to the leakage of JP-4 vapor into the cabin air, the pilot began inhaling an unreported concentration of jet fuel soon after takeoff that was relieved only when he switched his oxygen regulator to 100% oxygen. Beginning approximately 7 minutes following takeoff, the pilot reported becoming groggy, weak, and very sleepy, and that the normal engine noise had become exceptionally intense. Subsequent neurological examination upon landing revealed a slight staggering on walking, a slight slurring of speech, a "jovial" demeanor, a positive Romberg (nondirectional) quotient for postural body sway, mild muscular weakness, and a possibly reduced nociception to painful stimuli. During his emergency descent for landing, the pilot reported being lost, demonstrated an erratic landing roll, and did not follow the instructions of landing tower personnel. In follow-up studies, the pilot reported feeling "abnormal" for about 36 hours post exposure. Similarly, Porter (1990) reported the case of intoxication of two military aviators by inhalation of JP-5 fuel vapors. It was hypothesized that raw fuel may have entered the cockpit during aircraft roll maneuvers. One or both aviators experienced burning eyes, nausea, fatigue, impairing of eye-hand coordination, euphoria, and self-reported memory defects when their cockpit air became mixed with JP-5 vapor/aerosol. During the emergency landing approach, the student aviator was reported to be laughing and euphoric. Porter (1990) described acute effects of JP-5 exposure of students and pilots that included difficulty in recalling emergency procedures, flight plan information, and, in the extreme case, personal information (i.e., his wife's name).

Koshier (1999) reported that kerosene mid-range distillates (C_9 to C_{16}) can induce CNS depression characterized by ataxia, hypoactivity, and prostration when inhaled in high concentrations. The author further reported that there were no detectable changes in CNS function with moderate dermal exposure to kerosene middle distillates.

Since the 1992 through 1996 conversion from predominant use of JP-4 jet fuel to kerosene-based JP-8 by the USAF and U.S. Army, there have been increased self-

TABLE 4.2
Acute and Chronic Symptoms or Deficits from Repeated Exposure to Jet Fuel (MC 75, MC 77) in European Aircraft Manufacturing Workers, as Compared to Controls

Symptom*	Fuel-Exposed Group (%)	Control Group (%)
Acute Symptoms		
Fatigue	**43**	**NA**
Dizziness	**50**	**NA**
Headache	**23**	**NA**
Nausea	**13**	**NA**
Palpation/thoracic oppression	13	NA
Respiratory	10	NA
Chronic Symptoms		
Gastritis	50	33
Fatigue	[a]43	[a]3
Polyneuropathy	**40**	**20**
Depression; lack of initiative	**[a]33**	**[a]3**
Dizziness	**[a]33**	**[a]5**
Palpation/thoracic oppression	[a]30	[a]5
Sleep disturbances	**[a]30**	**[a]5**
Eye irritancy	**[a]30**	**[a]6**
Headache	**17**	**3**
Memory impairment	**17**	**5**
Irritability	**13**	**5**
Respiratory tract irritancy	10	3
Sweating	10	3
Neurological Examination		
Polyneuropathy	**45**	**0**
EEG - Alpha amplitude	*a*	*a*
Simple reaction time - Across successive blocks	*a*	*a*
Perceptual speed: Bourdon–Wiersma Vigilance Test	*a*	*a*
Reaction time: addition	**NS**	**NS**
Memory: recognition	**NS**	**NS**
Manual dexterity: Santa Ana	**NS**	**NS**
Memory (reproduction)	**NS**	**NS**
CPRS Subscale (of 39 Items)		
Inner tensions	*a*	*a*
Compulsive thoughts	*a*	*a*
Phobias	*a*	*a*
Lassitude	*a*	*a*

Continued

TABLE 4.2 (*Continued*)
Acute and Chronic Symptoms or Deficits from Repeated Exposure to Jet Fuel (MC 75, MC 77) in European Aircraft Manufacturing Workers, as Compared to Controls

Symptom*	Fuel-Exposed Group (%)	Control Group (%)
Reduced sleep	*a*	*a*
Muscular tension	*a*	*a*
Learning	*a*	*a*
	Additional	
Diagnosed neurasthenia 24/30 <0.0001		
Diagnosed anxiety or depression 23/30 <0.0008		
Self-reported fatigue 13/30 <0.0001		
Self-reported depression 10/30 <0.0003		
Self-reported dizziness 10/30 <0.0010		
Palpitations/chest oppression 10/30 <0.003		
Self-reported sleep disturbances 9/30 <0.003		
Eye irritation 9/30 <0.008		
Self-reported headache 5/30 <0.025		

Source: Adapted from Knave, B., Persson, H.E., Goldberg, J.M., et al. *Scand. J. Work Environ. Health*, 2: 152–164, 1976; Ritchie, G.D., Still, K.R., Alexander, W.K., et al. *J. Toxicol. Environ. Health. B Crit. Rev.*, 4: 223–312, 2001.

Note: NS Not significant. [a] Significant (< 0 .05) versus controls.

* CNS effects are boldfaced.

reported and medically diagnosed complaints from exposed personnel. While JP-8 or JP-5 exhibit higher flash points, lower vapor pressures, and increased handling safety as compared to JP-4, both JP-8 and JP-5 necessarily vaporize more slowly from skin, clothing, environmental surfaces, soil, and groundwater, and are more likely to be found in aerosolized versus vapor phase compared to JP-4 (Ritchie et al., 2001b, 2003). These characteristics of kerosene-based jet fuels may actually provide increased human dermal exposure to raw fuel as well as increased respiratory exposure to fuel in aerosol phase, as compared to JP-4. Fuel workers commonly complain of JP-8 odor persisting on the skin and in the saliva for more than 12 hours post exposure, and "sweating out" of jet fuel during unassigned weekend hours (Ullrich and Lyons, 2000; Ritchie et al., 2001b). Discarding fuel-soaked clothing by workers in the home environment may provide a basis for substantial inhalation exposure of family members to jet fuel vapor. Interestingly, similar complaints for JP-4 exposures have not been documented in the literature. Some of the JP-8-related complaints mimic those reported for repeated exposure to the European jet fuels MC75 and/or MC77, as listed previously. In a comparison between military workers occupationally exposed to high levels of JP-8 versus those exposed to low levels or unexposed to JP-8, there were significant differences ($p < 0.05$) in incidence of the following self-reported symptoms (Olsen et al., 1998; AFIERA, 2001). Those symptoms typically associated with CNS disorders are shown in boldfaced type:

1. **Dizziness**
2. **Imbalance**
3. **Walking difficulties**
4. **General weakness/fatigue**
5. **Difficulty in gripping objects**
6. **Numbness/tingling of limbs**
7. Itching skin
8. Blisters/skin rashes on hands and arms
9. Chemical allergy
10. Difficulty breathing
11. Chest tightness
12. Excessive sweating
13. **Trouble concentrating**
14. **Forgetfulness**
15. **Perception that "work is impacting health"**

Additionally, there were numerically greater incidences in self-reporting by the "high"-exposure group of the following symptoms (Olsen et al., 1998; AFIERA, 2001):

1. **Headache**
2. **Blurry vision**
3. **Tremors**
4. **Tearing eyes**
5. **Chronic pain/use of pain medication**
6. Heart palpitations
7. Scaly skin or weeping skin

An extensive medical records review for the same subjects, however, indicated that there were no significant differences between either male or female JP-8-exposed subjects versus controls (low exposure groups) for the mean number of health-care provider visits for skin disease, gastrointestinal concerns, sports-related injuries, workplace injuries, other injuries, respiratory conditions, neurological conditions, musculoskeletal conditions, cardiovascular conditions, or urogenital complaints. Total health-care provider visits were nearly identical for exposure groups, including 220 males and 45 females (AFIERA, 2001).

Olsen et al. (1998), evaluating a small number of military fuel workers during occupational exposure to JP-4 and then again following USAF conversion to JP-8, reported several significant decreases in neurobehavioral function in workers following the JP-8 conversion, as compared to unexposed controls. Using the MicroCog neurocognitive functioning battery (tests of attention/mental control, memory, reasoning/calculation, spatial processing, and reaction time), it was generally shown that unexposed workers scored numerically higher on all tests as compared to workers exposed occupationally to JP-4, then JP-8. On the Reaction Time index, a test of motor response speed, unexposed controls exhibited significantly faster reaction time scores than did JP-8 workers 6 months following the transition from JP-4 to JP-8.

Beginning in 2001, the USAF Office of the Surgeon General, in cooperation with the U.S. Navy, funded a large epidemiological study to identify possible health effects in military workers exposed subchronically or chronically to jet fuels. Approximately 340 jet fuel-exposed military personnel and matched controls at six environmentally diverse, domestic USAF bases (Davis Montham AFB, Arizona; Seymour Johnson AFB, North Carolina; Langley AFB, Virginia; Pope AFB, North Carolina; Little Rock AFB, Arizona; and Hurlbert Field, Florida) volunteered to submit to a large number of medical and physiological tests, including several tests of CNS function. Most commonly, the fuel-exposed workers were USAF fuel tank (cell) cleaners or jet aircraft mechanics, and were occupationally exposed to various concentrations of JP-8 on the flight line or in enclosed hangars during work weeks of up to 60 hours. Fuel workers reported that their fuel exposures ranged from approximately 90 days to careers exceeding 20 years. In a previously published study of the worst-case jet fuel exposure scenarios for fuel cell workers, Carlton and Smith (2000) measured a high 8-hour time-weighted average (TWA) of 1,304 mg/m^3 and a high short-term (15-minute average) exposure of 10,295 mg/m^3. Instantaneous sampling results indicated benzene exposures during fuel tank repair up to 49.1 mg/m^3. Fuel-exposed volunteers were tested before their occupational assignment after a 17- to 48-hour rest from jet fuel exposure, and then were retested approximately 20 to 60 minutes following completion of daily work assignments, commonly involving dermal and (or) inhalation exposure to jet fuel. In some cases, workers would report minimal inhalation exposure to JP-8, while in many cases workers reported prolonged exposure to very high concentrations of JP-8 aerosol combined with prolonged dermal exposure (e.g., fuel-soaked cotton work uniforms). Control subjects were volunteers who worked on one of the six USAF bases, but did not have occupational assignments involving direct exposure to jet fuels. It was shown that some of the control subjects exhibited detectable levels of JP-8 chemical components (i.e., benzene and naphtha) in the blood and breath that was assumed to result from breathing the air near the flight lines during daily work assignments or the operation of gasoline-powered personal vehicles.

As part of the AFIERA (2001) study, a subgroup of fuel-exposed and control subjects volunteered to complete eyeblink classical conditioning (EBCC) tests (Bekkedal et al., unpublished manuscript; McInturf et al., 2001). EBCC is a neurobehavioral method used to study the neurobiology of associative learning and memory; diagnose various neuropathologies (i.e., Alzheimer's disease, cerebrovascular dementia, Downs Syndrome, etc.); measure specific cerebellar, brainstem, or hippocampal deficits; and quantify effects from various pharmacological challenges or toxic exposures (e.g., lead), particularly as may impact the cholinergic or GABAergic neurotransmitter systems. While there are several EBCC protocols, a delay conditioning paradigm was used to evaluate possible acute and chronic jet fuel exposure effects. Delay conditioning occurs normally in subjects with intact brainstem, cerebellar, and thalamic systems. In EBCC delay conditioning, a neutral CS (conditioned stimulus, i.e., an auditory tone) occurs before but overlapping temporally with the occurrence of a noxious UCS (unconditioned stimulus, i.e., a corneal airpuff). The reflexive unconditioned response (UCR, i.e., an eyeblink) to the UCS becomes associated with, and ultimately occurs in response to the CS. This associated response is known as the CR (conditioned response), and is identified as a response topographically similar to the UCR that occurs in a specified temporal window following CS onset, as defined by the neural circuitry thought to subserve the association. Hence, responses that occur before the temporal window are defined as either non-stimulus contingent or abnormal, while those that occur beyond the temporal window are necessarily defined as either spontaneous or as UCRs (Woodruff-Pak and Thompson, 1988). First, it was demonstrated in animals adequately and humans minimally that specific regions of the brainstem, cerebellum (i.e., nucleus interpositus), and thalamus must be intact for delay EBCC to occur. Further, pharmacological manipulation of the cholinergic and GABAergic systems has been shown to influence EBCC.

Fifty-four USAF personnel exposed subchronically or chronically to JP-8 and 57 non-fuel-exposed matched controls were compared for acquisition, retention, and performance of the EBCC. Subjects were trained initially (acquisition) following a 17- to 48-hour rest from occupational responsibility and then were tested for retention approximately 1 hour after a 4- to 6-hour return to normal employment involving JP-8 exposure. It was shown at all six USAF bases tested that there were either statistically significant or numerical impairments in the acquisition of EBCC for the fuel-exposed subjects as compared to the controls. A scrutiny of the EBCC data showed that many of the deficits in EBCC could be attributed to the fact that the fuel-exposed personnel emitted CRs during the period before the temporal window established for "normal" responding. In other words, fuel-exposed subjects were emitting CR-like responses more quickly than the controls, possibly suggesting a persisting hyperactivation of the CNS. During the afternoon EBCC retention sessions, following 4 to 6 hours of occupational exposure to JP-8, there was only a minimal difference between the fuel-exposed subjects and the controls. In fact, at two of the six USAF bases tested, the retention of the fuel-exposed subjects numerically exceeded the retention of the controls, although during acquisition the performance of the controls had exceeded the performance of the fuel-exposed subjects. Again, this might suggest that learning the simple EBCC task was nearly identical for fuel-exposed and control subjects, although a possible hyperactivation of the CNS in the exposed subjects may have contributed to the reduced performance (percentage of properly emitted CRs) that was measured for this group.

Finally, the cognitive/neurobehavioral capacity of USAF workers with "high dose" occupational exposures to JP-8 was compared to that of workers with "low dose" or no JP-8 exposure (AFIERA, 2001). As with previously discussed testing, these military workers were evaluated 17 to 48 hours following the last occupational exposure to JP-8, and again following a 4- to 6-hour return to occupational responsibility. "High dose" workers exhibited a number of significant deficits on the U.S. Navy Behavioral Assessment and Research System/Global Assessment System for Humans (BARS/GASH) battery (Rossi et al., 1996), as compared to the "low dose" exposure group. Subjects classified as "high" JP-8 exposure exhibited significantly reduced performance (subchronic/chronic effect), relative to "low" exposure controls, on the following tests administered after a 17- to 48-hour rest from occupational exposure:

1. *Digit Span – Forward/Backward:* A series of numbers was presented sequentially on a computer screen, and the subject was asked to reproduce the sequence on a numbered keyboard in the same order or in reverse order. The length of the sequence increased until a criterion was met, or the subject consistently failed.
2. *Symbol Digit:* A coding test in which digits were paired with symbols in a 2×9 matrix of squares on a computer screen. A similar matrix occurring at the bottom of the screen contained the symbols, but not the digits. The subject was required to type in the digit that was previously paired with each of up to nine symbols.
3. *Tapping or Dual Tapping, Preferred Hand:* The subject was instructed to press (tap) a designated console button as rapidly as possible, first with the index finger of the preferred hand, then of the non-preferred hand (single tapping), then alternating between the index fingers of the preferred and non-preferred hands (dual tapping).

The same subject group exhibited significantly reduced performance ("acute" effect), as compared to the initial testing session, when retested after a 4- to 6-hour return to occupational responsibility (JP-8 exposure) on the following tests:

1. *Tapping, Preferred Hand:* The task is described above.
2. *Dual Tapping:* The task is described above.
3. *Delayed Matching-to-Sample (DMTS):* DMTS is a test requiring the subject to view a complex visual pattern, then chose the pattern observed initially from three choices, presented following a short, variable delay.

Deficits in the "high" exposure JP-8 group appeared to reflect both significant, acute, or subchronic/chronic reduction in higher cognitive capacity (Digit Span, Symbol Digit, Matching-to-Sample tests), emphasizing impaired short-term memory, and in simple motor skills (Tapping tests). In these studies, there was no evidence of a JP-8-induced improvement in task performance.

Tu et al. (2004) conducted a similar study of cognitive function of USAF personnel exposed chronically to jet fuels. The authors recruited volunteers at a domestic Air Guard Base where military reservists had worked with jet fuels for as many as 30 years. Breath analysis was used to provide an estimate of body burden as well as the recent inhalation and dermal exposure to JP-8. All individuals studied exhaled aromatic and aliphatic hydrocarbons that are found in JP-8. The highest breath concentration of "total" JP-8 was 11.5 mg/m^3. This breath concentration suggested that exposure to JP-8 at an Air Guard Base is typically much less than exposure observed at USAF bases (AFIERA, 2001). This reduction in exposure to JP-8 is attributed to enhanced safety practices and standard operating procedures carried out by Air Guard Base personnel. The base personnel who exhibited the highest exposures to JP-8 were fuel cell workers, fuel specialists, and controls who smoked cigarettes (who smoked outside and frequently downwind from the flight line). Although the average JP-8 body burdens of Air Guard Base personnel were generally lower than those of USAF base personnel, the neurocognitive functioning of these chronically fuel-exposed personnel was still lower than was measured in controls. JP-8-exposed individuals performed significantly worse than non-exposed, age- and education-matched individuals on 20 of 47 measures of information processing and other cognitive functions.

Bell et al. (2005) studied a number of Persian Gulf War veterans, many of whom complained of chronic neurocognitive deficits, fatigue, and musculoskeletal symptoms. Upon further studying this cohort, the authors determined that many of the complaints overlapped with civilian illnesses (e.g., chronic fatigue syndrome and fibromyalgia) thought to result from low-level exposures to environmental chemicals. The authors utilized a computerized visual divided attention test to identify neurobehavioral deficits in chronically unhealthy Gulf War veterans (n = 22, ill with low-level chemical intolerance [CI]; n = 24, ill without CI) and healthy Gulf War era veterans (n = 20). Many of the unhealthy veterans reported that they had multiple exposures to JP-8, other jet fuels, or industrial hydrocarbon solvents during their military deployments. Based on this observation, Bell et al. (2005) tested each veteran for cognitive deficits before and after each of three weekly, double blind, low-level JP-8 jet fuel or clean air sham exposure laboratory sessions, including acoustic startle stimuli. Unhealthy veterans receiving jet fuel had faster mean peripheral reaction times over sessions, compared with unhealthy veterans receiving sham clean air exposures. Unhealthy Gulf war veterans with CI exhibited faster post- versus pre-session mean central reaction times compared with unhealthy Persian Gulf War veterans without CI. Findings were controlled for psychological distress variables. These data on unhealthy Gulf veterans showed an acceleration of divided attention task performance over the course of repeated low-level JP-8 exposures. The present faster reaction times are consistent with some the other human and animal JP-8 dose-response patterns reviewed in this chapter. Together with previous research findings, the data suggest involvement of at least the CNS dopaminergic pathways in affected Persian Gulf War veterans. Bell et al. (2005) suggested that repeated jet fuel exposures may be involved in the diagnosis of Multiple Chemical Sensitivity syndrome in many military veterans.

Ocular System Effects in Humans

As previously indicated, self-reported health effects from occupational fuel workers may include tearing of the eyes and blurred vision (Ritchie et al., 2001a; AFIERA, 2001). Two aviators exposed to a "high level" of JP-5 reported a burning sensation in the eyes, while one experienced itching, watering eyes for 24 hours post exposure. One of the aviators was diagnosed with hyperemic conjunctiva that persisted for 4 days post exposure (Porter, 1990). Eye irritation was, however, not reported by six volunteers exposed for 15 minutes to 140 mg/m^3 deodorized kerosene vapor (Carpenter et al., 1976). There are no reports of more serious human eye injuries from repeated exposure to kerosene-based jet fuel vapor or aerosol.

Peripheral and Central Auditory System Effects in Humans

Morata et al. (1997a) reported a prevalence of hearing loss (42% to 50%) in petroleum workers exposed chronically to aromatics and paraffins that was 2.4 times greater than for controls. The results of acoustic reflex decay tests suggested a retrocochlear or central auditory pathway involvement in the losses observed in certain job categories. Morata at al. (1997b) similarly reported hearing loss (based on pure-tone audiometry and immittance audiometry testing) among 49% of rotogravure printing workers exposed to various levels of noise and an organic solvent mixture of toluene, ethyl acetate, and ethanol. The author concluded, based on urine testing, that chronic toluene exposure may be linked to hearing loss.

Odkvist et al. (1983, 1987) examined the effects of chronic exposure to European jet fuel (i.e., MC75 and MC77, containing a high percentage of unleaded gasoline) on audiological and vestibulo-oculomotor function in eight jet mechanics exposed from 15 to 41 years (mean = 25 years). Although there was no indication of any abnormal deficit in peripheral auditory mechanisms or thresholds, exposed workers exhibited significant deficit rates on Interrupted Speech Discrimination (38%) and Cortical Response to Frequency Glides (50%) tests. The former test, sensitive to lesions of the central auditory pathways and auditory cortex, evaluated the capacity of subjects to discriminate human speech interrupted at 4, 7, or 10 times per second. The latter test, highly sensitive to lesions of the cerebello-pontine and other auditory pathways, required the subject to identify frequency glides, in a 1,000-Hz tone, of 50 Hz or 200 Hz within 167 or 140 milliseconds, respectively. The outcome of vestibulo-oculomotor tests indicated significant results on Broad-Frequency Visual Suppression and Broad Frequency Smooth Pursuit tests, but reflected no deficit in the sensory organs. For the former test, subjects were required to suppress the vesbulo-oculomotor reflex in order to fixate a dot on a full visual field screen moving with a rotating chair. Unlike during the testing of solvent-exposed (positive control) workers, jet-fuel-exposed workers did not exhibit a significant deficit on the Visual Suppression test until it was made more difficult by using broad frequency pseudorandom swings. In the latter test, subjects were asked to follow a small target moving at a velocity of 10 or 25°/second across a monitor, with the dependent measure being the eye speed between saccades. Values below 8° and 18°/second, respectively, were considered significantly abnormal. This research appears to indicate the workers exposed chronically to jet fuel (mean = 25 years) may have exhibited subtle deficits in the higher-level inhibition (cerebellar, cortical, etc.) of brainstem functions, which may remain undetected without the use of complex testing batteries. It has been hypothesized (Odkvist et al., 1983) that deficits in the oculomotor system commonly observed in animals and humans following hydrocarbon solvent exposures may reflect a decreased inhibition of the vestibulo-oculomotor reflex, presumably exerted by the cerebellum.

Kaufman et al. (2005) conducted a study examining the effects of occupational exposure to jet fuel on hearing in military workers. Noise-exposed subjects working on a USAF base, with or without jet fuel exposure, underwent hearing tests. Work histories, recreational exposures, protective equipment, medical histories, alcohol, smoking, and demographics were identified by questionnaire. Jet fuel, solvent, and noise exposure data were collected from occupational records. Fuel exposure estimates were less than 34% of the OSHA Threshold Limit Values (TLVs). Subjects with 3 years of jet fuel exposure had a 70% increase in adjusted odds of hearing loss (Odds Ratio [OR] = 1.7; 95% Confidence Interval (CI) = 1.14 to 2.53) and the odds increased to 2.41 (95% CI = 1.04 to 5.57) for 12 yr of noise and fuel exposure. Consistent with the results of animal studies, these findings suggest that jet fuel has a chronic toxic effect on the auditory system.

Vestibular System Effects in Humans

Smith et al. (1997) reported the effects on postural balance/equilibrium in 27 USAF employees with 0.8 to 30 years (mean = 4.56 years) exposure to JP-8 (although some subjects also reported an exposure history to JP-4). Exposed subjects and matched controls were tested before the work shift (12- to 24-hour rest from exposure) and again after 4 to 6 hours of occupational exposure to JP-8. Eight-hour breathing zone samples were collected for each employee. Mean exposure levels for employees in all job categories exposed to JP-8 fuel were as follows: benzene (5.03 ± 1.4 ppm), toluene (6.11 ± 1.5 ppm), xylenes (6.04 ± 1.4 ppm), and naphthas (419.6 ± 108.9 ppm). Subjects were asked to stand erect on a thin pad containing sensitive postural sway-measuring instrumentation. Subjects were evaluated for the capacity to maintain postural equilibrium on a standard postural balance platform during each of four testing conditions: (1) eyes open:stable platform, (2) eyes closed:stable platform, (3) eyes open:standing on 4-inch foam, and (4) eyes closed:standing on 4-inch foam. Workers exposed for equal to or more than 9 months to JP-4/JP-8 exhibited significantly increased postural sway patterns, relative to controls, but only during the most difficult testing condition, in which the eyes were closed and the subject stood on a 4-inch thick section of packing foam. Performance deficits were correlated with breathing space and breath levels of benzene (5.03 ± 1.4 ppm), toluene (6.11 ± 1.5 ppm), and xylenes (6.04 ± 1.4 ppm), but not naphtha (491.6 ± 108.9 ppm). The reported deficits were subchronic/chronic and were not significantly modulated by the "acute" 4- to 6-hour JP-8 occupational exposures. The study authors noted that the strongest statistical association between sway length and benzene exposure, which may imply a subtle influence on vestibular/proprioception functionalities.

As a part of the previously discussed AFIERA study (2001) of JP-8 occupational exposure, the "high exposure" group, as compared to "low exposure" controls, exhibited a significant deficit on testing for postural equilibrium. There was a statistically significant difference between "high exposure" and "low exposure" personnel during only the "forward bending of the torso" condition. This testing condition is generally considered more difficult than the "eyes closed: standing on foam" testing condition (Smith et al., 1997), in which there were no significant differences between fuel-exposed and control subjects. It should be noted that workers in the AFIERA (2001) study were generally younger and reported less duration of exposure to JP-8 and no exposure history to JP-4, as compared to subjects tested in the previous study (Smith et al., 1997). Similar to the 1997 study result, deficits in postural equilibrium exhibited by workers with subchronic or chronic exposure to JP-8 were generally not exacerbated by "acute" exposure to JP-8. Persisting postural equilibrium deficits are known to reflect deficits in brainstem vestibular or proprioceptive control systems, but may additionally reflect deficits in peripheral proprioceptive mechanisms (Smith et al., 1997).

IN VITRO STUDIES

Lin et al. (2001) used cDNA nylon arrays (Atlas Rat 1.2 Array, Clontech Laboratories, Palo Alto, California) to measure changes in gene expression in whole brain tissue of rats exposed repeatedly to JP-8 under conditions that simulated possible real-world occupational exposure (1,000 mg/m^3 for 6 hours/day for 91 days) to JP-8 vapor. Rats were sacrificed following a rest period and neurobehavioral testing. Gene expression analysis of the exposure group compared to the control group revealed up-regulation or down-regulation of several genes, including glutathione-*S*-transferase Yb2 subunit (GST Yb2), cytochrome P450 IIIA1 (CYP3A1), glucose-dependent insulinotropic peptide (GIP), alpha1-proteinase inhibitor (alpha1-AT), polyubiquitin, GABA transporter 3 (GAT-3), and plasma membrane Ca^{2+}-transporting ATPase (brain isoform 2) (PMCA2). In a follow-up study, Lin et al. (2004) reported that the numerous changes in CNS gene expression associated with repeated exposure to JP-8 vapor can be divided into two main functional categories: (1) neurotransmitter signaling pathways and (2) stress response. These studies suggest that repeated exposure of rats to JP-8 vapor at human "real-world" concentrations results in persisting changes in gene expression in multiple CNS systems. These CNS systems appear to include those known to modulate at least some of the behaviors previously identified as affected by jet fuel exposures in animals and humans.

Edelfors and Ravn-Jonsen (1992), using rat CNS synaptosomal membranes, reported that exposure of the in vitro preparation to kerosene resulted in reduced activity of the enzyme (Ca^{2+}/Mg^{2+}) ATPase and subsequently increased levels of intracellular Ca^{2+}, resulting in increased neurotransmitter release.

There is one published study of the in vitro effects of acute exposure to JP-8 on nervous system tissue. Grant et al. (2000) investigated the in vitro cytotoxicity and electrophysiological effects of JP-8 on neuroblastoma × glioma (NG108-15) cell cultures, as well as on embryonic hippocampal neurons. Acute JP-8 (in 5% ethanol) exposure of the hippocampal neurons proved highly toxic ($IC_{50} < 2$ μg/ml) while, in contrast, the NG108-15 cells were much less sensitive. Electrophysiological examination of NG108-15 cells showed that administration of JP-8 at 1 μg/ml did not alter significantly any of the electrophysiological properties. However, exposure to JP-8 at 10 μg/ml during a current stimulus of +46 picoamperes decreased the amplitude of the action potential to 83% ± 7%, the rate of rise (dV/dtMAX) to 50% ± 8%, and the spiking rate to 25% ± 11% of the corresponding control levels. These results demonstrate that JP-8-induced cytotoxicity varies among cell types, and for the first time that CNS neurons may alter electrophysiological function without cell death in response to JP-8 exposure.

Additionally, there is one published study of the effects of hydrocarbon fuel exhaust on the rat brain. Microinjection of exhaust emissions, containing PAHs and nitro-PAHs, into the hippocampus or striatum induced significant lesions, with tissue loss and disappearance of immunoreactivity for glial fibrillary acidic protein (GFAP), tyrosine hydroxylase, and acetylcholinesterase (AChE) (Andersson et al., 1998).

NEUROTOXICITY FROM EXPOSURE TO JET FUEL EXHAUST

Literally all personnel working at military bases with aircraft, on military aircraft carriers, or at commercial airports experience jet fuel exhaust on a daily basis. When working in fuel handling or aircraft maintenance occupations, concurrent exposure to raw fuel and fuel combustion exhaust is generally unavoidable. Similarly, exposure of personnel living near military or commercial airports includes low levels of both raw fuel (vapor/aerosol) and combustion exhaust. A recent investigation of exhaust from military aircraft ground support equipment using JP-8, diesel fuel, or gasoline indicated particulate mass concentrations from 0.09 to 1.1 g/kg fuel, and emphasized that particle size distribution varied as a function of engine condition and engine load (Zielinska, et al., 2004). The following chemical components of kerosene-based jet fuel combustion exhaust have been identified:

1. *Inorganic gases:* CO, CO_2, NO_x, SO_x, formed from the reaction of nitrogen in the air, or carbon or sulfur in jet fuel with atmospheric oxygen during fuel combustion.
2. *Volatile organic compounds (VOCs):* Alkanes, cycloalkanes, alkenes, and aromatic hydrocarbons (including pentane; butane; 1,3-butadiene; benzene; toluene; ethylbenzenes; xylenes) derived from incomplete combustion of fuels.
3. *Raw fuel:* Up to 30% aerosolized uncombusted fuel, as a function of engine type, engine mechanical condition, and environmental temperatures.
4. *Oxygenated organics:* Carbonyl compounds, including formaldehyde, acetaldehyde, crotonaldehyde, acrolein, and benzaldehyde.
5. *Polycyclic aromatic hydrocarbons (PAHs):* Including anthracene, phenanthrene, fluoranthene, pyrene, chrysene, benzo[b,k]fluoranthene, cyclopental[c,d]pyrene, benzo[e]pyrene, benzo[a]pyrene, benzo[a]anthracene, indo[1,2,3-c,d]pyrene, benzo[g,h,i]perylene, perylene, coronene, and chrysene.
6. *Alcohols:* Including methanol and ethanol.
7. *Ozone:* Formed by the interaction of VOCs, NO_x, and sunlight.
8. *Particulate matter:* Elemental carbon, sulfate, and nitrate aerosols, both PM10 (particulate matter, diameter 10μm) and microfine particles, PM2.5 (Ritchie et al., 2001b; 2003).

The following provides a brief description of known neurotoxicity effects from acute or repeated exposure to jet fuel combustion exhaust.

- *Carbon monoxide:* CO is toxicologically important because it competes with O_2 for hemoglobin binding, resulting in the formation of carboxyhemoglobin. Hemoglobin CO saturation levels of 50% to 0% are associated with unconsciousness and death. Signs and symptoms of CO hemoglobin saturation approaching dangerous levels (20% to 30%) include headache, weakness, nausea, and dimness of vision (NML, 1996). Chronic exposure to low levels of CO produces medical symptoms associated with many other disease states, including chronic fatigue, headache, dizziness, nausea, and mental confusion, and is often mistaken for other illnesses (Knobeloc and Jackson, 1999). CNS neurons are particularly sensitive to the toxic effects of prolonged hypoxia produced by CO exposure (Gordon and Amdur, 1991).

- *Oxides of sulfur:* SO_x (SO, SO_2, SO_3) is produced during the combustion of JP-8 (Yost and Montalvo, 1995). Chronic occupational exposure to SO_X was associated with various health conditions including alteration of smell and taste, increased fatigue, neurotic and vegeto-asthenic nervous system disorders (ILO, 1983; Duffell, 1985).
- *Oxides of nitrogen:* NO and NO_2 are primarily respiratory and eye irritants with a reported threshold of 10 to 20 ppm (Sharp, 1978). Chronic exposure to low concentration of NO was reported to induce chronic irritation of the respiratory tract, cough, headache, and loss of appetite (NML, 1998a, 1998b).
- *Formaldehyde:* Formaldehyde can be formed by the incomplete combustion of kerosene-based jet fuels and was detected in the exhaust of military aircraft at concentrations of 0.86-2.78 ppm (50 meters behind the aircraft), as a function of the power setting of the engine, ambient temperature, and relative humidity (Kobayashi and Kikukawa, 2000). Formaldehyde is a concentration-dependent irritant of the eyes and mucous membranes at low-level exposures (Horvath et al., 1988). Occupational exposure to JP-8 exhaust was shown to irritate the eyes and respiratory tract (Miyamoto, 1986; Kobayashi and Kikukawa, 2000).
- *Elemental carbon:* Complete or partial combustion of kerosene-based jet fuels was shown to generate elemental carbon (carbon black). Exposure to elemental carbon was found to produce conjunctivitis, epithelial hyperplasia of the cornea, and eczematous inflammation of the eyelids (Sax, 1984).

CONCLUSIONS

Although hundreds of thousands of commercial and military workers are exposed to jet fuels for up to 8 hours/day, 5 days/week for extended periods of time, there have been remarkably few documented reports of serious persisting central or peripheral nervous system effects. Acute exposure to high concentrations of jet fuels (e.g., accidental spillages) may result in temporary irritancy of the ocular and respiratory tissues, disorientation, mental confusion, memory loss, and loss of equilibrium. Early studies of workers exposed chronically to European aircraft fuels thought to contain high levels of aromatic hydrocarbons and n-hexane, reported severe psychiatric, physical, cognitive, and emotional effects, as well as peripheral neuropathy. As these fuels are no longer manufactured, the results of the studies must be considered in context.

With chronic exposure to lower concentrations of JP-8, JP-5, or Jet A/A-1, there have been several reports of slightly impaired cognitive processing function, one report of modulated learning of a classically conditioned paradigm, and several reports of impaired vestibular function. In some cases, acute or subchronic exposures to jet fuels have been shown to "activate" the CNS, while longer-term exposures have been shown to result in subtle but persisting deficits on complex postural equilibrium, sensory discrimination, or higher cognitive tasks. In most cases, when chronically exposed workers are evaluated on simple cognitive tests, there are no detectable deficits. However, when the same workers are challenged by more complex tests, significant deficits are observed. One reason that subtle deficits are not commonly reported is that affected jet fuel workers may seldom have the opportunity to be evaluated for higher-level CNS deficits.

In animals, the range of reported CNS effects from jet fuel exposure is similar to those observed in humans. In most cases, acute or subchronic exposure to jet fuels has resulted in minimal health and cognitive effects. Typically, animals exposed to jet fuels exhibit dramatically increased locomotor activity, occasional autonomic signs, and deficits in learning or recall of moderate to complex learning tasks. There is at least one report suggesting that exposure to low levels of JP-8 may actually improve learning or performance of operant tasks of moderate complexity (neurobehavioral "hormesis"). One explanation for the cognitive deficits observed in animals with high-level fuel exposures is that the CNS is activated to the point of inducing inattentive behaviors.

There is evidence from several animal studies that jet fuels may perturb at least two of the major CNS neurotransmitter systems, specifically dopamine and serotonin, for many months following the cessation of fuel exposure. Additionally, several in vitro studies suggest that exposure to jet fuel or jet fuel exhaust components can damage or destroy cells from the several brain areas tested. It has been shown that exposure to jet fuels results in perturbation of a number of genes thought to be involved in normal CNS function. These neurohistopathological and genetic changes in the CNS are sufficient to possibly explain many of the fuel-induced changes in neurobehavior reported in both humans and animals. Further, direct exposure to fuels containing ethylbenzene and toluene, in the presence of loud noise, has been shown to induce significant cochlear hair cell damage in rats.

Although jet fuels typically consist of more than 200 different hydrocarbons and up to six chemical additives, there is increasing evidence that significant human and animal toxicity generally results from exposure to specific smaller-chain fractions, including benzene, toluene, xylenes, n-hexane, and ethylbenzene, and may vary substantially with the routes of exposure.

In summary, chronic exposure to jet fuel may induce subtle, but persisting changes in a number of human and animal CNS systems and may be responsible for small, but significant reductions in cognitive capacity. It would seem obvious that better controlled studies of human jet fuel workers are needed, especially those with tests capable of measuring subtle changes in cognitive processing, orientation, and memory systems.

REFERENCES

AFIERA (Air Force Institute for Environment, Safety and Occupational Health Risk Analysis). *JP-8 Final Risk Assessment.* The Institute of Environmental and Human Health, Texas Tech University, Lubbock, TX, 2001.

Akamaguna, A.L., and Odita, J.C. Radiology of kerosene poisoning in young children. *Ann. Trop. Paediatr.*, 3: 85–88, 1983.

Aldy, D., Siregar, R., and Siregar, H. Accidental poisoning in children with special reference to kerosene poisoning. *Paediatr. Indones.*, 18: 45–50, 1978.

Andersson, K., Fuxe, K., Nilsen, O.G., et al. Production of discrete changes in dopamine and noradrenaline levels and turnover in various parts of the rat brain following exposure to xylene, ortho-, meta-, and para-xylene, and ethylbenzene. *Toxicol. Appl. Pharmacol.,* 60: 535–548, 1981.

Andersson, K., Nilsen, O.G., Toftgard, et al. Increased amine turnover in several hypothalamic noradrenaline nerve terminal systems and changes in prolactin secretion in the male rat by exposure to various concentrations of toluene. *Neurotoxicology,* 4: 43–55, 1983.

Andrews, L.S., Lee, E.W., Witmer, C., et al. Effects of toluene on metabolism, disposition, and hematopoietic toxicity of (^{3}H) benzene. *Biochem. Pharmacol.,* 26: 293–300, 1977.

Baldwin, C.M., Figueredo, A.J., Wright, L.S., et al. Repeated aerosol-vapor JP-8 jet fuel exposure affects neurobehavior and neurotransmitter levels in a rat model. *J. Toxicol. Environ. Health A*, 70: 1203–1213, 2007.

Baldwin, C.M., Houston, F.P., Podgornik, M.N., et al. Effects of aerosol-vapor JP-8 jet fuel on the functional observational batter and learning and memory in the rat. *Arc. Environ. Health,* 56: 216–226, 2001.

Bell, I.R., Brooks, A.J., Baldwin, C.M., et al. JP-8 jet fuel exposure and divided attention test performance in 1991 Gulf War veterans. *Aviat. Space Environ. Med.,* 76: 1136–1144, 2005.

Bogo, V., Young, R.W., Hill, T.A., et al. Neurobehavioral toxicology of petroleum- and shale-derived jet propulsion fuel No. 5. In *Advances in Modern Environmental Toxicology: Applied Toxicology of Petroleum Hydrocarbons*. MacFarland, H.N., Holdsworth, C.E., MacGregor, J.A., et al. (Eds.), Vol. VI, Princeton NJ: Princeton Scientific Publishers, pp. 17–32, 1984.

Campo, P., Loquet, G., Blachère, et al. Toluene and styrene intoxication route in the rat cochlea. *Neurotoxicol. Teratol.*, 21: 427–344, 1999.

Carlton G.N., and Smith, L.B. Exposures to jet fuel and benzene during aircraft fuel tank repair in the U.S. Air Force. *Appl. Occup. Environ. Hyg.,* 15: 485–491, 2000.

Carpenter, C.P., Kinkead, E.R., Geary, D.L. Jr., et al. Petroleum hydrocarbon toxicity studies: I. Methodology. *Toxicol Appl. Pharmacol.,* 32: 246–262, 1975.

Carpenter, C.P., Geary, D.L., Myers, R.C., et al. Petroleum hydrocarbon toxicity studies. XI. Animal and human response to vapors of deodorized kerosene. *Toxicol. Appl. Pharmacol.,* 36: 443–456, 1976.

Crofton K.M., Lassiter, T.L., and Rebert, C.S. Solvent-induced ototoxicity in rats: An atypical selective mid-frequency hearing deficit. *Hear. Res.,* 80: 25–30, 1994.

Davies, N.E. Jet fuel intoxication. *Aerosp. Med.,* 35: 481–482 1964.

Dudin, A.A., Rambaud-Cousson, A., Thalji, A., Jubeh, I.I., et al. Accidental kerosene ingestion: A 3-year prospective study. *Ann. Trop. Paediatr.*, 11: 155–161, 1991.

Duffell, G.M. Pulmonotoxicity: Toxic effects in the lungs. In *Industrial Toxicology. Safety and Health Applications in the Workplace*, P.L. Williams and J.L. Burson (Eds.). New York: Van Nostrand-Reinhold, pp. 162–197, 1985.

Fechter, L.D., Gearhart, C., Fulton, S., Campbell J., et al. Promotion of noise-induced cochlear injury by toluene and ethylbenzene in the rat. *Toxicol. Sci.*, 98: 542–551, 1990.

Fechter, L.D., Gearhart, C., Fulton, S., Campbell, J., et al. JP-8 jet fuel can promote auditory impairment resulting from subsequent noise exposure in rats. *Toxicol. Sci.,* 98: 510–525, 2007.

Foo, S.C., Jeyaratnam, J., and Koh, D. Chronic neurobehavioural effects of toluene. *Brit. J. Ind. Med.,* 47: 480–484, 1990.

Fornazzari, L, Wilkinson, D.A., Kapur, B.M., et al. Cerebellar, cortical and functional impairment in toluene abusers. *Acta. Neurol. Scand.,* 67: 319–329, 1983.

Gamberale, F., Annwall, G., and Hultengren, M. Exposure to xylene and ethylbenzene. III. Effects on central nervous system functions. *Scand. J. Work Environ. Health,* 4: 204–211, 1978.

Glowa, J.R. Some effects of sub-acute exposure to toluene on schedule-controlled behavior. *Neurobehav. Toxicol. Teratol.*, 3: 463–465, 1981.

Gordon, T., and Amdur, M.O. Responses of the respiratory system to toxic agents. In *Casarett &Doull's Toxicology: The Basic Science of Poisons, 4th ed.*, M.O. Amdur, J. Doull, and C.D. Klaassen (Eds.), New York: McGraw-Hill, pp. 383–407, 2001.

Gralewicz, S., Wiaderna, D., and Tomas, T. Development of spontaneous, age-related non-convulsive seizure electrocortical activity and radial-maze learning after exposure to m-xylene in rats. *Int. J. Occup. Med. Environ. Health*, 8: 347–360, 1995.

Gralewicz, S., and Wiaderna, D. Behavioral effects following subacute inhalation exposure to *m*-xylene or trimethylbenzene in the rat: A comparative study. *Neurotoxicology,* 22: 79–89, 2001.

Grant, G.M., Shaffer, K.M., Kao, W.Y., et al. Investigation of *in vitro* toxicity of jet fuels JP-8 and Jet A. *Drug Chem. Toxicol.*, 23: 279–291, 2000.

Hanninen, H., Eskelinen, L., Husman, K., et al. Behavioral effects of long-term exposure to a mixture of organic solvents. *Scand. J. Work Environ. Health,* 4: 240–255, 1976.

Holm, S., Norbäck, D., Frenning, B., et al. Hydrocarbon exposure from handling jet fuel at some Swedish aircraft units. *Scand. J. Work Environ. Health,* 13: 438–444, 1987.

Horvath, E.P., Jr., Anderson, H., Jr., Pierce, et al. Effects of formaldehyde on the mucous membranes and lungs. A study of an industrial population. *J. Am. Med. Assoc.*, 259: 701–707, 1988.

Hougaard, K.S., Andersen, M.B., Hansen, A.M., et al. Effects of prenatal exposure to chronic mild stress and toluene in rats. *Neurotoxicol. Teratol.*, 27: 153–167, 2005.

Hougaard, K.S., Hass, U., Lund, S.P., et al. Effects of prenatal exposure to toluene on postnatal development and behavior in rats. *Neurotoxicol. Teratol.*, 21: 241–50, 1999.

Ikeda, T., and Miyake, H. Decreased learning in rats following repeated exposure to toluene: Preliminary report. *Toxicol. Lett.,* 1: 235–239, 1978.

ILO (International Labour Office). *Encyclopedia of Occupational Health and Safety*, Vols. I and II. Geneva, Switzerland: International Labour Office, 1983.

Kaufman, L.R., LeMasters, G.K., Olsen, D.M., et al. Effects of concurrent noise and jet fuel exposure on hearing loss. *J. Occup. Environ. Med.,* 47: 212–218, 2005.

Kinkead, E.R., Salins, S.A., and Wolfe, R.E. Acute irritation and sensitization potential of JP-8 jet fuel. *J. Am. College Toxicol.*, 11: 700, 1992a.

Kinkead, E.R., Wolfe, R.E., and Salins, S.A. Acute irritation and sensitization potential of petroleum-derived JP-5 jet fuel. *J. Am. College Toxicol.*, 11: 706, 1992b.

Knave, B., Mindus, P., and Struwe, G. Neurasthenic symptoms in workers occupationally exposed to jet fuel. *Acta Psychiatr. Scand.,* 60: 39–49, 1979.

Knave, B., Olson, B.A., Elofsson, S., Gamberale, et al. Long-term exposure to jet fuel. II. A cross-sectional epidemiologic investigation on occupationally exposed industrial workers with special reference to the nervous system. *Scand. J. Work Environ. Health,* 4: 19–45, 1978.

Knave, B., Persson, H.E., Goldberg, J.M., et al. Long-term exposure to jet fuel: an investigation on occupationally exposed workers with special reference to the nervous system. *Scand. J. Work Environ. Health,* 2: 152–164, 1976.

Knobeloch, L., and Jackson, R. Recognition of chronic carbon monoxide poisoning. *Western Med. J.*, 98: 26–29, 2001.

Kobayashi, A., and Kikukawa, A. Increased formaldehyde in jet engine exhaust with changes to JP-8, lower temperature, and lower humidity irritates eyes and respiratory tract. *Aviat. Space Environ. Med.*, 71: 396–399, 2000.

Koschier, F.J. Toxicity of middle distillates from dermal exposure. *Drug Chem. Toxicol.*, 22: 155–164, 1999.

Lataye, R., Campo, P., and Loquet, G. Toluene ototoxicity in rats: assessment of the frequency of hearing deficit by electrocochleography. *Neurotoxicol. Teratol.*, 21: 267–276, 1999.

Lazar, R.B, Ho, S.U., Melen, O., et al. Multifocal central nervous system damage caused by toluene abuse. *Neurology,* 33: 1337–1340, 1983.

Lin, B., Ritchie, G.D., Rossi, J. 3rd, et al. Identification of target genes responsive to JP-8 exposure in the rat central nervous system. *Toxicol. Indus. Health,* 17: 262–269, 2001.

Lin, B., Ritchie, G.D., Rossi, J. 3rd, et al. Gene expression profiles in the rat central nervous system induced by JP-8 jet fuel vapor exposure. *Neurosci. Lett.,* 363: 233–238, 2004.

Lucas, G.N. Kerosene oil poisoning in children: A hospital-based prospective study in Sri Lanka. *Indian J. Pediatr.*, 61: 683–687, 1994.

Mahdi, A.H. Kerosene poisoning in children in Riyadh. *J. Trop. Pediatr.*, 34: 316–318, 1988.

Majeed, H.A., Bassyouni, H., Kalaawy, M., et al. Kerosene poisoning in children: A clinico-radiological study of 205 cases. *Ann. Trop. Paediatr.*, 1: 123–130, 1981.

Mattie, D.R., Alden, C.L., Newell, T.K., et al. A 90-d continuous vapor inhalation toxicity study of JP-8 jet fuel followed by 20 or 21 months of recovery in Fischer 244 rats and C57BL/6 mice. *Toxicol. Pathol.,* 19: 77–87, 1991.

Mattie, D.R., Marit, G.B., Flemming, C.D., et al. The effects of JP-8 jet fuel on male Sprague-Dawley rats after a 90-day exposure by oral gavage. *Toxicol. Indus. Health,* 11: 423–435, 1995.

McGuire, S., Bostad, E., Smith, L., et al. Increased immunoreactivity of glutathione-*S*-transferase in the retina of Swiss Webster mice following inhalation of JP8 + 100 aerosol. *Arch Toxicol.*, 74(4-5): 276–280, 2000.

McInturf, S.M., Bekkedal, M.Y., Ritchie, G.D., et al. Effects of repeated JP-8 jet fuel exposure on eyeblink conditioning in humans. *The Toxicologist,* 60: 555, 2001.

McWilliams, M.L., Chen, G.D., and Fechter, L.D. Low-level toluene disrupts auditory function in guinea pigs. *Toxicol. Appl. Pharmacol.,* 167: 18–29, 2000.

Mindus, P., Struwe, G., and Gullberg, B. A CPRS subscale to assess mental symptoms in workers exposed to jet fuel-some methodological considerations. *Acta Psychiatr. Scand. Suppl.,* 271: 53–62, 1978.

Miyake, H., Ikeda, T., Maehara, N., et al. Slow learning in rats due to long-term inhalation of toluene. *Neurobehav. Toxicol. Teratol.,* 5: 541–548, 1983.

Miyamoto, Y. Eye and respiratory irritants in jet engine exhaust. *Aviat. Space Environ. Med.*, 57: 1104–1108, 1986.

Morata, T.C., Engel, T., Durão, A., et al. Hearing loss from combined exposures among petroleum refinery workers. *Scand. Audiol.,* 26: 141–149,1997a.

Morata, T.C., Fiorini, A.C., Fischer, F.M., et al. Toluene-induced hearing loss among rotogravure printing workers. *Scand. J. Work Environ. Health,* 23: 289–298, 1997.

Moser, V.C., and MacPhail, R.C. International validation of a neurobehavioral screening battery: the IPCS/WHO collaborative study. *Toxicol Lett.*, 64-65(Spec. No.): 217–223, 1992.

NML (National Library of Medicine). 1996. HSDB (Hazardous Substances Database). http://toxnet.nlm. nih.gov/ cgi-bin/sis/htmlgen:HSDB.htm, *Carbon monoxide*. CASRN: 630-08-0, 1996.

NML (National Library of Medicine). HSDB (Hazardous Substances Database). http://toxnet.nlm. nih.gov/cgi-bin/sis/htmlgen:HSDB.htm. *Nitric oxide*. CASRN: 10102-43-9, 1998a.

NML (National Library of Medicine). HSDB (Hazardous Substances Database). http://toxnet.nlm. nih.gov/cgi-bin/sis/htmlgen:HSDB.htm, 1988b. *Nitrogen dioxide*. CASRN: 10102-44-0. 1998b.

Nordholm, A.F., Rossi, J. 3rd, Ritchie, G.D., et al. Repeated exposure of rats to JP-4 vapor induces changes in neurobehavioral capacity and 5-HT/5-HIAA levels. *J. Toxicol. Environ. Health A,* 56: 471–499, 1999.

Odkvist, L.M., Arlinger, S.D., Edling, C., et al. Audiological and vestibulo-oculomotor findings in workers exposed to solvents and jet fuel. *Scand. Audiol.*, 16: 75–81, 1987.

Odkvist, L.M., Larsby, B., Tham, R., et al. Vestibulo-oculomotor disturbances caused by industrial solvents. *Otolaryngol. Head. Neck. Surg.*, 159: 326–338, 1983.

Olson, B.A., Gamberale, F., and Iregren, A. Co-exposure to toluene and *p*-xylene in man: central nervous functions. *Brit. J. Ind. Med.*, 42: 117–122, 1985.

Olsen, D.M., Mattie, D.R., Gould, W.D., et al. A pilot study of occupational assessment of Air Force personnel exposure to jet fuel before and after conversion to JP-8. Air Force Research Lab, Wright-Patterson AFB, OH Human Effectiveness Directorate. Interim Report 1, 1998.

Pleil, J.D., Smith, L.B., and Zelnick S.D. Personal exposure to JP-8 jet fuel vapors and exhaust at air force bases. *Environ. Health Perspect.*, 108: 183–192, 2000.

Porter, H.O. Aviators intoxicated by inhalation of JP-5 fuel vapors. *Aviat. Space Environ. Med.*, 61: 654–656, 1990.

Ritchie, G.D., Rossi, J. 3rd, Nordholm, A.F., et al. Effects of repeated exposure to JP-8 jet fuel vapor on learning of simple and difficult operant tasks by rats. *J. Toxicol. Environ. Health A*, 64: 385–415, 2001a.

Ritchie, G.D., Still, K.R., Alexander, W.K., et al. A review of the neurotoxicity risk of selected hydrocarbon fuels. *J. Toxicol. Environ. Health. B Crit. Rev.*, 4: 223–312, 2001b.

Ritchie, G.D., Still, K., Rossi, J. 3rd, et al. Biological and health effects of exposure to kerosene-based jet fuels and performance additives. *J. Toxicol. Environ. Health B Crit. Rev.*, 357–451, 2003.

Rosenberg, N.L., Spitz, M.C., Filley, C.M., et al. Central nervous system effects of chronic toluene abuse–clinical, brainstem evoked response and magnetic resonance imaging studies. *Neurotoxicol. Teratol.*, 10: 489–495, 1988.

Rossi, J. 3rd, Nordholm, A.F., Ritchie, G.D., et al. Effects of repeated exposure of rats to JP-5 or JP-8 jet fuel vapor on neurobehavioral capacity and neurotransmitter levels. *J. Toxicol. Environ. Health A*, 63: 397–428, 2001.

Rossi, J. 3rd, Ritchie, G.D., Macys, D.A., et al. An overview of the development, validation, and application of neurobehavioral and neuromolecular toxicity assessment batteries: potential applications to combustion toxicology. *Toxicology*, 115: 107–117, 1996.

Smith, L.B., Bhattacharya, A., Lemasters, G., et al. Effect of chronic low-level exposure to jet fuel on postural balance of US Air Force personnel. *J. Occup. Environ. Med.*, 39; 623–632, 1997.

Savolainen, K., Kekoni, J., Riihimaki, V., et al. Immediate effects of *m*-xylene on the human central nervous system. *Arch. Toxicol. Suppl.*, 7: 412–417, 1984.

Struwe, G., Knave, B., and Mindus, P. Neuropsychiatric symptoms in workers occupationally exposed to jet fuel—combined epidemiological and casuistic study. *Acta Psychiatr. Scand. Suppl.*, 303: 55–67, 1983.

Ullrich, S.E., and Lyons, H.J. Mechanisms involved in the immunotoxicity induced by dermal application of JP-8 jet fuel. *Toxicol Sci.*, 58: 290–298, 2000.

Wada, H. Toluene and temporal discrimination in rats: effects on accuracy, discriminability, and time estimation. *Neurotoxicol. Teratol.*, 21: 709–718, 1999.

WHO (World Health Organization) Working Group. *Chronic Effects of Organic Solvents on the Central Nervous System and Diagnostic Criteria*. WHO Regional Office, Copenhagen, and the Nordic Council of Ministers, Oslo, Norway, 1985.

Wolfe, R.E., Kinkead, E.R., Feldman, M.L., et al. Acute Toxicity Evaluation of JP-8 Jet Fuel and JP-8 Jet Fuel Containing Additives. Mantech Environmental Technology, Inc. Final Report ADA318722, November 1996.

Woodruff-Pak, D.S., and Thompson, R.F. Classical conditioning of the eyelid response in the delay paradigm in adults aged 18-83 years. *Psych. Aging*, 3: 219–229, 1988.

Yin, S.N., Li, G.L., Hu, Y.T., et al. Symptoms and signs of workers exposed to benzene, toluene or the combination. *Indus. Health*, 25: 113–130, 1987.

Yost, D.M., and Montalvo, D.A. Comparison of diesel exhaust emissions using JP-8 and low-sulfur diesel fuel. *TFLRF-308*. San Antonio, TX: Southwest Research Institute, 1995.

Zielinska, B., Sagebiel, J., Arnott, W.P., et al. Phase and size distribution of polycyclic aromatic hydrocarbons in diesel and gasoline vehicle emissions. *Environ. Sci. Technol.,* 38: 2557–2567, 2004.

5 Differential Protein Expression Following JP-8 Jet Fuel Exposure

Frank A. Witzmann and
Mark L. Witten

CONTENTS

INTRODUCTION

This chapter summarizes the collaborative efforts of two laboratories in assessing the toxicological effects of exposure to JP-8 jet fuel using several routes of exposure, and analyzing protein expression in various organ targets by proteomic techniques. JP-8 aerosol and vapor exposures included several rodent models where the effects on lung, liver, kidney, or testis tissues were evaluated. Additionally, in vitro exposures to JP-8 were conducted using alveolar type II epithelial cell cultures. While many insights into the effects of JP-8 exposure were gained through these studies, the limitations of the analytical approach also became self-evident. While two-dimensional gel electrophoresis (2-DE) is a powerful tool in the assessment of differential protein expression and is well-suited for toxicologic studies, the sensitivity of the approach clearly limited the scope of the analyses and many low-expression members of the target-tissue proteomes eluded clear interpretation. This statement is not to minimize the importance of the results, but rather to draw attention to recent developments in sample preparation for all manners of proteomic analyses, as described at the end of this chapter.

JP-8 continues to be the dominant military and civilian (Jet A) aviation fuel, although synthetic fuels (i.e., aliphatic hydrocarbon fuel S-8) are being developed, used, and their toxicity assessed [36, 50]. JP-8 is a kerosene-like, complex mixture containing over 228 hydrocarbon constituents, often including performance additives. Because of its high flash point and low vapor pressure, JP-8 has characteristics that reduce the risk of fire and explosion but increase its availability for (dermal and inhalation) exposure, as compared to other previously used hydrocarbon fuels. Consequently, individuals on or near flight facilities and elsewhere where the fuel is used are likely to be exposed [68, 69]. The most toxicologically significant human exposure routes seem to be dermal contact to neat fuel, JP-8 vapor or aerosol inhalation, and inhalation of combusted fuel by-products. The effects of these exposures have been shown to include neurobehavioral [56] (e.g., fatigue, depressed mood, lack of initiative, dizziness, sleep disorders, abnormal electroencephalograms, shortened attention spans, and decreased sensorimotor speed); liver function [15] (e.g., hepatic drug metabolism system induction); and immune system dysfunction [37, 52] (e.g., altered white blood cell [WBC] counts and lymphocytic DNA damage).

The outcome of several military exposure studies may have inspired an extensive toxicologic evaluation of JP-8 effects in animal models [32, 53]. The results of these studies suggest toxicologically significant effects in a wide range of organ systems in contrast to JP-4, the predecessor jet fuel [7]. A number of target organs and tissues have been associated with JP-8 exposure, including the immune system [27–29, 41, 51, 70], lungs [31, 59, 64, 72], skin [11, 13, 40, 60, 63], liver [24], and the central nervous system [4, 43, 55, 61]. Despite the studies cited above, the molecular mechanisms underlying the acute and persistent effects of JP-8 observed in rodents were generally not elucidated, and the human risk associated with JP-8 exposure remains the subject of debate. To support this evaluation, our laboratories investigated JP-8 effects at the protein level. We conducted a number of studies in which the proteomes of various JP-8 organ and tissue targets were analyzed by two-dimensional electrophoresis (2-DE) and label-free quantitative mass spectrometry (LFQMS). Figure 5.1 illustrates the analytical aspects of this dual research platform. All of the initial proteomic analyses of JP-8 effects were conducted using 2-DE, and thus are presented first. At the end of the chapter, the results of the more recent LFQMS-based experiments are presented.

THE 2-DE GEL-BASED PROTEOMIC APPROACH

In 2-DE, complex protein mixtures are solubilized, denatured, and subjected to orthogonal separation methods: first by protein charge via isoelectric focusing (IEF), and then by mass in sodium dodecyl sulfate polyacrylamide gel electrophoresis (SDS-PAGE). The final product of 2-DE separation is essentially an in-gel array of proteins, each assuming a coordinate position corresponding to the unique combination of isoelectric point (pI) and mass (MW) [26]. Resulting 2D protein patterns

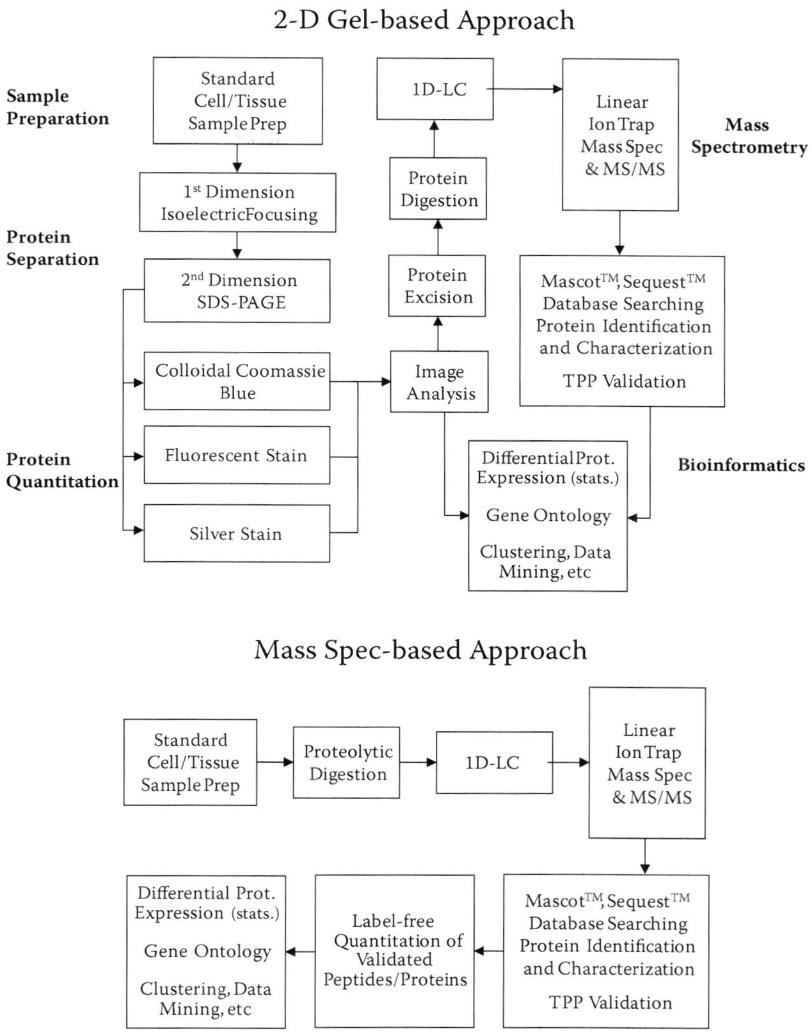

FIGURE 5.1 Complementary gel-based and gel-free components of proteomic platform for protein quantitation and identification in differential expression used in the studies described in this chapter.

are visualized by any of a number of methods, including visible and fluorescent dyes, silver stains, or autoradiography. Typically, scanned gel images are analyzed by any of a number of ever-improving 2D gel analysis software packages. It is here that both the strengths and weaknesses of this approach become evident. Protein expression comparisons (e.g., differential expression) are easily made, as differences in protein spot density are readily detectable, can be quantified robustly, and compared statistically. However, because the dynamic range of protein expression in most whole cell or tissue lysates is vast, only the most abundant proteins from 2D gels can be analyzed, and many proteins are overlooked. Furthermore, proteins with pIs at extreme pH and significant hydrophobicity are typically absent from 2D gel patterns. Typically, only approximately 2,000 of the most abundant proteins in a particular cell or tissue can be reliably separated and identified.

Despite its shortcomings, 2D gel-based experimentation remains an exceptional approach for assessing differential protein expression. 2-DE not only enables the researcher to separate, detect, and quantify the proteins, but it is also a preparative technique. Proteins separated on a 2D gel are isolated at their unique pI and MW coordinates and therefore can be physically excised, identified, and characterized using a variety of mass spectrometric techniques. This approach enables the researcher to (1) determine relative expression levels of cellular proteins; (2) identify the resolved proteins; (3) characterize modifications such as phosphorylation, glycosylation, proteolytic cleavage, etc.; and (4) discover unpredicted or hypothetical gene products.

ELECTROPHORESIS, IMAGE ANALYSIS, AND PROTEIN IDENTIFICATION

Whole target tissue pieces or specific tissue regions were solubilized using a standard 2-DE lysis buffer containing urea, a nonionic detergent, a reducing agent, and carrier ampholytes for nucleic acid precipitation and buffering, and separated by 2-DE using a 20-gel platform, essentially as described in [75]. The resulting slab gels were stained using a colloidal Coomassie Blue G-250 procedure [10] and were scanned and analyzed using PDQuest™ software. Statistical comparisons between mean individual protein expressions were conducted by calculation of Student-*t* after data export to Excel software.

Protein spots of interest were cut from the stained gels and processed for tryptic digestion, peptide recovery, and mass analysis using matrix-assisted laser desorption/ionization-time-of-flight mass spectrometry (MALDI-TOF MS) [81]. After acquisition of the peptide mass spectra, proteins were identified by manually searching the ProFound™ peptide mass database.

More recently, owing to the high false-discovery rates associated with peptide mass fingerprinting, proteins cut from gels have been identified using tandem mass spectrometry. In this case, tryptic peptide samples are injected into a Thermo Scientific LTQ linear ion trap mass spectrometer via an auto-sampler coupled to a nanoHPLC apparatus. Peptides are ionized by nanoelectrospray, and data are collected in a "Triple-Play" (MS scan, Zoom scan, and MS/MS scan). The acquired mass spectral data are then searched against a FASTA format database (corresponding to the species used in the experiment) assembled using gene annotations publicly available from PIR (Protein Information Resource, http://pir.georgetown.edu) using the SEQUEST (v. 28 rev. 12) program in Bioworks (v. 3.3). To avoid false-positive identifications, the searched peptides and proteins are validated via PeptideProphet and ProteinProphet in the Trans-Proteomic Pipeline (TPP) (http://tools.proteomecenter.org/software.php), and only those proteins with greater than 90% confidence (containing multiple peptides with greater than 90% confidence) are considered positive identifications.

SHORT-TERM JP-8 AEROSOL EXPOSURE

LUNG PROTEOMICS

Given the proclivity for inhalation of JP-8 fuel vapor and aerosols, it was logical to investigate the lungs as a probable target organ. The development of nose-only inhalation exposure protocols in rodents enabled in-depth evaluation of systemic injury due to highly controlled JP-8 aerosol exposures [32, 58]. Using this model in Fischer 344 rats, apoptosis of lung epithelial cells was observed as a result of 1 hour/day JP-8 exposure for 7, 28, and 56 days at both low (approximately 500 mg/m^3) and high (approximately 1,000 mg/m^3) JP-8 levels. At JP-8 levels (50 mg/m^3) well below the current inhalation safety standard, morphological lung injury has been observed [59]. In general, lung injury in the form of increased epithelial permeability, interstitial edema, alveolar septal damage, alveolar epithelial type II cell injury, and accumulation of inflammatory cells all demonstrate the histological and morphological damage in the pulmonary system as a direct effect of chronic JP-8 exposure.

Our initial proteomic study [73] investigated differential expression of cytoplasmic proteins, isolated from whole lung homogenates, following acute exposure to varying levels of 7-day, 1 hour/day JP-8 aerosol exposure in Swiss-Webster mice. When studied by image analysis, these cytoplasmic 2D gel patterns contained an average of approximately 800 completely matched protein spots, of which 30 were found to be significantly up-regulated and 135 down-regulated ($p < 0.05$) by 2,500 mg/m^3 JP-8 jet fuel exposure. The lower exposure (1,000 mg/m^3) had a lesser effect, as 21 proteins were up-regulated and 99 down-regulated ($p \leq 0.05$). At these relatively high exposure levels, the dominant trend in protein expression was toward down-regulation, nearly four-to-one. Tryptic digestion and mass spectrometric analysis (MALDI-MS and ESI-MS/MS), with identification of some of these differentially expressed proteins cut from the gels, revealed that JP-8 was effective in modulating protein expression in four general areas (see Table 5.1).

High-dose JP-8 exposure decreased the expression of cytoplasmic proteins hsc70, Hop, transitional endoplasmic reticulum ATPase (TER ATPase), and Nedd5 (septin). Hsc70 (heat shock cognate, hsp73) is a cytoplasmic/nuclear homologue of hsp70 involved in the folding, translocation, and trafficking of newly synthesized proteins [19]. Because hsc70 is actually resolved as three spots, a true measure of hsc70 expression requires the summation of each charge variant. When this is done, a nearly 20% decline in hsc70 expression was calculated. If this observation truly reflects its total cellular down-regulation, then the implications for impaired protein synthesis and protein misfolding are clear, and the notion is supported by the nearly identical percent decline in Hop expression. Hop (hsc70/hsp90-organizing protein) is required by hsc70 (and hsp70) for protein folding activity [38]. If one follows this line of reasoning further, the potential increase in the number of misfolded proteins that might result from JP-8 intoxication leads quite naturally to the more classic response to toxic stress found in the apparent up-regulation of hsp84 (murine hsp90-β). This molecular chaperone carries out generalized functions in protein folding and, as mentioned above, an interaction with hsp70 via Hop.

TABLE 5.1
Summary of Mouse Lung Cytoplasmic Proteins Altered by Exposure to 2,500 mg/m³ JP-8 Aerosol, and Identified by MALDI-MS or ESI-MS/MS

Function	Protein Identification	% Change[a]	MW	pI
Protein processing	Transitional endoplasmic reticulum ATPase (TER ATPase)	−60	89,308.3	5.14
	Heat shock cognate 70 protein (hsc70)[#]	−4	70,837.4	5.38
	Heat shock protein hsp 84 (murine hsp90-beta)	+36	83,194.1	4.97
	hsc70 (charge variant)	−56	70,837.4	5.38
Cell ultrastructure	Nedd5 protein	−27	41,525.8	6.10
	Nedd5 protein	−27	41,525.8	6.10
	Ulip2 protein (dihydropyrimidinase related protein-2)[b]	+4	62,170.8	5.95
	Ulip2 protein (charge variant)	−38	62,170.8	5.95
	Laminin receptor	−46	32,719.1	4.74
Toxic stress response	Heat shock 70 protein (hsc70)	−4	70,837.4	5.38
	Heat shock protein hsp 84 (murine hsp90-beta)	+36	83,194.1	4.97
	Heat shock 70 protein (hsc70)	−56	70,837.4	5.38
	Hop (hsc70/hsp90-organizing protein; extendin)	−18	62,582.5	6.40
	Thioether S-methyltransferase (TEMT)	+8	29,460.0	6.00
	1-cys peroxiredoxin	−20	24,826.8	5.98
	Thiol-specific antioxidant	−58	21,778.8	5.20
	Glutathione S-transferase	+66	27,497.8	6.91
CO_2/Acid-Base Balance	Carbonic anhydrase II	+49	29,068.7	6.49
	Carbonic anhydrase I	+84	28,320.6	6.44
Other	Annexin V (lipocortin V; endonexin II; calphobindin I)[b]	−4	35,732.7	5.03
	45-kDa secretory protein	+4	46,027.4	5.64

Source: Adapted from Witzmann, F.A., Bauer, M.D., Fieno, A.M., et al. *Electrophoresis,* 20, 3659, 1999a.

[a] Relative to sham exposed mice; [b] ID confirmed by ESI-MS/MS.

Additional impairment in protein processing was suggested by a decline in total transitional endoplasmic reticular (TER) ATPase and Nedd5. TER ATPase is required for the vesicular transport of proteins and lipids, such as phosphatidylcholine (a major constituent of pulmonary surfactant) and cholesterol, to the Golgi apparatus [83]. Furthermore, the expression of Nedd5 protein, a septin (cytoskeletal Rho-like GTPase) that organizes submembranous structures and is involved in vesicle trafficking, was decreased by JP-8. Together with the altered expression discussed above, these protein alterations strongly imply a JP-8 mediated disturbance of post-translational protein and lipid processing in the lung and may underlie the appearance of vacuolated type II alveolar cells [73] as evidence of JP-8-mediated disruption of surfactant processing.

Previously, JP-8 was shown to cause ultrastructural damage preferentially in Clara cells and type II alveolar epithelium [31, 59]. Previous examination of JP-8 effects on lung ultrastructure revealed breaks in the alveolar-capillary membrane and increased epithelial permeability, replacement of type I by type II alveolar epithelia, areas of microvillar and epithelial loss, and extravasation and hemorrhage [32]. The decline in total Ulip2 expression, along with Nedd5 and laminin receptor protein, provides additional molecular correspondence between the observations mentioned above. Ulip2, also known as unc-33–like phosphoprotein and dihydropyrimidinase related protein-2 (DRP-2), is prominent in neuronal cells and ubiquitously expressed, except in liver. Decreased Ulip2 expression has been directly related to microvillous atrophy [2] and so its down-regulation documented via 2-DE-based proteomics may explain the JP-8 mediated loss of microvilli observed in type II alveolar cells [32].

Chemical exposure is generally associated with an increased expression of specific detoxification systems, bioactivation systems, and stress response proteins that are specific to the target tissue. The lung in general, and type II alveolar cells specifically, are no exception. Glutathione-*S*-transferases (GST) represent a class of mostly cytoplasmic detoxification enzymes that catalyze the conjugation of xenobiotics and their electrophilic metabolites with reduced glutathione. Thus they constitute an important defense mechanism against chemical toxicity/carcinogenicity as well as serving as markers of toxic exposure in the lung [21].

Exposure to JP-8 (2,500 mg/m^3) resulted in a 66% increase in the expression of GSTP1 over the unexposed group (35% by 1,000 mg/m^3), confirming JP-8 lung toxicity. Further confirmation was provided by increased expression of hsp84 (murine hsp90), a constitutive cytoplasmic protein induced in response to various biological stresses as well as chemical exposure in the lung [3]. The moderate induction (8%, $p < 0.01$) was observed in one of the more abundant cytoplasmic proteins resolved. This reduction was seen in thioether *S*-methyltransferase, an important enzyme in the metabolism of sulfur- and selenium-containing compounds [47]. JP-8 exposure also reduced the expression of thiol-specific antioxidant thioredoxin peroxidase 1 by more than 50%. Thioredoxin peroxidase 1 catalyzes the removal of thiyl radicals before they generate more reactive radicals, thereby protecting biomolecules from oxidative damage [82]. The observed down-regulation may increase the risk of oxidative damage to the lung during JP-8 and other toxicant exposure.

The protein alterations occurring during JP-8 exposure scenarios that may most closely reflect the differences between normal lung physiology and JP-8-related functional injury are the induction of two carbonic anhydrase forms. Carbonic anhydrases (CA) catalyze the equilibration of carbon dioxide and carbonic acid, thus playing important roles in the lung, for example, carbon dioxide exchange, lung fluid balance and secretion, acid-base balance, and equilibration of ventilation/perfusion ratios. Both CAI (84%) and CAII (49%) were up-regulated in JP-8-exposed mice. It is well-known that alveolar type II cells are the principal source for CAII activity in rat lung [20] while CAI is dominant in erythrocytes. The 84% increase in CAI expression is most likely the result of erythrocyte infiltration that accompanies jet fuel-mediated lung damage [73].

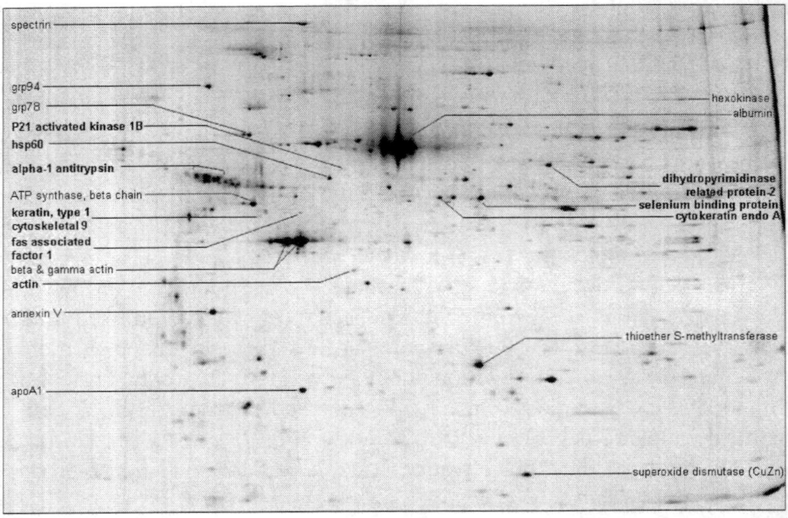

FIGURE 5.2 2-DE gel image of Swiss-Webster mouse lung showing 20 proteins identified by peptide mass fingerprinting. Whole lung was solubilized and separated on an immobilized pH gradient strip (pH 3–10), followed by SDS-PAGE separation on an 11% to 17% acrylamide gradient slab gel stained with colloidal Coomassie blue.

MODERATE CONCENTRATION/DURATION JP-8 EXPOSURES

Lung Proteomics

To analyze a more occupationally relevant dose of aerosolized JP-8 (between the occupational safety standard of 350 mg/m^3 [inhalation] and 50 mg/m^3 where the injury threshold for JP-8 has been observed in mice [59]) a nose-only inhalation study was conducted in mice. Male Swiss-Webster mice were exposed for 1 hour/day to aerosolized JP-8 jet fuel at concentrations of 250, 1,000, or 2,500 mg/m^3 for 7 days [16]. Whole lung samples were prepared and separated by 2-DE as described earlier, and the protein expression in exposed and non-exposed samples compared. A representative gel image is shown in Figure 5.2, depicting roughly 1,000 matched and analyzed protein spots.

Image analysis revealed exposure-mediated expression changes in 41 proteins (or 4% of the detectable proteome). In direct contrast to the higher-level JP-8 exposures described earlier in this chapter, far fewer proteins (41) were affected by the lower dose of 250 mg/m^3, although 31 proteins were up-regulated and only 10 were down-regulated. Of the altered proteins, several identified proteins were found to be significant markers of JP-8-induced stress on lung epithelial cells (see Table 5.2). For example, the decrease in α1-antitrypsin, and with it decreased antiproteinase activity, has significant implications in potential damage to lung ultrastructure [14]. In what may be a related response, lung proteins involved in cell remodeling activities, including actin, keratin type 1 cytoskeleton 9, desmin fragments, and cytokeratin endo A, were all up-regulated. Furthermore, a generalized stress response to the

TABLE 5.2
Effect of 250 mg/m³ JP-8 Aerosol Exposure on Protein Expression in Various Whole Lung Proteins Resolved on 2D Gels

Protein	Control	JP-8	p Value
Actin	1,038.3	1,546.7	0.001
α1-Antitrypsin	2,869.6	2,296.2	0.02
Cytokeratin endo A	1,007.9	1,365.2	0.03
Desmin fragment	1,02.3	137.4	0.03
Dihydropyrimidinase related protein-2	1,278.0	1,681.3	0.04
fas-associated factor 1	349.9	503.6	0.05
Hexokinase, type 1	1,131.6	714.0	0.05
hsp60	1,295.2	1,753.1	0.007
Keratin, type 1 cytoskeletal 9	190.4	266.7	0.02
P21 activated kinase 1B	198.5	278.5	0.0004
Selenium binding protein 1	2,335.5	2,698.8	0.05

Source: Adapted from Drake, M.G., Witzmann, F.A., Hyde, J., et al. *Toxicology,* 191, 199, 2003.
Note: Proteins were identified by peptide mass fingerprinting; values are mean integrated density obtained via PDQuest™ Gel Analysis Software; probabilities calculated by Student's *t*-test.

intoxication was indicated by the up-regulation of the mitochondrial stress protein hsp60 and the apoptotic mediator Fas-associated factor 1.

The overall response to the 250-mg/m³ exposure was, as expected, lower in magnitude than the higher exposure level, yet the protein expression changes were generally opposite in the direction observed. The lowered response to the 250-mg/m³ exposure may reflect an adaptive or compensatory response to the JP-8 perturbation and was, thus, reflected as protein up-regulation and mobilization. This is in contrast to the high-dose exposures where significant injury processes may already be in effect, and the more widespread protein down-regulation manifests the cellular injury responses coinciding with histological evidence of JP-8 tissue effects.

Renal Proteomics

The kidney is essential in electrolyte and water balance, acid-base homeostasis, recovery and excretion of compounds, as well as many other physiological functions. Consequently, a very large fraction of the total cardiac output is delivered to the kidney to meet these demands. The kidney also serves as a major route for xenobiotic excretion, particularly of drugs and their metabolites. Like the liver, the kidney is a site for substantial xenobiotic metabolism [44]. For these reasons, the kidney is a prime target for a variety of xenobiotics and a particularly appropriate proteome to explore for the effects of JP-8.

As in the lung proteome experiments described previously, 2-DE of cytoplasmic proteins prepared from whole kidney homogenates of Swiss-Webster mice exposed to 1,000 mg/m^3 aerosolized JP-8 for 1 hour/day for 7 days [16] revealed nearly 1,000 proteins that were matched across 12 individual patterns ($n = 6$). When between-group statistics were calculated and protein expression compared [74], it was apparent that JP-8 exposure had significantly altered ($p < 0.05$) the expression of 56 proteins spots (21 up-regulated and 35 down-regulated); 26 of these protein spots (representing 22 individual proteins) were identified by peptide mass fingerprinting, and these proteins, listed in Table 5.3, can be categorized functionally as follows: (1) ultrastructural abnormalities, (2) altered protein processing, (3) metabolic effects, and (4) stress protein/detoxification system response.

Similar to JP-8's pulmonary effects, alterations in ultrastructural proteins and those involved in various aspects of protein synthesis and processing were observed. The expression of ultrastructural proteins tropomyosin 4 and high mobility group 1 protein (HMG1) decreased with JP-8 exposure, while ezrin increased. Ezrin belongs to a family of plasma membrane-cytoskeleton linking, actin binding proteins (ezrin-radixin-moesin family) involved in signal transduction, growth control, cell-cell adhesion, and microvilli formation. In the kidney, ezrin plays a role in the cytoskeletal organization, such as actin filament reassembly accompanying podocyte injury and may serve as a marker for podocytes in normal and injured glomeruli [35]. The apparent up-regulation of ezrin may thus play a role in the JP-8-mediated response to glomerular injury or brush-border damage, and signify the onset of the repair processes. In a related manner, two low abundant isoforms of another structural element, tropomyosin (TM, a cytoskeletal protein believed to bind to actins to stiffen and stabilize the filaments), were affected differently. TM1 up-regulation (4%, but not significantly increased) suggests a trend toward accumulation of actin bundles such as those found in filipodia or stress fiber elements. However, in view of the relative unresponsiveness of TM1 to JP-8, down-regulation of TM4 (–11%, statistically significant effect) is more interesting. TM isoform expression is known to be highly cell and compartment specific. In kidney cells, TM1 has been shown to reside on stress fibers in tubuloepithelial cells, while TM4 is found at adhesion sites near the apical region of some epithelial cells, but is absent from others. While an explanation for the TM alterations is unclear, these changes suggest JP-8-mediated modification of ultrastructural or adhesion dynamics in rodents. Whether or not these changes are indicative of tubular or glomerular epithelial or mesangial cell injury remains to be determined.

Another renal protein alteration with ultrastructural implications concerns the protein identified as the HMG1 amphoterin. HMG1 proteins are normally associated with the nucleus where they bind preferentially to single-stranded DNA, unwind double-stranded DNA, and are involved in DNA compaction and transcriptional control. HMG1 detection in the kidney cytoplasm and its apparent down-regulation by JP-8 exposure may be linked to its association with the plasma membrane of filipodia in process-growing cells (e.g., glomerular epithelial cells or podocytes of the renal filtration apparatus), as well as the extension of cytoplasmic processes in developing cells [45]. Together, the protein alterations indicate a change in the ultrastructural proteome of the kidneys that may have significant functional consequences.

TABLE 5.3
Summary of Mouse Kidney Cytoplasmic Proteins, Altered by Exposure to 1,000-mg/m³ Aerosolized JP-8 for 1 hour/day for 5 days, and identified by MALDI-MS or ES-MS/MS

Function	Protein Identification	% Change[a]	MW	pI
Protein processing	Heat shock 70 protein (hsc70)	−8	70,837.4	5.38
	Heat shock 70 protein (hsc70) (acidic charge variant)#	−14	70,837.4	<5.38
	Rab GDP dissociation inhibitor beta#	+11	50,537.4	5.93
	Elongation factor 2	−12	95,284.6	6.41
	Elongation factor 2 (acidic charge variant)	+5	95,284.6	<6.41
	Aminopeptidase#	−11	52,607.3	5.77
	hsp70	+1	70,837.4	5.38
	hsp60	0	58,870.2	5.48
	grp75	+3	73,448.6	5.72
Cell ultrastructure	Ezrin (P81, cytovillin, villin-2)	+12	69,346.1	5.83
	Tropomyosin (embryonic fibroblast isoform; TM-4)	−11	28,509.8	4.66
	High mobility group 1 protein (amphoterin)	−11	24,863.9	5.75
	Annexin VI (lipocortin protein)#	−14	75,886.7	5.34
	Tropomyosin (beta, TM-1)	+4	32,944.9	4.61
Toxic stress response	Heat shock 70 protein (hsc70)	−8	70,837.4	5.38
	Heat shock 70 protein (hsc70) (acidic charge variant)#	−14	70,837.4	<5.38
	Heat shock protein 70 (hsp70)	+1	70,837.4	5.38
	hsp60 protein	0	58,870.2	5.48
	grp75	+3	73,448.6	5.72
	Thioether S-methyltransferase (TEMTase)	−10	29,460.0	6.00
	Amine N-sulfotransferase	+11	35,099.5	5.44
	Cu/Zn superoxide dismutase (SOD)	−10%	15,761.5	6.03
	hsp20	−11%	19,000.0	5.65
Metabolic	alpha-enolase	−23%	47,125.0	6.37
	alpha-enolase (acidic charge variant -2)	−10%	47,125.0	<<6.37
	alpha-enolase (acidic charge variant -1)	−12%	47,125.0	<6.37
	lactate dehydrogenase	+10%	36,441.1	5.70
	calbindin	−11%	29,862.9	4.71
	pyruvate carboxylase	+32%	129,685.3	6.25
	carbonic anhydrase II	−10%	29,091.7	6.53
	phosphoglycerate kinase 1	−30%	44,536.7	7.54

Source: Adapted from Witzmann, F.A., Bauer, M.D., Fieno, A.M., et al. *Electrophoresis,* 21, 976, 2000a.

[a] Relative to sham exposed mice.

As in the lung (see Table 5.1), a decline in renal protein processing and vesicular trafficking activities associated with JP-8 exposure was observed in the kidney. Whereas a different set of proteins was affected in the kidney, a similar protein processing effect is suggested by decreased aminopeptidase and increased Rab GDP-dissociation inhibitor beta (Rab GDI-2) expression. Aminopeptidase is an abundant renal exopeptidase that plays a substantial role in the normal disposition of a variety of proteins. Aminopeptidase regulates cell proliferation, adhesion, cell signaling, cell activation, differentiation, and cell-cell communication by activating precursor proteins or inactivating specific cellular proteins. The decreased aminopeptidase expression suggests reduced processing and turnover of intracellular proteins, further implying impaired cell growth and viability observed with JP-8 exposure.

In a related subcellular system, renal vesicular trafficking activities involve small GTPases of the Rab subfamily whose membrane associations and dissociations are regulated by GDP-dissociation inhibitor proteins such as Rab GDI-beta. The expression of this renal cytoplasmic protein was increased by JP-8 exposure, a response that could conceivably hinder protein secretion by sequestering membrane-associated Rab GTPases in the cytoplasm [12, 17]. Because Rab GDI complexes are also required for vesicular transport from the endoplasmic reticulum (ER) to the Golgi stack, the increase in Rab GDI-beta could imply decreased vesicular transport of newly synthesized proteins to the Golgi apparatus. This response is supported by the observed decrease in annexin VI, a protein that promotes interactions between membranes that are destined to undergo fusion.

It should also be noted that the change in EF-beta expression (Table 5.3) is a qualitative one; that is, the total EF-beta expression (e.g., sum of all charge forms) was unaffected by JP-8 exposure. As noted in the original publication of this observation [74], a slight but significant decrease in charge modification index (CMI) calculated for elongation factor 2 (EF-beta) indicates a decreased chemical modification (fewer charge variants). CMI, a quantitative expression of the extent of protein post-translational modification [1], suggests that JP-8 exposure is associated with decreased EF-beta phosphorylation. This, in turn, implies that EF-beta activation was therefore [8] stimulated by JP-8 exposure, and protein elongation was enhanced. The true nature of EF-beta chemical modification and the consequence of the implied changes (decreased phosphorylation or ADP-ribosylation) remain to be determined.

Enolase, a cytoplasmic glycolytic enzyme, resides in cortical renal epithelia and is phosphorylated in vivo. Several charge forms of alpha-enolase resulting from this post-translational modification were identified by peptide mass fingerprinting (Table 5.3). JP-8 exposure decreased total alpha-enolase expression (sum of all charge variants) by 12% ($p < 0.05$) but had no effect on protein charge modification [74]. Because the key role of alpha-enolase is glycolysis, it seems that JP-8 exposure via the lung may have decreased glycolytic capacity in the kidney. This is reinforced by a 30% decrease in phosphoglycerate kinase expression, another major glycolytic enzyme, preceding by one additional step, the enolase-catalyzed reaction.

Coincident with this apparent decline in glycolytic capacity is the moderate up-regulation of lactate dehydrogenase (LDH) and pyruvate carboxylase (PC) (a mitochondrial matrix protein)—adaptations suggesting an increased gluconeogenic state in kidney cells following JP-8 exposure. This particular LDH form, LDH-B or H chain, catalyzes the conversion of lactate to pyruvate (perhaps yielding reducing equivalents for detoxification systems) for carboxylation in the mitochondria by PC. Once again, the appearance of a mitochondrial protein in the cytoplasmic fraction is unusual. This may be due to mitochondrial damage prior to, or during fractionation, or it is the newly synthesized fraction isolated in the cytosol prior to mitochondrial import. Nonetheless, the metabolic enzyme alterations observed in this experiment suggest a JP-8-mediated shift from glucose breakdown to gluconeogenesis in the kidney. The underlying reasons for this shift are unclear, but the implications for chronic exposure are noteworthy.

Stress proteins normally are viewed as sensitive indicators of toxic exposure [66], as seen in the lung (see Table 5.1) in response to JP-8. In this experiment, four prominent stress proteins were identified in the kidney cytoplasm; mitochondrial stress proteins grp75 and hsp60 (presumably as contaminants introduced from ruptured mitochondria), and cytoplasmic hsc70 (constitutive) and hsp70 (inducible). Surprisingly, the expression of all these proteins was not significantly affected by JP-8 exposure.

However, an equally relevant member of the phase II detoxification system, amine *N*-sulfotransferase (ST3A1), was detected in the kidney cytoplasm, and its expression was increased by JP-8 exposure. Although the expression of sulfotransferases is thought to be under hormonal control, their regulation is poorly understood. In contrast, two prominent detoxification enzymes—thioether S-methyltransferase and superoxide dismutase—were down-regulated by JP-8. Unlike the lung, where JP-8 induced glutathione *S*-transferase, the decreased expression of two important detoxifiers in the kidneys is puzzling.

EFFECTS OF LONG-TERM JP-8 VAPOR EXPOSURE

Increasing evidence suggests that repeated exposure to low levels of certain volatile organic chemical mixtures (e.g., JP-8) may be involved in the etiology of several poorly defined diagnostic conditions [53], with symptoms quite unique from the single dose or acute responses generally observed. Of utmost concern are occupational exposures suffered by individuals whose long-term service involves low-level, yet regular JP-8 exposure. Additionally, the potential persistence of the potential effects of JP-8 is relevant to those who are no longer occupationally exposed, but were chronically exposed at some point. Chronic exposure to JP-8 is thus an additional paradigm in which proteomics is applicable and has been applied in relevant animal models. This section of the chapter addresses two studies [75, 76] that examined the differential protein expression in male reproductive toxicity after prolonged exposure to JP-8 vapor, and possible persistent effects on the kidney and liver after long-term JP-8 vapor exposure and subsequent recovery. It was hypothesized that molecular markers of vapor exposure in rats that may occur in the absence of easily observable health consequences might prove useful in predicting human health effects from career-long exposures.

TESTIS PROTEOMICS

Of the few investigations that have studied the reproductive effects of jet fuel exposure, only [49] reported that repeated exposure of male rats to JP-8 vapor induced

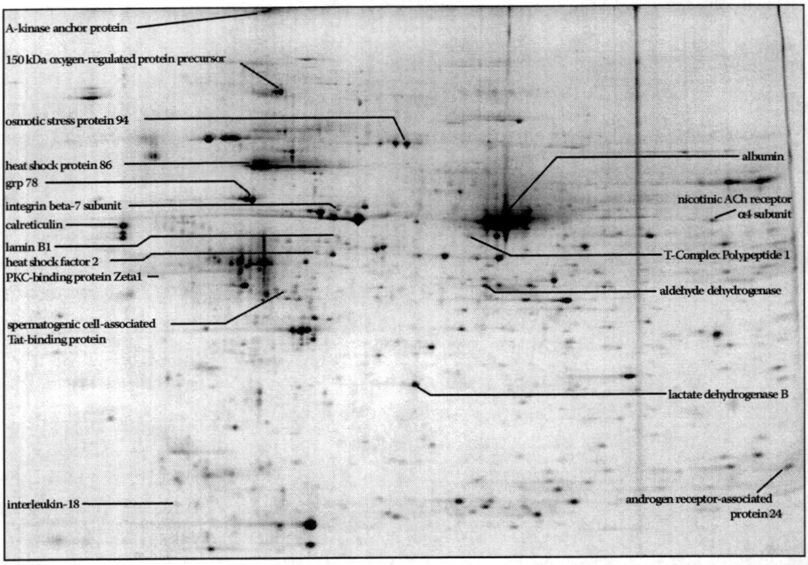

FIGURE 5.3 2-DE gel image of normal, untreated Sprague-Dawley rat testis showing 18 proteins identified by peptide mass fingerprinting whose expression was altered by JP-8 exposure. The testis was solubilized and separated by tube-gel IEF, pH gradient 4–8, followed by SDS-PAGE separation on an 11% to 17% acrylamide gradient slab gel stained with colloidal Coomassie blue.

changes in spermatogenesis or sperm motility without concomitant changes in testis histopathology. As a follow-up study, we analyzed frozen samples from the Price et al. investigation where Sprague-Dawley rats had been exposed to room air control conditions or atmospheres containing 0, 250, 500, or 1,000 mg/m^3 JP-8 vapor (n = 5) for 6 hours/days (typical work shift) for 91 consecutive days. As in our previous studies, 2D gel-based proteomics was used to analyze protein expression, as we attempted to identify JP-8-mediated protein alterations that might explain the motility changes and identify adaptations to exposure not distinguished in the histopathology. A representative testis 2-DE gel pattern is shown in Figure 5.3.

Image analysis indicated that the expression of 77 of 1,320 matched proteins (5.8%) was significantly ($p < 0.05$) altered by exposure to JP-8 vapor. JP-8 vapor at 250 mg/m^3 resulted in fivefold more up-regulated proteins than did exposure to the highest dose (1,000 mg/m^3). In contrast, exposure to the highest dose level resulted in twice as many down-regulated proteins as the lowest dose exposure. In fact, proteomic analyses of individual protein expression, plotted as a function of JP-8 vapor levels, seldom reflected a strict linear dose relationship [75]. As shown in Table 5.4, only 4 of the 19 identified proteins that were significantly altered by exposure increased expression as a function of dose. These were Hsp86, nicotinic acetylcholine receptor alpha subunit, serum albumin, and T-complex protein 1. Likewise, no identified protein underwent a statistically significant linear decrease in expression as a function of JP-8 dose exposure. In the remaining 15 cases of significantly altered expression

TABLE 5.4
Identified Proteins from Whole Testis and the Effect of Prolonged JP-8 Jet Fuel Exposure on Mean Protein Expression

Name	MW	pI	Control	250 mg/m³	P	500 mg/m³	P	1000 mg/m³	P
150-kDa Oxygen-regulated protein precursor	111.5	5.1	2,405	2,225	0.2	1,966	0.05	2,158	0.6
78 kd Glucose-regulated protein precursor (grp 78)	72.5	5.1	3,193	4,777	0.06	3,949	0.2	3,620	0.5
A-kinase anchor protein	256.9	5.0	1,772	2,547	0.05	1,784	1.0	1,184	0.5
Aldehyde dehydrogenase, mitochondrial (ALDH1)	48.7	6.1	1,580	3,330	0.03	1,771	0.8	1,743	0.8
Androgen receptor-associated protein 24	24.6	7.0	2,609	2501	0.7	2,366	0.6	3,640	0.01
Calreticulin	48.2	4.3	5,077	5,787	0.02	6,003	0.08	5,707	0.05
Heat shock factor 2	57.8	4.8	741	936	0.01	852	0.3	895	0.2
Heat shock protein 86	85.2	4.9	2,278	1,034	0.3	2,987	0.05	2,657	0.7
Heat shock protein 86	85.2	4.9	3,103	4,460	0.2	4,794	0.2	5,342	0.05
Integrin beta-7 subunit	71.7	5.4	279	604	0.02	632	0.01	472	0.2
Interleukin-18	22.5	4.9	467	556	0.04	595	0.007	496	0.5
Lactate dehydrogenase B	36.9	5.7	4,467	5298	0.03	4,660	0.9	5,033	0.4
Lamin B1	66.8	5.2	574	816	0.002	597	0.8	846	0.2
Nicotinic acetylcholine receptor alpha 4 subunit	71.3	6.7	264	658	0.4	971	0.1	2,020	0.001
Osmotic stress protein 94 (heat shock 70-related protein APG-1)	95.2	5.5	2,403	2,767	0.06	2,501	0.7	2,680	0.4
Protein kinase C-binding protein Zeta1	45.5	4.3	171	256	0.01	255	0.002	155	0.8
Serum albumin	68.7	6.4	21,471	18,429	0.5	29,682	0.2	41,175	0.05
Spermatogenic cell/sperm-associated Tat-binding protein	49.8	5.0	350	427	0.04	461	0.007	387	0.3
T-complex polypeptide 1	60.9	5.9	545	632	0.2	583	0.5	789	0.05

TABLE 5.5
Summary of Sprague-Dawley Liver and Kidney Proteins Whose Expression Remained Altered after Recovery from Exposure to JP-8 Jet Fuel Vapor

Name	% Change[a]	MW	pI
Lamin A	−50	74,323.7	6.54
Lamin A (charge variant)	+575		
10-Formyltetrahydrofolate dehydrogenase	+315	99,127.1	5.79
10-Formyltetrahydrofolate dehydrogenase (charge variant)	−36		
Glutathione-*S*-transferase	+35	27,497.8	6.91
Glutathione-*S*-transferase (charge variant)	+57		

Source: Adapted from Witzmann, F.A., Carpenter, R.L., Ritchie, G.D., et al. *Electrophoresis,* 21, 2138, 2000.

[a] % change in expression relative to average integrated density for each corresponding protein in the control group.

across all doses, there was no clear linear relationship between JP-8 dose (250 to 1,000 mg/m^3) and the change from baseline protein expression. This suggests that when assessing human risk of complex mixtures like JP-8, researcher should (1) consider data from both low- and high-dose exposures, (2) not assume linear extrapolation among dose-response effects for a complex mixture exposed by inhalation, and (3) avoid excluding nonlinear dose-response data from a toxicological profile.

An analysis of the functional significance of the JP-8-mediated protein expression alterations (and absence of effect) listed in Table 5.5 is complex, as the exact role of many of these proteins in testis function is conjectural. In the 250-mg/m^3 exposure group, there was a significant twofold increase ($p < 0.03$) in mitochondrial aldehyde dehydrogenase (ALDH) expression, a change absent from both the moderate- and high-exposure groups. ALDH is commonly expressed in a number of organ systems, including the testis, where it is located principally in the Leydig cell and is known to increase in response to the presence of alcohols or glycols [46]. Ethylene glycol monomethylether (EGME), a structural analog of DiEGME (diethylene glycol monomethylether), the anti-icing additive in JP-8, is metabolized by ALDH and possibly lactate dehydrogenase (LDH) to the major metabolite methoxyacetic acid (MAA). EGME or MAA has been widely implicated in male reproductive system toxicity, including disruption of spermatogenesis [5].

In view of the temperature sensitivity of spermatogenesis, it is not surprising that the major stress proteins (hsp and grp) have reasonably high constitutive expression in the testis of rodents. These include 70-kDa hsp-1, hsp-2, and hsp-3; 70-kDa testis-specific hsp (hsp70t); 70-kDa heat shock cognate protein (hsc70); 75-kDa glucose-regulated protein (grp75); 78-kDa glucose-regulated protein (grp78); and heat shock 70-related APG-1 (osmotic stress protein 94), as well as regulators of the stress response including heat shock factors 1 and 2. Several proteins related to this stress protein family responded to JP-8 exposure. Increased expression of heat shock factor 2 (low dose), total hsp70t (sum of all charge variants) (high dose), Grp78 (low dose), and Hsp86 (moderate and high doses) all suggest a generalized cellular stress response at those individual exposure levels. In addition, T-complex polypeptide 1 (TCP-1) was up-regulated at high-dose exposure. TCP-1 is classified as one of the Type II molecular chaperones, the variety that has a specific role in the folding of newly synthesized tubulins and actins. All the stress respondent proteins mentioned above may play a role in the recovery of cells from possible protein damage inflicted by JP-8, assisting in the folding of proteins that are actively synthesized and (or) renatured as a consequence of prolonged exposure to the volatile components of JP-8. The slight (+22%) but significant increase in Tat binding protein-1 (TBP-1) also may be related to this response. TBP-1 is abundant in the testis and shows particular expression in spermatogonia, spermatocytes, spermatids, and epididymal sperm. Its function is linked to the 26S proteasome and appears to be involved in numerous germ cell activities, including ATP-dependent degradation of ubiquinated proteins [57].

The interleukin-1 system is present in testis and may be involved indirectly in spermatogenesis. However, interleukin-18 is associated with responses in tissue injury and has been shown [64] to induce apoptosis in response to JP-8 exposure. Although induction of apoptosis was not investigated directly in that study, it should be noted that in addition to interleukin-18, lamin B is also associated with the induction of apoptosis [65]. This raises the possibility that JP-8 exposure, which has already been shown to induce apoptosis in the lung [6], may also have induced the beginning stages of apoptosis in the exposed testis.

Lamin B1 is a PKC binding protein, but nothing is known regarding this activity in the testis. However, JP-8 caused changes in the expression of another PKC binding protein, Zeta1, including a +50% increase at low and moderate exposures. PKC binding proteins are abundant in the testis, starting with the onset of spermiogenesis with high expression of specific mRNA in the elongating spermatids. Changes in the expression of these proteins following JP-8 exposure raise the possibility that subtle effects were taking place in the Sertoli cell cytoplasm or in the developing germ cells, both of which depend on stable microtubule formations for normal spermatogenesis. Another up-regulated (+50%) cytoskeleton-related protein, A-kinase anchor protein (AKAP) is expressed in high levels in the testis and is closely associated with the microtubule elements of the sperm flagellum. AKAPs may also play a role in targeting type II PKA for cAMP-responsive peroxisomal events, so its altered expression certainly warrants further investigation of potential long-term effects of JP-8 on spermatozoa formation, sperm maturation, and sperm motility.

Integrins are ubiquitously expressed on the surface of cells and enable cell-to-cell communication. The identification of the β7 integrin in this study is somewhat surprising as this integrin type is localized in the lymphocyte, where it is involved in homing. Several integrin subtypes are found in the testis, but integrin β7 has not been described previously. With barely detectable expression in control testis, its up-regulation at low and moderate JP-8 exposures may be related to an injury-related increase in the local lymphocyte population. Despite a reasonably high Z-score of 1.39 (92%) calculated by the Profound peptide mass database search engine, no sequence analysis by MS/MS was conducted to identify it unequivocally. The

Z-score is an indicator of the quality of the search result and reflects the distance to the population mean in units of standard deviation. It also corresponds to the percentile of the search in the random peptide mass match population, as shown above. Even at the 92nd percentile, it is possible that β7 may have been misidentified and is actually one of the normally resident integrins. An additional explanation for the integrin expression changes lies in one of the more subtle histological observations that the lumen of certain seminiferous tubules was dilated after JP-8 exposure. It is well established that integrins are involved in mechanical stress to cells and tissues and, thus, stretching of the seminiferous epithelium may have altered the expression of the integrin family of genes in the testis.

The observation that repeated low-dose JP-8 vapor exposure resulted in far more proteins with increased expression than did exposure to the highest dose, and that exposure to the highest dose level resulted in significantly reduced expression of twice as many proteins as the low dose exposure (Table 5.4), is difficult to explain, but not unusual. This phenomenon has been observed consistently with JP-8 vapor and aerosol exposures in other organs, including the brain [56], as described in this chapter, and its explanation requires additional study. Whether the alterations are related to injury versus adaptive or repair mechanisms as mentioned earlier, or perhaps due to a more complex hormetic [9] response, remains to be determined.

PERSISTING EFFECTS OF JP-8 VAPOR EXPOSURE, FOLLOWING A PERIOD OF RECOVERY

To assess the persistence of JP-8-related effects, once exposure to the fuel had ceased, we applied 2-D gel-based proteomic analysis methods to examine protein expression in the kidney and liver from rats exposed subchronically to JP-8 vapor (1,000 mg/m^3) 6 hours/day, 5 days/week, for 6 consecutive weeks, and then allowed to recover [76]. Given our understanding of the moderate (although significant) effects of acute and prolonged JP-8 exposure on the proteome of various rodent target tissues, we hypothesized that little if any significant variation in protein expression should be observed after 82 days of recovery from six consecutive weeks of 1,000-mg/m^3 JP-8 vapor exposure. Despite expectations against such an effect, 6 weeks of JP-8 vapor exposure followed by 82 days of recovery had significant effects on the expression of several biologically important proteins in the liver and kidney.

Gel image analysis and statistical testing revealed significant ($p < 0.001$) expression alterations in three liver and four kidney protein spots following recovery. As in the other studies, these protein spots were cut from replicate gels, and subjected to tryptic digestion and MALDI-MS peptide mass fingerprinting. Six of the seven gel cutouts produced searchable masses and these resulted in the identification of three proteins: lamin A in the liver, and 10-formyltetrahydrofolate dehydrogenase and glutathione *S*-transferase (GST) in the kidney (Table 5.6).

Of the two protein spots independently identified as lamin A, the one with more alkaline pI (considered the unmodified form) [19] decreased (–50%) by treatment conditions while the other, more negatively charged (modified) form increased by

TABLE 5.6
Differential Expression and Charge Modification Index (CMI) in Kidney Proteins after Recovery from Extended JP-8 Jet Fuel Exposure

	Control		JP-8	
Protein	Total Expression	CMI	Total Expression	CMI
Lamin A	10,736 ± 1,578	−0.09 ± 0.01	11,434 ± 1,272	−0.58 ± 0.01[a]
10-Formyltetrahydrofolate dehydrogenase	2,779 ± 455	−0.87 ± 0.04	3,131 ± 149	−0.53 ± 0.01[a]
Glutathione-*S*-transferase	5,356 ± 252	−0.42 ± 0.01	7,580 ± 1,583[b]	−0.56 ± 0.12[b]

Source: Adapted from Witzmann, F.A., Carpenter, R.L., Ritchie, G.D., et al. *Electrophoresis,* 21, 2138, 2000.

[a] P < 0.001, control versus JP-8; [b] Total expression reflects sum of both charge variants; values are means ± SEM; *P < 0.05, control versus JP-8; [#] P < 0.001, control versus JP-8.

575%. Total expression of lamin A calculated by adding the integrated densities of the two charge forms (Table 5.7) was unaffected by JP-8 exposure and recovery. Following exposure and recovery, it seems that there is a shift in charge from the main form, which declines in expression, to the charge variant, which increases in expression by nearly sixfold. These individual expression changes in lamin A demonstrate a significant ($p < 0.01$) and persisting modulation in the protein charge modification index (CMI) between groups (Table 5.7), and suggest a considerable post-translational modification not observed in controls. While the nature of this modification was not assessed and thus remains unknown, thus rendering the functional significance of this finding conjectural, it would appear that this result suggests some long-term, post-exposure hepatic effect and (or) evidence of significant and continued hepatic regeneration from JP-8-mediated consequences.

The nuclear lamina of hepatocytes and many other cell types, a meshwork of intermediate filaments closely apposed to the inner nuclear membrane, is believed to provide a framework important for nuclear envelope integrity and interphase chromatin organization. Its assembly state is regulated in the cytoplasm by

TABLE 5.7
Proteins Altered by JP-8 Exposure (≥1.5-fold; p ≤ 0.05) of 1,269 Proteins Identified and Quantified

	0.1 μg/ml	0.4 μg/ml	2.0 μg/ml
1 hour	0 ↑ 0 ↓	0 ↑ 2 ↓	0 ↑ 3 ↓
6 hour	0 ↑ 36 ↓	55 ↑ 185 ↓	69 ↑ [a]649 ↓

[a] Cell viability nearly 0% (see Figure 5.4).

phosphorylation of its protein components, minimally including lamins A, B, and C. Isoprenylation of the lamins is essential for their proper membrane anchoring and functionality. Normal and regenerating hepatocytes, biliary epithelial cells (ductal and ductular cells), and hepatocellular carcinoma cells invariably expressed both A-type lamins and lamin B. Enhanced prenylation of proteins is observed during rat liver regeneration and may be either an additional or alternative explanation for our observations. It should also be noted that current mass spectrometric (MS) capabilities make the characterization of such modifications highly feasible [28, 62], so that the prefractionation of charge forms in the IEF dimension, as accomplished in this example, can and should be exploited in experiments of this type.

A pair of kidney proteins was identified as glutathione s-transferase (homolog), without designation to the GST class. According to previous electrophoretic results [77], the coordinate position of these proteins matches those identified as GSTP1 (π class). The GSTs are a multi-gene family of enzymes responsible for the detoxification or, infrequently, the activation of a wide range of xenobiotics. The cytoplasmic family of GST consists of several families (α, μ, π, δ, and θ) with catalytically active GST formed as homo- and hetero-dimers from subunits within each family. GST substrates include environmental carcinogens, environmental toxicants, pesticides, pharmaceuticals, and endogenous molecules. GST catalyzes the nucleophilic activity of glutathione on electrophilic substrates, decreasing their toxic effects on cellular macromolecules. Therefore, the high catalytic activity of GST toward a wide diversity of endogenous and exogenous reactive species is thought to be protective against oxidative tissue damage.

Like lamin A, the expression of the two charge variants of GST is altered by JP-8 exposure and recovery (Table 5.6), as is total GST (Table 5.7). CMI calculation (Table 5.7) confirmed the chemical modification of the GST homolog, a 33% increase in CMI. Although the nature of the charge shift was not determined, it may be due to either increased deamidation of glutamine or asparagine residues, or the phosphorylation of serine, tyrosine, or threonine residues. Because GST elevation is frequently observed and interpreted as indicative of toxic exposure or effect [73, 78, 79], these data suggest that GST is indeed a good marker of jet fuel intoxication, even 82 d after the last exposure.

10-Formyltetrahydrofolate dehydrogenase (10-FTHF DH) was resolved as two charge variants. As Table 5.6 indicates, expression of the main, unmodified form increased by 315%, while the more acidic charge variant decreased by 36%. This right-ward charge shift is the reverse of that seen in lamin A and GST, where the extent of post-translational modification was increased. Table 5.7 demonstrates that total 10-FTHF DH expression was unaltered, but a significant ($p < 0.001$) decline in charge modification resulted after exposure and recovery. The modification implied (CMI of −0.87) in control 10-FTHF DH may be the result of known in vivo covalent binding of tetrahydrofolate polyglutamates [71]. The decline in expression of the modified charge variant and the reduction of CMI to −0.53 following JP-8 exposure and recovery may be a reduction in the number of covalently bound folate polyglutamates, formation of phosphorates, or chemical adduction. The effect this difference might have on enzyme activity in this study is unknown.

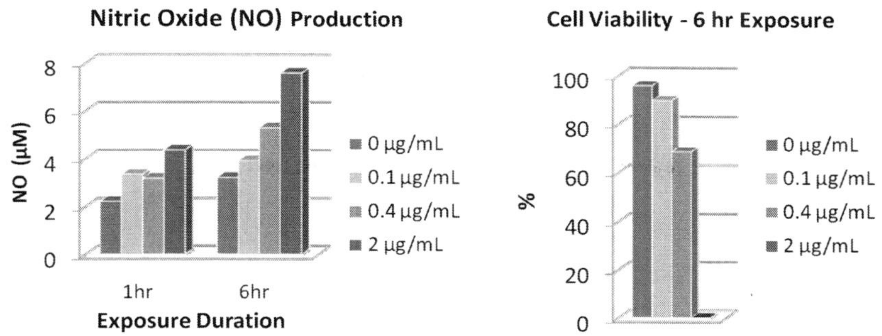

FIGURE 5.4 Effect of in vitro JP-8 exposure on rat alveolar type II cells on nitric oxide production and cell viability.

Label-Free Quantitative Mass Spectrometry (LFQMS): Alternative to 2-DE

As an alternative to electrophoretic analysis of differential protein expression, a study was undertaken to assess the effect of acute JP-8 exposure on rat alveolar type II epithelial cells, at occupationally relevant levels [80]. An equally important aim was to assess the utility of a novel isotope label-free proteomics approach (LFQMS) to quantify JP-8-mediated alterations in cellular protein expression. Rat alveolar type II cells (1×10^6 cells/well) were cultured in six-well plates and exposed to JP-8 jet fuel (0.1 μg/ml, 0.4 μg/ml, 2.0 μg/ml, vehicle control, and media control) for 1 and 6 hours. Cells from each of six wells per treatment group (n = 6) were solubilized in situ and proteins trypsinized. Tryptic peptides from each sample were analyzed serially via linear ion-trap LC-MS/MS. Proteins were identified using SEQUEST and X!Tandem, while the assessment of relative protein expression was carried out based on chromatographic alignment and peptide quantification via integration of extracted ion chromatograms [33]. In this approach, the integral volume under each selected peptide peak was measured, normalized, and compared for relative abundance.

As Figure 5.4 illustrates, JP-8 exposure increased nitric oxide production, an injury response indicator, in both 1-hour and 6-hour exposures, and most notably at the higher dose and longer duration. However, despite injury induction, JP-8 exposure did not affect cell viability at any exposure level at 1 hour, while the viability of 6-hour cells significantly decreased at 0.4 μg/ml, with zero cell viability observed at the 6-hour, 2.0-μg/ml exposure.

LFQMS identified and provided relative quantification of 2,461 proteins. Of these, 1,269 proteins were identified with high confidence, based on the quality of the amino acid sequence identification (peptide ID confidence, e.g., 90% to 100%) and whether one or more sequences were identified (single or multiple sequences). Proteins identified with 75% to 89% confidence, whether or not multiple peptides were used in the process, were not used in further data interpretation. Of the 1269 proteins (median % Coefficient of Variation (%CV) was 9.09%), 608 proteins had at least one significant change between groups. Relevant group comparisons are presented in Tables 5.7, 5.8, and 5.9. The differential protein expression data obtained

TABLE 5.8
Alveolar Type II Epithelial Cell Protein Groups Down-Regulated by JP-8 Exposure

Protein	Members	Mean Fold Change	Mean No. of Peptide Sequences
14-3-3 gamma (zeta/delta, epsilon, eta, theta)	4	−2.00	6
26S Protease regulatory subunit (6B, 7, 13)	3	−1.85	2
40S Ribosomal protein	19	−1.87	3
60S Acidic ribosomal protein (P1, P2a, P2b)	3	−1.60	5
60S Ribosomal protein	10	−1.92	3
Actin or actin-related	24	−1.86	15
AHNAK nucleoprotein (desmoyokin)	8	−1.90	28
Aldose reductase	4	−1.68	7
Collagen/collagen-related	14	−1.84	8
Dynein and spectrin	5	−1.77	2
Eukaryotic translation elongation factors	13	−2.28	8
Glyceraldehyde-3-phosphate dehydrogenase	14	−2.02	4
GTP-binding proteins (Ras, Rab, Ran, etc.)	17	−1.82	3
Heat shock proteins	21	−1.84	8
Pyruvate kinase	7	−1.98	14
Ribosomal protein (other)	11	−1.87	2
Splicing factors	10	−1.71	2
Tubulin	18	−2.09	11
Ubiquitin related	10	−1.74	2

Note: Groups include proteins with three or more isoforms detected or related proteins; average-fold change for 6-hours control versus 6-hour, 2.0-μg/ml JP-8 exposure.

and analyzed in this way indicate that in vitro JP-8 jet fuel exposure, at a level considered to be low (≤2 μg/ml), had virtually no effect after 1 hour. Likewise, cell culture duration had little effect on protein expression in unexposed controls. However, after 6 hours exposure, significant differences in protein expression were observed. Most of these changes were apparent down-regulations, and the most severe effects were observed at 2 μg/ml, where 57% of the measured proteins were altered by ≥1.5-fold, corresponding to the viability data. Decreased cell viability corresponded to significant down-regulation of proteins involved in all manner of cell activity, but predominantly via decreased translational and protein synthetic machinery. These results are consistent with the gene expression alterations (as measured by microarray) in mice exposed for 1 hour to 1,000 mg/m^3 JP-8 and organs sampled at 6 hours where JP-8 exposure induced the loss of both 18S and 28S RNA in exposed cells (due to loss of ribosomal proteins) that occurred between 6 and 24 hours after in vivo exposure [30]. These results suggest that under these culture conditions, JP-8 causes cell death by mechanisms that remain to be determined. The study also demonstrates the unique utility of LFQMS.

TABLE 5.9
Alveolar Type II Epithelial Cell Proteins Down-Regulated ($p \leq 0.05$) at 6 Hours by 0.1 μg/ml JP-8

IPI	Gene ID	RGD	Annotation
IPI00191142.1	81773	RGD:621024	40S ribosomal protein S10
IPI00191142.1	81773	RGD:621024	40S ribosomal protein S10 (17-kDa protein)
IPI00373448.3	81773	RGD:621024	40S ribosomal protein S10 (18-kDa protein)
IPI00366014.3	684988	RGD:621027	40S ribosomal protein S13
IPI00392390.2	83789	RGD:619887	40S ribosomal protein S2
IPI00214582.1	29258	RGD:61907	40S ribosomal protein S7
IPI00358285.2	290678	RGD:1559708	Ribosomal protein S10
IPI00203523.1	360572	RGD:1304897	60S ribosomal protein L23a
IPI00204703.5	29345	RGD:69302	47-kDa heat shock protein (serpin H1 precursor; collagen-binding protein)
IPI00324539.6	286758	RGD:628763	Aquaporin-11
IPI00569197.2	501389	RGD:1588387	Chromosome 9 open reading frame 36 (LOC501389)
IPI00388632.3	307492	RGD:1310695	Dead end homolog 1 (52-kDa protein)
IPI00371802.3	304719	RGD:1560340	Excision repair cross-complementing rodent repair deficiency, complementation group 4
IPI00200757.1	25661	RGD:2624	Fibronectin precursor, isoform 1
IPI00368017.4	691044	RGD:1359558	GTPase activating protein testicular GAP1
IPI00210090.3	117280	RGD:620372	Heterogeneous nuclear ribonucleoprotein U (SP120)
IPI00211813.1	79433	RGD:71000	Myosin heavy chain (Neuronal)
IPI00360637.3	310178	RGD:1307193	Myosin-10 (Myosin heavy chain, nonmuscle IIb) (cellular myosin heavy chain, type B)
IPI00558912.1	24582	RGD:3136	Myosin-11
IPI00367479.4	308572	RGD:1306821	Myosin-14
IPI00209113.3	25745	RGD:3140	Myosin-9
IPI00370406.2	679306	RGD:1623873	NMDA receptor-regulated gene 2, isoform 2
IPI00394392.3	56768	RGD:69406	Piccolo (presynaptic cytomatrix protein)
IPI00464535.1	246303	RGD:619907	Plasminogen activator inhibitor 1 RNA-binding protein, isoform 2)
IPI00359590.5	308918	RGD:1565383	Protein tyrosine phosphatase, receptor-type, F interacting protein, binding protein 2
IPI00202873.1	24718	RGD:3553	Reelin precursor, isoform 1
IPI00203078.1	116475	RGD:620744	Selective LIM binding factor, rat homolog
IPI00191416.1	79111	RGD:708535	Solute carrier family 27 (fatty acid transporter), member 5 (bile acyl-CoA synthetase)
IPI00779007.1	310360	RGD:1566344	Translation elongation factor 1-alpha 1 (50-kDa protein)
IPI00195372.1	171361	RGD:67387	Translation elongation factor 1-alpha 1
IPI00568311.1	171361	RGD:67387	Translation elongation factor 1-alpha 1C (46-kDa protein)

Continued

TABLE 5.9 (*Continued*)
Alveolar Type II Epithelial Cell Proteins Down-Regulated ($p \leq 0.05$) at 6 Hours by 0.1 μg/ml JP-8

IPI	Gene ID	RGD	Annotation
IPI00195372.1	171361	RGD:67387	Translation elongation factor EF-1, subunit alpha 1 (50-kDa protein)
IPI00325281.1	24799	RGD:3781	Translation elongation factor EF-1, subunit alpha 2 (45-kDa protein)
IPI00568311.1	171361	RGD:67387	Translation elongation factor Tu
IPI00372810.4	292148	RGD:1307269	Translation initiation factor 3, subunit 10 (theta) (ZH12 protein)
IPI00373703.2	301079	RGD:1311745	Uncharacterized protein KIAA1143 homolog

OVERALL CONCLUSIONS

The various studies investigating possible JP-8 toxicity reviewed in this chapter have resulted in at least two significant accomplishments. First, it was shown that rodent in vivo and in vitro models of JP-8 toxicity, where occupational exposures were simulated and protein expression was analyzed by 2D gel-based or MS-based proteomics, consistently reflected significant changes in protein expression. Depending on the species, exposure conditions, and target tissues, the protein alterations were quantitative (up- or down-regulation) and (or) qualitative (post-translational); and based on protein identification, the functional consequences of the altered proteomes could be postulated. There would appear to be little doubt, if the rodent model of JP-8 toxicity is considered adequate for human risk, that acute exposure to high levels or repeated (occupational) exposure to lower levels of jet fuels must be expected to induce acute or persisting changes in protein systems that are biologically significant.

Second, the powerful utility of 2D gel-based and MS-based approaches in analyzing differential protein expression, across many samples/dose groups, was demonstrated in these studies. Using a large-scale, highly parallel approach to 2D gel electrophoresis of numerous samples, the resulting gel-gel uniformity enabled accurate comparisons of protein expression. The detection of statistically significant protein up- and down-regulation, and accurate, quantitative assessment of the chemical modification of specific proteins through the calculation of CMI, are capabilities possibly unique to the 2-DE approach.

As these results also clearly demonstrate, only a fraction of the entire target tissue proteomes was analyzed. It must be considered that many important protein alterations associated with JP-8 exposure have yet to be identified.

However, before the ultimate analytical power of these experimental approaches can be realized and fully appreciated, traditional toxicological uses of these systems must be modified and improved. Methods such as 2D-DIGE [28, 39, 67] and other developing quantitative mass spectrometry techniques [22, 23, 25, 33, 63, 80] show great promise and are emerging as preferred tools of toxicoproteomics.

REFERENCES

[1] Anderson, N.L., Copple, D.C., Bendele, R.A., et al. Covalent protein modifications and gene expression changes in rodent liver following administration of methapyrilene: A study using two-dimensional electrophoresis. *Fundam. Appl. Toxicol.,* 18, 570, 2002.

[2] Assman, B., Hoffmann, G.F., Wagner, L., et al. Dihydropyrimidinase deficiency and congenital microvillous atrophy: Coincidence or genetic relation? *J. Inherit. Metab. Dis.,* 20, 681, 1997.

[3] Bagghi, D., Bhattacharya, G., and Stohs, S.J. In vitro and in vivo induction of heat shock (stress) protein (Hsp) gene expression by selected pesticides. *Toxicology,* 112, 57, 1996.

[4] Baldwin, C.M., Houston, F.P., Podgornik, M.N., et al. Effects of aerosol-vapor JP-8 jet fuel on the functional observational battery, and learning and memory in the rat. *Arch. Environ. Health,* 56, 216, 2001.

[5] Berndtdson, W.E., and Foote, R.H. Disruption of spermatogenesis in rabbits consuming ethylene glycol monomethyl ether. *Reprod. Toxicol.,* 11, 29, 1997.

[6] Boulares, A.H., Contreras, F.J., Espinoza, L.A., et al. Roles of oxidative stress and glutathione depletion in JP-8 jet fuel-induced apoptosis in rat lung epithelial cells. *Toxicol. Appl. Pharmacol.,* 180, 92, 2002.

[7] Bruner, R.H., Kinkead, E.R., O'Neill, T.P., et al. The toxicologic and oncogenic potential of JP-4 jet fuel vapors in rats and mice: 12-month intermittent inhalation exposures. *Fundam. Appl. Toxicol.,* 20, 97, 1993.

[8] Buss, W.C., Stepanek, J., and Queen, S.A. Association of tissue-specific changes in translation elongation after cyclosporin with changes in elongation factor 2 phosphorylation. *Biochem. Pharmacol.,* 48, 1459, 1994.

[9] Calabrese, E.J., and Baldwin, L.A. Hormesis: the dose-response revolution. *Annu. Rev. Pharmacol. Toxicol.,* 43, 175, 2003.

[10] Candiano, G., Bruschi, M., Musante, L., et al. Blue silver: a very sensitive colloidal Coomassie G-250 staining for proteome analysis. *Electrophoresis,* 25, 1327, 2004.

[11] Chao, Y.C., and Nylander-French, L.A. Determination of keratin protein in a tape-stripped skin sample from jet fuel exposed skin. *Ann. Occup. Hyg.,* 48, 65, 2004.

[12] Chen, W., Feng, Y., Chen, D., et al. Rab11 is required for trans-Golgi network-to-plasma membrane transport and a preferential target for GDP dissociation inhibitor. *Mol. Biol. Cell.,* 9, 3241, 1998.

[13] Chou, C.C., Riviere, J.E., and Monteiro-Riviere, N.A. The cytotoxicity of jet fuel aromatic hydrocarbons and dose-related interleukin-8 release from human epidermal keratinocytes. *Arch. Toxicol.,* 77, 384, 2003.

[14] Coakley, R.J., Taggart, C., O'Neill, S., et al. Alpha1-antitrypsin deficiency: biological answers to clinical questions. *Am. J. Med. Sci.,* 321, 33, 2001.

[15] Dossing, M., Loft, S., and Schroeder, E. Jet fuel and liver function. *Scand. J. Work. Environ. Health,* 11, 433, 1985.

[16] Drake, M.G., Witzmann, F.A., Hyde, J., et al. JP-8 jet fuel exposure alters protein expression in the lung. *Toxicology,* 191, 199, 2003.

[17] Elazar, Z., Mayer, T., and Rothman, J.E. Removal of Rab GTP-binding proteins from Golgi membranes by GDP dissociation inhibitor inhibits inter-cisternal transport in the Golgi stacks. *J. Biol. Chem.,* 269, 794, 1994.

[18] Emmett, M.R. Determination of post-translational modifications of proteins by high-sensitivity, high-resolution Fourier transform ion cyclotron resonance mass spectrometry. *J. Chromatogr. A,* 1013, 203, 2003.

[19] Fink, A.L. Chaperone-mediated protein folding. *Physiol. Rev.,* 79, 425, 1999.

[20] Fleming, R.E., Moxley, M.A., Waheed, A., et al. Carbonic anhydrase II expression in rat type II pneumocytes. *Am. J. Respir. Cell. Mol. Biol.,* 10, 499, 1994.

[21] Forkert, P.G., D'Costa, D., and El-Mestrah, M. Expression and inducibility of alpha, pi, and mu glutathione S-transferase protein and mRNA in murine lung. *Am. J. Respir. Cell. Mol. Biol.,* 20, 143, 1999.

[22] Gerber, S.A., Rush, J., Stemman, O., et al. Absolute quantification of proteins and phosphoproteins from cell lysates by tandem MS. *Proc. Natl. Acad. Sci.,* 100, 6940, 2003.

[23] Gluckmann, M., Fella, K., Waidelich, D., et al. Prevalidation of potential protein biomarkers in toxicology using iTRAQ reagent technology. *Proteomics,* 7, 1564, 2007.

[24] Grant, G.M., Jackman, S.M., Kolako, C.J., et al. JP-8 jet fuel-induced DNA damage in H4IIE rat hepatoma cells. *Mutat. Res.,* 490, 67, 2001.

[25] Gygi, S.P., Rist, B., Gerber, S.A., et al. Quantitative analysis of complex protein mixtures using isotope-coded affinity tags. *Nat. Biotechnol.,* 17, 994, 1999.

[26] Hamdan, M., and Righetti, P.G. Assessment of protein expression by means of 2-D gel electrophoresis with and without mass spectrometry. *Mass. Spectrom. Rev.,* 22, 272, 2003.

[27] Harris, D.T., Sakiestewa, D., Robledo, R.F., et al. Effects of short-term JP-8 jet fuel exposure on cell-mediated immunity. *Toxicol. Indus. Health,* 16, 78, 2000.

[28] Harris, D.T., Sakiestewa, D., Titone, D., et al. Jet fuel-induced immunotoxicity. *Toxicol. Indus. Health,* 16(7-8), 261–265, 2001.

[29] Harris, D.T., Sakiestewa, D., Titone, D., et al. JP-8 jet fuel exposure results in immediate immunotoxicity, which is cumulative over time. *Toxicol. Indus. Health,* 18, 77, 2002.

[30] Harris, D.T., and Witten, M.L. Immunotoxicity of aerosolized JP-8 jet fuel exposure and its prevention by substance P. *Toxicol. Sci.,* 84, 4, 2005.

[31] Hays, A.M., Lantz, R.C., and Witten, M.L. Correlation between in vivo and in vitro pulmonary responses to Jet Propulsion Fuel-8 using precision-cut lung slices and a dynamic organ culture system. *Toxicol. Pathol.,* 31, 200, 2003.

[32] Hays, A.M., Parliman, G., Pfaff, J.K., et al. Changes in lung permeability correlate with lung histology in a chronic exposure model. *Toxicol. Indus. Health,* 11, 325, 1995.

[33] Higgs, R.E., Knierman, M.D., Gelfanova, V., et al. Comprehensive label-free method for the relative quantification of proteins from biological samples. *J. Proteome Res.,* 4, 1442, 2005.

[34] http://gltrs.grc.nasa.gov/reports/2006/tm-2006-214365.pdf.

[35] Hugo, C., Nangaku, M., Shankland, S.J., et al. The plasma membrane-actin linking protein, ezrin, is a glomerular epithelial cell marker in glomerulogenesis, in the adult kidney and in glomerular injury. *Kidney Int.,* 54, 1934, 1998.

[36] Inman, A.O., Monteiro-Riviere, N.A., and Riviere, J.E. Inhibition of jet fuel aliphatic hydrocarbon induced toxicity in human epidermal keratinocytes. *J. Appl. Toxicol.,* 28, 543, 2008.

[37] Jackman, S.M., Grant, G.M., Kolanko, C.J., et al. DNA damage assessment by comet assay of human lymphocytes exposed to jet propulsion fuels. *Environ. Mol. Mutagen.,* 40, 18, 2002.

[38] Johnson, B.D., Schumacher, R.J., Ross, E.D., et al. Hop modulates Hsp70/Hsp90 interactions in protein folding. *J. Biol. Chem.,* 273, 3679, 1998.

[39] Jung, E.J., Avliyakulov, N.K., Boontheung, P., et al. Pro-oxidative DEP chemicals induce heat shock proteins and an unfolding protein response in a bronchial epithelial cell line as determined by DIGE analysis. *Proteomics,* 7, 3906, 2007.

[40] Kanillannan, N., Locke, B.R., and Singh, M. Effect of jet fuels on the skin morphology and irritation in hairless rats. *Toxicology,* 175, 35, 2002.

[41] Keil, D.E., Warren, D.A., Jenny, M.J., et al. Immunological function in mice exposed to JP-8 jet fuel in utero. *Toxicol. Sci.,* 76, 347, 2003.

[42] Lee, Y.H., Boeslsterli, U.A., Lin, Q., et al. Proteomics profiling of hepatic mitochondria in heterozygous Sod2+/– mice, an animal model of discreet mitochondrial oxidative stress. *Proteomics,* 8, 555, 2008.

[43] Lin, B., Ritchie, G D., Rossi, J., 3rd, et al. Identification of target genes responsive to JP-8 exposure in the rat central nervous system. *Toxicol. Indus. Health,* 17, 262, 2001.
[44] Lock, E.A., and Reed, C.J. Xenobiotic metabolizing enzymes of the kidney. *Toxicol. Pathol.,* 26, 18, 1998.
[45] Merenmies, J., Pihlaskari, R., Laitinen, J., et al. 30-kDa heparin-binding protein of brain (amphoterin) involved in neurite outgrowth. Amino acid sequence and localization in the filopodia of the advancing plasma membrane. *J. Biol. Chem.,* 266, 16722, 1991.
[46] Messiha, F.S. Ethanol and hypobarometric simulated high altitude: A gonadal-hepatic toxicity study in the male rat. *J. Toxicol. Environ. Health,* 10, 247, 1982.
[47] Mozier, N.M., McConnell, K.P. and Hoffman, J.L. S-adenosyl-L-methionine:thioether S-methyltransferase, a new enzyme in sulfur and selenium metabolism. *J. Biol. Chem.,* 263, 4527, 1988.
[48] Pfaff, J.K., Tollinger, B.J., Lantz, R.C., et al. Neutral endopeptidase (NEP) and its role in pathological pulmonary change with inhalation exposure to JP-8 jet fuel. *Toxicol. Indus. Health,* 12, 93, 1996.
[49] Price, W.A., Briggs, G.B., Grasman, K.A., et al. Evaluation of reproductive toxicity from exposure of male rats to jet propulsion fuel JP-8 vapor. *The Toxicologist,* 60, 251, 2001.
[50] Ramos, G., Limon-Flores, A.Y., and Ullrich, S.E. Dermal exposure to jet fuel suppresses delayed-type hypersensitivity: A critical role for aromatic hydrocarbons. *Toxicol. Sci.,* 100, 415, 2007.
[51] Ramos, G., Nghiem, D.X., Walterscheid, J.P., et al. Dermal application of jet fuel suppresses secondary immune reactions. *Toxicol. Appl. Pharmacol.,* 180, 136, 2002.
[52] Rhodes, A.G., Lemasters, G.K., Lockey, J.E., et al. The effects of jet fuel on immune cells of fuel system maintenance workers. *J. Occup. Environ. Med.,* 45, 79, 2003.
[53] Richardson, M.R., Hong, D., Sturek, M., et al. Quantitative mass spectrometry to determine effects of diabetic dyslipidemia and exercise on the plasma low density lipoproteome. *FASEB J.,* 21, A855, 2007.
[54] Ritchie, G., Still, K., Rossi, J. 3rd, et al. Biological and health effects of exposure to kerosene-based jet fuels and performance additives. *J. Toxicol. Environ. Health B Crit. Rev.,* 6, 357, 2003.
[55] Ritchie, G.D., Rossi, J., 3rd, Nordholm, A., et al. Effects of repeated exposure to JP-8 jet fuel vapor on learning of simple and difficult operant tasks by rats. *J. Toxicol. Environ. Health A,* 64, 385, 2001a.
[56] Ritchie, G.D., Still, K.R., Alexander, W.K., et al. A review of the neurotoxicity risk of selected hydrocarbon fuels. *J. Toxicol. Environ. Health B Crit. Rev.,* 4, 223, 2001b.
[57] Rivkin, E., Cullinan, E.B., Tres, L.L., et al. A protein associated with the manchette during rat spermiogenesis is encoded by a gene of the TBP-1-like subfamily with highly conserved ATPase and protease domains. *Mol. Reprod. Dev.,* 48, 77, 1997.
[58] Robledo, R.F., and Witten, M.L. Acute pulmonary response to inhaled JP-8 jet fuel aerosol in mice. *Inhalat. Toxicol.,* 10, 531, 1998.
[59] Robledo, R. F., Young, R.S., Lantz, R.C., et al. Short-term pulmonary response to inhaled JP-8 jet fuel aerosol in mice. *Toxicol. Pathol.,* 28, 656, 2000.
[60] Rosenthal, D.S., Simbulan-Rosenthal, C.M., Liu, W.F. Mechanisms of JP-8 jet fuel cell toxicity. II. Induction of necrosis in skin fibroblasts and keratinocytes and modulation of levels of Bcl-2 family members. *Toxicol. Appl. Pharmacol.,* 171, 107, 2001.
[61] Rossi, J., 3rd, Nordholm, A.F., and Carpenter, R.L. Effects of repeated exposure of rats to JP-5 or JP-8 jet fuel vapor on neurobehavioral capacity and neurotransmitter levels. *J. Toxicol. Environ. Health A,* 63, 397, 2001.

[62] Seo, J., and Lee, K.J. Post-translational modifications and their biological functions: Proteomic analysis and systematic approaches. *J. Biochem. Mol. Biol.,* 37, 35, 2004.
[63] Singh, S., Zhao, K., and Singh, J. In vivo percutaneous absorption, skin barrier perturbation, and irritation from JP-8 jet fuel components. *Drug Chem. Toxicol.,* 26, 135, 2003.
[64] Stoica, B.A., Boulares, A.H., Rosenthal, D.S., et al. Mechanisms of JP-8 jet fuel toxicity. I. Induction of apoptosis in rat lung epithelial cells. *Toxicol. Appl. Pharmacol.,* 171, 94, 2001.
[65] Swamy, M.V., Cooma, I., Reddy, B.S., et al. Lamin B, caspase-3 activity, and apoptosis induction by a combination of HMG-CoA reductase inhibitor and COX-2 inhibitors: A novel approach in developing effective chemopreventive regimens. *Int. J. Oncol.,* 20, 753, 2002.
[66] Timbrell, J.A. Biomarkers in toxicology. *Toxicology,* 129, 1, 1998.
[67] Tonge, R., Shaw, J., Middleton, B., et al. Validation and development of fluorescence two-dimensional differential gel electrophoresis proteomics technology. *Proteomics,* 1, 377, 2001.
[68] Tu, R.H., Mitchell, C.S., Kay, G.G., et al. Human exposure to the jet fuel, JP-8. *Aviat. Space Environ. Med.,* 75, 49, 2004.
[69] Tunicliffe, W.S., O'Hickey, S.P., Fletcher, T.J., et al. Pulmonary function and respiratory symptoms in a population of airport workers. *Occup. Environ. Med.,* 56, 118, 1999.
[70] Ullrich, S.E. Dermal application of JP-8 jet fuel induces immune suppression. *Toxicol. Sci.,* 52, 61, 1999.
[71] Wagner, C., Briggs, W.T., Horne, D.W., et al. 10-Formyltetrahydrofolate dehydrogenase: Identification of the natural folate ligand, covalent labeling, and partial tryptic digestion. *Arch. Biochem. Biophys.,* 316, 141, 1995.
[72] Wang, S., Young, R.S., Sun, N.N., et al. In vitro cytokine release from rat type II pneumocytes and alveolar macrophages following exposure to JP-8 jet fuel in co-culture. *Toxicology,* 173, 211, 2002.
[73] Witzmann, F.A., Bauer, M.D., Fieno, A.M., et al. Proteomic analysis of simulated occupational jet fuel exposure in the lung. *Electrophoresis,* 20, 3659, 1999a.
[74] Witzmann, F.A., Bauer, M.D., Fieno, A.M., et al. Proteomic analysis of the renal effects of simulated occupational jet fuel exposure. *Electrophoresis,* 21, 976, 2000a.
[75] Witzmann, F.A., Bobb, A., Briggs, G.B., et al. Analysis of rat testicular protein expression following 91-day exposure to JP-8 jet fuel vapor. *Proteomics,* 3, 1016, 2003a.
[76] Witzmann, F.A., Carpenter, R.L., Ritchie, G.D., et al. Toxicity of chemical mixtures: proteomic analysis of persisting liver and kidney protein alterations induced by repeated exposure of rats to JP-8 jet fuel vapor. *Electrophoresis,* 21, 2138, 2000b.
[77] Witzmann, F.A., Daggett, D.A., Fultz, C.D., et al. Glutathione S-transferases: Two-dimensional electrophoretic protein markers of lead exposure. *Electrophoresis,* 19, 1332, 1998a.
[78] Witzmann, F.A., Fultz, C.D., Grant, R.A., et al. Differential expression of cytosolic proteins in the rat kidney cortex and medulla: Preliminary proteomics. *Electrophoresis,* 19, 2491, 1998b.
[79] Witzmann, F.A., Fultz, C.D., Grant, R.A., et al. Regional protein alterations in rat kidneys induced by lead exposure. *Electrophoresis,* 20, 943, 1999b.
[80] Witzmann, F.A., Lee, K., Wang, M., et al. Pulmonary effects of JP-8 jet fuel exposure — Label-free quantitative analysis of protein expression in alveolar type II epithelial cells using LC/MS. *Toxicol. Sci.,* 96, 102, 2007.
[81] Witzmann, F.A., Li, J., Strother, W.N., et al. Innate differences in protein expression in the nucleus accumbens and hippocampus of inbred alcohol-preferring and nonpreferring rats. *Proteomics,* 3, 1335, 2003b.

[82] Yim, M.B., Chae, H.Z., Rhee, S.G., et al. On the protective mechanism of the thiol-specific antioxidant enzyme against the oxidative damage of biomacromolecules. *J. Biol. Chem.,* 269, 1621, 1994.
[83] Zhang, L., Ashendel, C.L., Becker, G.W., et al. Isolation and characterization of the principal ATPase associated with transitional endoplasmic reticulum of rat liver. *J. Cell. Biol.,* 127, 1871, 1994.

6 Immune Modulation by Dermal Exposure to Jet Fuel

Gerardo Ramos and Stephen E. Ullrich

CONTENTS

INTRODUCTION

Over the years, military jet fuel has evolved to meet modern equipment needs, provide increased safety in fires and crashes, and provide better combat survivability. In the early 1990s, the United States Air Force began a gradual transformation to a new jet fuel. Jet Propulsion (JP)-4 was replaced by JP-8. JP-8 was refined to have a higher flash point, lower vapor pressure, and a lower freezing point than JP-4, to provide a fuel that was less combustible and more explosion proof, more resistant to evaporation during storage, and provide a fuel that performs well at the higher altitudes required during military operations. JP-8 is essentially the same fuel used by commercial airlines (Jet A), but it is supplemented with an anti-corrosive agent, an anti-icing agent, and an anti-static agent to meet the military's performance specifications. JP-8 is a multi-use fuel used to fuel jet and turboprop aircraft, helicopters, and Navy ships. The Army and Marines use JP-8 to fuel tanks and fighting vehicles and trucks that can run on diesel. In the field, JP-8 is also used to fuel portable heating and air-conditioning units. It is estimated that over 2 million people a year are exposed to 60 billion gallons of jet fuel (Jet A or JP-8) (Ritchie et al., 2003). Jet fuel exposure is a major chemical exposure problem for military and civilian personnel who maintain jet engines, handle fuel, perform fuel tank maintenance, and work on the flight line. Those living on bases with increased flight traffic report that the smell of jet fuel combustion lingers in the air during peak hours (Gerardo Ramos, personal experience), making jet fuel exposure a major potential chemical risk exposure problem for military personnel regardless of whether or not they work on the flight line.

Standard toxicological screening during its development suggested that JP-8 caused minimal adverse effects. Gastric gavage with high doses (oral LD_{50} = 16 g/kg body weight) resulted in decreased body weights, gastric and perianal irritation, and elevated liver enzyme levels (although livers showed normal histology on necropsy). Little to no other morbidity or morbidity was reported (Mattie et al., 1995). Exposing JP-8 to pregnant rats did not induce fetal malformation (Cooper and Mattie, 1996). JP-8 was not irritating to the eyes and, at best, it was a weak sensitizer after dermal exposure (Kinkead, Salins, and Wolfe, 1992; Kanikkannan et al., 2000).

Subsequent studies indicated that JP-8 did have some toxic effects. Anecdotal reports from personnel at the Air Force bases that changed over to JP-8 suggested increased health problems, especially in personnel whose jobs included fuel handling. Chief among these were nausea, headaches, fatigue, blocked nasal passages, ear infections, and skin irritation. Laboratory studies demonstrated that exposure to JP-8 vapors, the most prominent route of JP-8 exposure, altered the postural sway response of human volunteers, suggesting an effect on neurological function (Smith et al., 1997). In addition, inhaling JP-8 reduced pulmonary function in vivo (Hays et al., 1995; Pfaff et al., 1995). Immune function was also suppressed in rats and mice exposed to aerosolized JP-8. Short-term exposure to low-dose JP-8 (once a day for 7 days; 100 mg/m^3) suppressed cell-mediated immune reactions, which persisted for up to 4 weeks post exposure (Harris et al., 1997a; Harris et al., 1997b; Harris et al., 2000). Indeed, Harris and colleagues suggested that immune function is more sensitive to JP-8-induced damage because immunotoxicity is generally induced with lower doses, and is found before toxic effects are found in other organ systems (Harris et al., 1997b). Suppression of antibody responses in mice born to jet fuel exposed dams, in the face of absolutely normal development of the pups (Keil et al., 2003), supports the concept that the immune system is more susceptible to the toxic effects of JP-8.

These studies clearly demonstrate that inhaling jet fuel aerosols and vapors can modulate immune function. They also indicate that the immune system is more susceptible than other organ systems to the toxic effects of jet fuel. Because the second major route of jet fuel exposure is via the skin, and because aviation personnel who work with fuel often experience long-term exposure due to contact with fuel-saturated clothing (Chao et al., 2006), the focus of this chapter is to review the immunological changes that occur following dermal exposure to jet fuel.

IMMUNE MODULATION FOLLOWING DERMAL APPLICATION OF JET FUEL

Using a mouse model of dermal exposure, we noted that applying JP-8 to the skin induced immune suppression (Ullrich, 1999). Our initial experiments were designed to characterize what types of immune reactions were affected by dermal application of jet fuel. We observed the following. First, applying jet fuel to the skin preferentially affects cell-mediated immune reactions. We observed that delayed type hypersensitivity (DTH), contact hypersensitivity (CHS), and T cell proliferation, but not antibody production, were suppressed by JP-8 treatment (Ullrich, 1999; Ullrich and Lyons, 2000). Both primary and secondary immune reactions were suppressed by jet fuel application (Ramos et al., 2002). In early experiments we used one application of a relatively large amount of JP-8 (200 to 240 mg/mouse); however, we also noted that repeated exposure to small doses of jet fuel (40 to 80 mg over a 4- to 5-day period) would also induce immune suppression (Ramos et al., 2002). Doses of JP-8 that suppressed in mice compared well to dermal exposures doses reported in Air Force personnel. Using a comparison of body surface area, we estimated that applying a single dose of 300 μl of JP-8 to a 20-gram mouse would be roughly equal to applying 100 ml of JP-8 to a 6-foot-tall human (Ullrich, 1999). Similarly, we noted significant immune suppression when as little as 50 μl undiluted JP-8 was applied to a mouse over the course of 4 to 5 days. This would be approximately equal to a human dermal exposure of 16 to 20 ml/day. After a single exposure to JP-8, the immune suppression persists for approximately 3 weeks (Ullrich and Lyons, 2000).

Second, activation of cytokine production by JP-8-treatment appears to play an important role in the activation of immune suppressive pathways. Blocking prostaglandin E_2 (PGE_2) production with a selective cyclooxygenase-2 (COX-2) inhibitor, or neutralizing interleukin (IL)-10 activity with a monoclonal antibody, blocked JP-8-induced immune suppression (Ullrich and Lyons, 2000).

Third, in regard to immune suppression induced following dermal application of jet fuel, we noted that there was no difference between military jet fuel and the fuel used by commercial airlines. As mentioned above, JP-8 is Jet A, supplemented with an anti-corrosive agent (DCI-450), an anti-icing agent (diethylene glycol monomethylether), and an anti-static agent (Statis 450). Early on, attention focused on the additive package as the agent(s) responsible for inducing immune suppression. Experimental results, however, indicated that this was not the case. The dose-response curve for immune suppression induced by JP-8 and Jet A are identical (Figure 6.1) (Ramos et al., 2002). Also, the mechanisms involved are similar. Both platelet-activating factor (PAF) and PGE_2 are involved in the immune suppression induced by JP-8 and Jet A, in that injecting a selective cyclooxygenase-2 (COX-2) inhibitor, or selective PAF receptor antagonists into JP-8 or Jet-A-treated mice blocked immune suppression (Ramos et al., 2004; Ramos et al., 2002). While these findings indicated that the base kerosene fuel—and not the additive package—induces immune suppression, it left unanswered what chemical or class of chemicals in jet fuel causes the immunotoxicity (see section entitled "The Aromatic Compounds in Jet Fuel Drive Immune Suppression").

MECHANISMS UNDERLYING JET-FUEL-INDUCED IMMUNOTOXICITY

In our initial experiments, we observed systemic immune suppression following dermal jet fuel exposure. That is to say, when we applied the jet fuel to the dorsal skin of a C3H/HeN mouse, and then applied a contact allergen to the unexposed

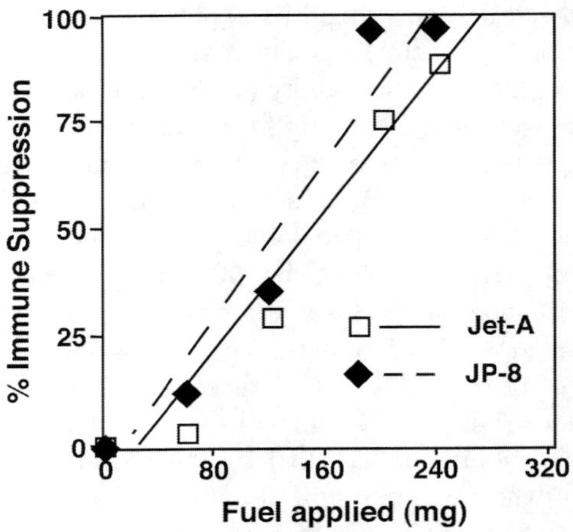

FIGURE 6.1 The dose-response curves for JP-8 and Jet A induced immune suppression are identical. Different concentrations of Jet A (□) and JP-8 (◆) were applied to the dorsal skin of C3H/HeN mice. The data are expressed as % immune suppression versus the dose of jet fuel applied. (*Source:* Adapted from data published by Ramos, G., D.X. Nghiem, J.P. Walterscheid, and S.E. Ullrich. *Toxicol. Appl. Pharmacol.,* 180 (2): 136–144, 2002.)

ventral skin, we suppressed the CHS reaction. Similarly, applying JP-8 to the dorsal skin of a mouse suppressed the induction of DTH to microbial antigens, such as *Candida albicans* or *Borrelia burgdorferi* injected into the subcutaneous space at a distant, untreated site (Ullrich, 1999; Ramos et al., 2002). The systemic nature of the immune suppression, coupled with the fact that primarily T cell-mediated immune reactions were suppressed (Ullrich and Lyons, 2000) drew our attention to cytokines and biological response modifiers that suppressed cell-mediated immune reactions. We found that dermal jet fuel exposure activated the production of IL-10 (Ullrich, 1999), a cytokine known to suppress cell-mediated immune reactions (Moore et al., 2001). Injecting JP-8-treated mice with monoclonal anti-IL-10 blocked the induction of immune suppression (Ullrich and Lyons, 2000). Moreover, injecting JP-8-treated mice with recombinant IL-12, a cytokine that counteracts the activity of IL-10 (Schmitt, Walterscheid, and Ullrich, 2000), blocked the induction of immune suppression (Ullrich and Lyons, 2000). Treating cultured keratinocytes with JP-8 activated the production of PGE_2, (Ramos et al., 2004) and injecting JP-8-treated mice with a selective COX-2 inhibitor (Seibert et al., 1994) blocked JP-8-induced immune suppression (Ullrich and Lyons, 2000).

Cytokines and Immune-Modulatory Factors Activate Jet Fuel-Induced Immune Suppression

An early step in jet-fuel-induced immune suppression is the production of PGE_2 by JP-8-treated keratinocytes (Ramos et al., 2004). A critical step in PGE_2 secretion, and perhaps one of the earliest steps in the cascade of events leading to immune suppression is the secretion of the lipid mediator of inflammation, PAF.

As its name implies, PAF activates a wide variety of cells, including platelets, monocytes, mast cells, and polymorphonuclear leukocytes. In addition to activating platelets, it also activates monocytes, mast cells, and polymorphonuclear leukocytes. PAF plays a role in cell communication. Cells that are responsive to PAF express a seven transmembrane spanning G-coupled protein receptor. Binding of PAF to its receptor activates a variety of intracellular events, such as increased calcium flux, activation of mitogen-activated protein kinase pathways, activation of phospholipase (PLA_2), and transcriptional activation of a variety of genes, including COX-2 (Ishii and Shimizu, 2000) and IL-10 (Walterscheid, Ullrich, and Nghiem, 2002). PAF is secreted in response to oxidative stress, and is secreted by epidermal cells almost immediately following skin trauma (Alappatt et al., 2000). Because PAF up-regulates the production of PGE_2 (Pei et al., 1998), we tested the hypothesis that JP-8-induced PAF activates cytokine production and initiates immune suppression. We pretreated jet-fuel-treated mice with a series of PAF receptor antagonists. Injecting the PAF receptor antagonists totally reversed jet-fuel-induced immune suppression. We noted reversal of immune suppression regardless of whether Jet A or JP-8 was used. At the cellular level, we observed that pretreating JP-8 or Jet-A-treated keratinocytes with a PAF receptor antagonist blocked PGE_2 secretion. We also found that treating jet-fuel-exposed mice with antioxidants, such as Vitamin C, Vitamin E, and beta-hydroxyl toluene (BHT), blocked immune suppression (Ramos et al., 2004). Because jet fuel treatment induces oxidative stress (Rogers et al., 2001), which has been associated with increased PAF production (Alappatt et al., 2000), these findings support the hypothesis that PAF production is an essential and early step in the cascade of effects leading to JP-8- and Jet-A-induced immune suppression. We propose that jet-fuel-induced PAF induces keratinocytes to secrete PGE_2, which then activates a cascade of events, including IL-10 secretion as described previously (Shreedhar et al., 1998), that ultimately suppresses cell mediated immune reactions, such as DTH, CHS, and T cell proliferation.

The molecular pathway leading from JP-8 exposure to COX-2 up-regulation and PGE_2 synthesis is not entirely clear. We decided to examine a potential role for nuclear factor kappa beta (NF-κβ). We focused on NF-κβ for two reasons:

1. Oxidative stress activates NF-κβ (Boulares et al., 2002).
2. PAF activates NK-κβ (Ko et al., 2002).

Using a fluorescent indicator dye, we confirmed that treating keratinocytes with JP-8 induced reactive oxygen species (ROS). Fluorescence was greatly diminished in JP-8-treated keratinocytes over-expressing genes encoding either catalase or superoxide dismutase. The expression of COX-2 was also significantly suppressed in catalase and superoxide dismutase over-expressing keratinocytes. Treating wild-type keratinocytes with N-acetylcysteine also suppressed COX-2 expression, suggesting a link between ROS production and COX-2 expression. We noted that application of JP-8 to murine skin activated NF-κβ (as measured by an electorphoretic mobility shift assay); and because this transcriptional activator can turn on COX-2, we decided to try to block NF-κβ activity and determine what effect this had on COX-2 expression and the induction of immune suppression. When JP-8-treated keratinocytes were treated with a selective inhibitor of NF-κβ activity (parthenolide), NF-κβ activation, COX-2 expression, and COX-2 transcription were all suppressed. Similarly, when small interfering RNA specific for the p65 subunit of NF-κβ was added to the JP-8-treated keratinocytes, COX-2 expression decreased to background. Further, when parthenolide was injected into JP-8-treated mice, immune suppression was abrogated. These data indicate that applying jet fuel to the skin causes the production of ROS, which in turn activates NF-κβ, and the up-regulation of COX-2, which results in PGE_2 production and the induction of immune suppression (Ramos, Limón-Flores, and Ullrich, 2009).

The Aromatic Compounds in Jet Fuel Drive Immune Suppression

An important issue concerning jet-fuel-induced toxicity is identifying the chemical, or family of chemicals, present in jet fuel that activates immune suppression. Jet fuel is made up of over 260 different aliphatic and aromatic compounds (C_6 to C_{17}) (Ritchie et al., 2003). Initially, those interested in determining what components in jet-fuel-induced immune suppression focused their attention on the additive package. However, because the dose-response curves for JP-8- and Jet-A-induced immune suppression are identical (Figure 6.1), and the mechanism of actions are similar (Ramos et al., 2002; Ramos et al. 2004), it appears that the immunosuppressive properties of jet fuel are inherent to the base kerosene fuel and not due to the three additives that refiners add to Jet A to make it into JP-8.

To experimentally address this question, we took advantage of the fact that the U.S. Air Force is currently flight-testing a synthetic jet fuel (hereafter called S-8) that can be produced from coal or natural gas using the Fischer-Tropsch (FT) reaction. Because S-8 is devoid of aromatic hydrocarbons (www.syntroleum. com/tech_specifications.aspx), testing its immunosuppressive properties provided an excellent opportunity to determine the role of the aromatic and aliphatic components of jet fuel in the induction of immune suppression. The literature suggests that there is a difference in the penetration (McDougal et al., 2000; Riviere et al., 1999) and toxicity (Chou, Riviere, and Monteiro-Riviere, 2003; Rogers et al., 2004) of aromatics and aliphatics, and that certain aromatics are immune suppressive (Abadin, Chou, and Llados, 2007). With this in mind, we began a series of studies to determine if applying S-8 to the skin of mice induces immune suppression. We found that applying S-8 to the skin, at all doses tested, did not induce immune suppression. Also, unlike JP-8, applying S-8 to the skin did not up-regulate COX-2 expression. To confirm the role of the aromatic compounds in activating immune suppression, we added to S-8 a cocktail of the seven most common aromatic compounds found in jet fuel (i.e., benzene, toluene, ethylbenzene, xylene, trimethylbenzene, cyclohexylbenzene, and dimethylnaphthalene) at the same concentration found in JP-8. When we applied the supplemented S-8 to the skin, we noted up-regulation of COX-2 expression and suppressed DTH. We also found that if the cocktail of the aromatic compounds was added directly to the skin (i.e., not diluted in S-8), it induced immune suppression. Finally, when the mice treated with the supplemented S-8 or the aromatic compounds alone, were injected with either a PAF receptor antagonist or a COX-2 inhibitor, no immune suppression was noted (Ramos, Limón-Flores, and Ullrich, 2007). These experiments suggest that the aromatic compounds in jet fuel induce immune suppression, and they do it using a mechanism identical to that already described for JP-8 and Jet A.

Our findings add to the observations made by McDougal et al. (2000) and Riviere et al. (1999) concerning differences in the deposition and toxicity of aliphatic and aromatic hydrocarbons after dermal exposure. Aromatic compounds rapidly penetrate through the skin, whereas aliphatic compounds are absorbed by the skin and have an increased residence time in the epidermis. Also, the aromatics compounds are more potent in inducing keratinocyte cell death. Aromatic cytotoxicity in this context has been reported to be a function of the number of side chains and the number of rings, in which a more complex ring structure correlates with increased cytotoxic potential (Chou, Riviere, and Monteiro-Riviere, 2003; Rogers et al., 2004). The aliphatic hydrocarbons found in jet fuel are responsible for dermal irritation (i.e., erythema, edema, and increased epidermal thickening), and are more potent at inducing the secretion of pro-inflammatory cytokines by keratinocytes (Muhammad et al., 2005). Our findings add to this body of work by demonstrating a differential effect of aromatic and aliphatic hydrocarbons on the induction of immune suppression.

ARE HUMANS AT RISK FOLLOWING DERMAL EXPOSURE TO JET FUEL?

One question that is bound to come up during a discussion of the dermal immunotoxicity of jet fuel is whether JP-8 induces immune suppression in humans. Initial studies by McDougal et al. (2000) suggest that even holding ones hands in JP-8 for 8 hours would not be sufficient to induce systemic toxicity because of the low amount of fuel that can be absorbed through the skin. Of course this analysis assumes that direct interaction of jet fuel components with target cells, which in the case of the immune system would presumably occur in the lymph node, is required. On the other hand, measurement of jet fuel metabolites in the urine of fuel cell maintenance workers indicates that dermal exposure to jet fuel results in the presence of naphthalene breakdown products (2-naphthols) in the urine, indicating that dermal exposure to jet fuel can lead to systemic exposure (Chao et al., 2006). Subsequent research in Nylander-French's laboratory (Kim, Andersen, and Nylander-French, 2006) studied the penetration of aromatic and aliphatic compounds of JP-8 in humans by exposing volunteers to 1 ml JP-8 on a surface area of 20 cm^2 for 30 minutes. Tape stripping was done to gain an estimation of penetration of the aromatic and aliphatic components of jet fuel through the skin, and blood samples were used to estimate permeability coefficients. A rank order of permeability through the skin (naphthalene > 1-methylnaphthalene = 2-methylnaphthalene > decane > dodecane > undecane) was determined. The authors also concluded that the rank order permeability coefficients for the compounds tested were similar to those obtained using rat and pig skin. These data suggest that the skin provides a significant route for human jet fuel exposure because jet fuel components can be found in the serum after dermal exposure.

Our findings in mice suggest that the induction of immune suppression may occur by an indirect mechanism. We know from the literature that the aromatic compounds in jet fuel penetrate the skin and induce toxic reactions (Riviere et al., 1999; McDougal et al., 2000). They induce keratinocytes to release PAF and other immune modulatory cytokines (PGE_2 and IL-10) that may "carry" the immune suppressive signal out of the skin to the immune system. Moreover, we recently reported that applying JP-8 to mast-cell-deficient mice does not induce immune suppression (Limón-Flores et al., 2009). We found that when mast-cell-deficient mice were reconstituted with bone-marrow-derived mast cells, immune suppression was restored. Immune suppression was not restored when the mast-cell-deficient mice were reconstituted with PGE_2-/ bone-marrow-derived mast cells, indicating that PGE_2 is the mast cell mediator that drives immune suppression. Moreover, we found that applying JP-8 to the skin of mice induced that migration of mast cells from the skin to the draining lymph nodes, an observation reminiscent of what happens when another dermal immunotoxin (ultraviolet radiation) is applied to the skin (Byrne, Limón-Flores, and Ullrich, 2008). Although some chemical compounds may penetrate to the lymph nodes and interact with immune elements residing there, our data with mast-cell-deficient mice suggest that this is not sufficient to activate immunosuppression. Because the skin of these mice is normal in all respects, one may assume that the aromatic hydrocarbons found in jet fuel are penetrating through the skin of mast-cell-deficient mice and reaching immune cells in the lymph nodes. Because we see no immune suppression in mice deficient in dermal mast cells, we suggest that an indirect mechanism is involved.

Additional data to support this scenario come from the observation that as little as 10.7 mg (12 μl) of a cocktail that contains the seven most prevalent aromatic compounds in JP-8 was able to induce immune suppression, whereas 227 mg (300 μl) of S-8 failed to induce immune suppression (Ramos, Limón-Flores, and Ullrich, 2007). McDougal's findings would indicate that this small amount of chemicals would not be sufficient to induce systemic toxicity (McDougal et al., 2000), yet we note the induction of systemic immune suppression. We suggest that the immune suppressive signal is generated in the skin and transmitted to the immune system by biological mediators and migrating mast cells, and that it does not take large internal doses of jet fuel—or its breakdown products—to activate immune suppression. This raises an important question: When studying the immunosuppression that occurs after a chemical toxicant is applied to the skin, where should we place our focus—on the skin or on the effect of the chemical directly on the immune organs?

IMMUNE MODULATION BY OTHER ROUTES OF EXPOSURE

The most prominent route of human exposure to jet fuel is via the respiratory tract. Inhaling jet fuel vapors and aerosol does modulate immune function, but because this topic is reviewed in detail elsewhere in this book, it is not addressed here.

A series of studies has been performed using oral gavage to administer JP-8. As pointed out in the National Research Council's report on the toxicity of JP-8,

"Although of questionable relevance for determining or assessing a personal exposure limit for JP-8, oral-toxicity studies with fuels have the advantage of administering the sample without prior fractionation of components" (National Research Council, 2003). That is to say, when oral gavage is used, there is little chance for evaporation of the more volatile components of jet fuel that may occur with the dermal or pulmonary routes of exposure. Keil and colleagues diluted JP-8 in olive oil and used oral gavage to administer JP-8 (1,000 to 2,000 mg/kg/day) over 7 days. Both B6C3F1 and DBA/2 mice were used, and in both strains, IgM production, as measured by the Jerne plaque-forming assay, was significantly suppressed. The authors also noted decreased thymus weight and cellularity, and an increase in liver weights in the JP-8-treated mice (Dudley et al., 2001).

In a subsequent publication (Keil et al., 2004), the same group examined hematological and immunological changes in B6C3F1 mice after a 14-day JP-8 exposure. There were no overt signs of toxicity, and no significant weight loss. Organ cellularity was decreased in the thymus, but not in the spleen. Peripheral white blood cell counts and differential cell counts were not markedly altered. There was no appreciable effect on T and B cell mitogen-induced proliferation, nor was Natural Killer cell function affected in the JP-8-treated mice. When antibody formation was assessed using the Jerne plaque-forming assay, a significant suppression of plaque-forming cells was found. However, when antigen-specific serum IgM was measured by two separate assays (ELISA and anti-hemagglutination), there was no difference in the antibody response between the JP-8-treated mice and the vehicle-treated controls. The non-concordance between these results is puzzling, especially in light of the fact that all three assays measure the production of antigen-specific IgM .

The results from these studies suggest that unlike dermal and pulmonary exposure to JP-8, where cell-mediated immune reactions are particularly sensitive to the effects of JP-8 (Harris et al., 2000; Ullrich and Lyons, 2000), oral exposure results in a suppression of the humoral response. Further data to support the suppression of antibody formation come from a study that examined immune alterations induced by in utero jet fuel exposure. Daily gavage (1,000 to 2,000 mg/kg/day) was used to introduce JP-8 to B6C3F1 pregnant dams days 6 through 15 of gestation. Immune function was measured 3 and 8 weeks after birth. At 3 weeks, there was no appreciable change in body, thymus spleen, and liver weights. There was a significant, but relatively marginal decrease in mitogen-induced B cell proliferation in mice exposed to 2,000 mg/kg/day of JP-8. At 8 weeks of age, the decrease in B cell proliferation was not observed. However, at 8 weeks of age, a significant and substantial decrease in the IgM plaque-forming response was observed in mice treated in utero with either 1,000 or 2,000 mg/kg/day of JP-8. On the other hand, no change in the response to challenge with *Listeria monocytogenes*, or modulation in tumor incidence in mice injected with the B16F10 melanoma cell line, was observed in mice treated with JP-8 in utero (Keil et al., 2003). These data again suggest that administration of JP-8 by gavage preferentially affects the humoral arm of the immune response. However, it is not clear how these findings can be translated to human risk assessment because oral exposure to JP-8 is not a common route of human exposure.

SUMMARY AND CONCLUSIONS

Exposure to JP-8 jet fuel represents the single largest potential for chemical exposure for U.S. military personnel (Ritchie et al., 2003). Dermal exposure to JP-8, suppresses the immune response in experimental animal models. Similarly, dermal exposure to commercial jet fuel (Jet A) induces immune suppression when applied to the skin of mice, and the dose-response curves for Jet A and JP-8, as well as the mechanisms of immune suppression, are, in every aspect tested, identical. The induction of immune suppression depends on a cascade of events, including the production of immune regulatory inflammatory mediators (PAF), reactive oxygen species and immune regulatory cytokines (IL-10), and the activation of COX-2 in the skin with subsequent production of the immunomodulatory factor PGE_2 by dermal mast cells. For the dermal route of exposure, the aromatic hydrocarbons in jet fuel are responsible for inducing immune suppression.

Although the term "immune suppression" is used throughout this review, the immune modulation induced following dermal exposure to jet fuel should be more formally described as an immune deviation. We find that cell-mediated immune reactions, such as DTH, CHS, and T cell proliferation, are suppressed by dermal application of jet fuel, whereas antigen-specific antibody formation *in vivo* (IgM, IgG_1, and IgG_{2b}) is not. The suppression of cell-mediated reactions is associated with the production of immune regulatory factors, such as IL-10 and PGE_2, which are known to suppress these reactions. This is in marked contrast to what is observed following oral administration of jet fuel, which appears to depress the humoral immune response but leaves cell-mediated immune reactions intact. However, the fact that only one out of three measurements of antibody secretion was suppressed following oral exposure to jet fuel raises concern about the overall ability of JP-8 to suppress the humoral immune response (Keil et al., 2004). Regardless, it is that clear immune toxicity is induced at doses that are significantly lower than those needed to induce toxic effects on other organ systems (Harris et al., 1997b; Keil et al., 2003), thus indicating that the immune system is particularly susceptible to the toxic effects of jet fuel.

We find that jet fuel administration can suppress both the induction of DTH, when the jet fuel is applied to the skin prior to immunization (Ullrich, 1999), or the elicitation of DTH in a previously immunized animal (Ramos et al., 2002). We suggest that the ability of dermal jet fuel exposure to suppress secondary immune reactions may have the potential to cause the most harm in exposed populations. Perhaps the most successful public health campaign of the past century was the widespread use of vaccination to control infectious disease. By suppressing the recall response to microbial antigens, dermal jet fuel exposure may have the potential to reduce vaccine efficacy. This may be particularly important in military operations in which personnel are often immunized against infectious agents not normally encountered in their home country prior to overseas deployment. This may also be of particular concern during combat operations, in which stress and fatigue, coupled with accidental jet fuel exposure, may contribute to a weakened immune response.

The two most prevalent routes of jet fuel exposure are the skin and the respiratory tract. Interestingly, the mechanisms underlying the immune suppression that occurs after pulmonary and dermal exposure are very similar. Cell-mediated immune reactions are preferentially suppressed (Harris et al., 2000; Ullrich and Lyons, 2000), a short exposure to JP-8 can result in long-term effects (Harris et al., 1997b; Ullrich and Lyons, 2000), and immune modulatory cytokines such as IL-10 and PGE_2 are involved in mediating the immune suppression (Ramos et al., 2002; Ullrich and Lyons, 2000; Harris et al., 2007). The similar mechanism suggests that agents such as Substance P (Harris et al., 1997c), selective COX-2 inhibitors (Ramos et al., 2002), antioxidants such as Vitamins C and E (Ramos et al., 2004), PAF receptor antagonists (Ramos et al., 2004), and anti-IL-10 or IL-10 antagonists such as IL-12, (Ullrich and Lyons, 2000) may all have some utility in blocking the immune suppression that occurs by either route.

Almost all the data demonstrating immune modulation by JP-8 and Jet A comes from studies employing mice as the experimental model (Harris et al., 2007; Ramos, Limón-Flores, and Ullrich, 2007). On the other hand, much of the data concerning the penetration of jet fuel through the skin comes from experiments using rat (McDougal et al., 2000) or pig skin (Riviere et al., 1999). A recent report using rats indicates no immune suppression when Jet A was applied to the skin during a 28-day dermal exposure protocol. In rats, dermal exposure to Jet A did not suppress T cell proliferation or Natural Killer cell function, nor did dermal exposure suppress antibody formation as measured by the Jerne plaque cell-forming assay (Mann et al., 2008). The authors questioned "[w]hether the differences observed between this study and previous mouse studies is due to a species difference in the immunological responses to jet fuel or the result of other factors." A careful examination of the methodology used by Mann et al., and that used by most of the investigations using mice, indicates one major difference that may contribute to the different experimental outcomes. To alleviate the skin toxicity that occurs during chronic jet fuel exposure, Mann and colleagues diluted the Jet A in a mineral oil vehicle. This is a standard technique to reduce the skin toxicity associated with chronic exposure and has been used extensively in studies to determine whether jet fuel and other petroleum middle distillates are carcinogenic. It is important to note, however, that Nessel et al. (1999) found that skin irritation and inflammation were markedly reduced and the tumor-promoting activity of jet fuel was almost absent when the fuel was applied to the skin using a mineral oil vehicle. Because our previous studies indicated that inflammatory products, such as PAF and PGE_2, are critical for activating the cascade of events that leads to immune suppression, we wondered if the real difference between our results and those reported by Mann et al. was the use of a mineral oil vehicle, which was dampening inflammation. To address this question, we performed the following experiment: 80 mg JP-8 was applied to C3H/HeN mice on 4 successive days as described previously (Ramos et al., 2002). Alternatively, 80 mg JP-8 was diluted 1:1 in mineral oil and this mixture was applied to the dorsal skin of the mice on 4 successive days. Three hours after the last JP-8 treatment, the mice were immunized and we measured the effect that jet fuel treatment had on the induction of DTH (Figure 6.2A). As described previously, multiple applications of small amounts of JP-8 induced immune suppression (85% immune suppression; $p < 0.01$ versus the positive control). However, when the jet fuel was diluted in mineral oil, no immune suppression was noted. We also

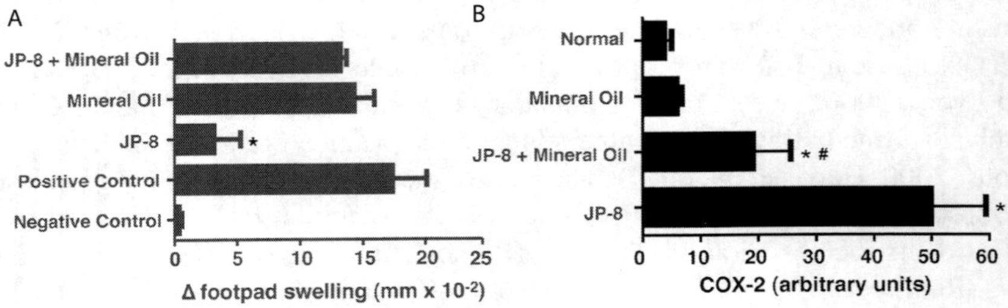

FIGURE 6.2 Applying JP-8 in mineral oil prevents the induction of immune suppression. (A) JP-8 (80 mg) neat or diluted in an equal volume of mineral oil (100 μl) was applied to the skin of C3H/HeN mice over 4 successive days. At 3 hours after the last exposure, the mice were immunized with *Candida albicans* and the effect that jet fuel treatment had on DTH was measured. Compared to the positive control (mice that were immunized and challenged, but not treated with jet fuel), JP-8 treatment causes a significant immune suppression (* $p < 0.001$). Applying the same concentration of jet fuel diluted in mineral oil prevented the induction of immune suppression. Mineral oil alone did not affect the magnitude of the immune response. (B) JP-8-induced COX-2 expression is prevented when the jet fuel is diluted in mineral oil. Skin samples were isolated from normal mice, mice treated with JP-8 (80 mg over 4 successive days), or mice treated with JP-8 diluted in mineral oil (80 mg diluted in an equal volume of mineral oil, 4 successive days). At 6 hours after the last treatment, the skin was removed and real-time PCR was used to measure COX-2 expression. The data are expressed as arbitrary units relative to the amount of GAPDH mRNA. * Indicates a statistically significant difference ($p < 0.001$) from the normal control. # Indicates a statistically significant differences ($p < 0.001$) from the JP-8 only group. Differences between control and experimental groups were determined with a one-way ANOVA, followed by Student-Newman-Keuls multiple comparison test.

measured the expression of COX-2 mRNA in the jet-fuel-treated skin using the real-time polymerase chain reaction (PCR) assay (Figure 6.2B). As described previously (Ramos, Limón-Flores, and Ullrich, 2007), applying undiluted JP-8 to the skin of mice up-regulated the expression of COX-2 mRNA, but when the JP-8 was diluted in mineral oil, the up-regulation of COX-2 was suppressed (* $p < 0.001$ versus normal control; # $P < 0.001$ versus JP-8 only group). These data indicate that diluting the jet fuel in mineral oil renders it non-immunosuppressive, in part by dampening its ability to activate COX-2 expression in the skin. We are not sure of the exact mechanism, but it is important to recall McDougal's comment that "the relative affinity of a chemical for the skin versus the affinity of the chemical for the vehicle will determine whether the chemical will have a tendency to stay in the vehicle or be driven into the skin" (McDougal et al., 2000). The data presented in Figure 6.2 indicate that mineral oil will ameliorate the immunotoxicity of jet fuel applied to the skin of mice, suggesting that the major reason for the lack of immunotoxicity in Mann's study is the choice of vehicle (i.e., "the result of other factors"). Although it is clear that mineral oil was employed by Mann and colleagues to comply with ethical animal use issues (i.e., to prevent the skin erosion that can occur during chronic jet fuel exposure), it is also readily apparent that Air Force and civilian aviation personnel are exposed to undiluted jet fuel, and are not normally exposed to fuel diluted in mineral oil or any other vehicle. We suggest that care must be taken to avoid over-analyzing results from animal experiments that do not closely mimic human exposure scenarios.

Along the same line, Selgrade (2007) has recently asked if rodent immunotoxicity data is relevant for extrapolating risk to humans. She reviewed data concerning immune suppression caused by agents such as cigarette smoke, arsenic, and polychlorinated biphenyls; she concluded that immune suppression in rodents is predictive of immune suppression in humans and that there is a link between developmental exposure to toxicants and increased risk of infections and neoplastic disease in humans. Most of the data demonstrating that JP-8 and Jet A induce immune suppression comes from animal studies. Although we agree with Selgrade and suggest that the results from the animal studies may be used to predict that jet fuel could be an immunosuppressive agent in humans, the lack of studies demonstrating an effect of jet fuel, either positive or negative, on the human immune system is a critical gap in our knowledge. Studies to directly address this question are needed and represent the next important step in determining jet fuel toxicity. Furthermore, in our opinion, simply measuring fluctuations in the numbers of circulating immune cells following jet fuel exposure is not a sufficient measure of immune status (Rhodes et al., 2003). *Studies are needed to measure the effect of jet fuel on the function of the human immune system!* Although these studies may be expensive to carry out, they are not impossible to do. We suggest that a page can be taken from our colleagues in the dermatology community. To demonstrate that sunscreens could effectively block UV-induced immune suppression of secondary immune reactions, assessment of DTH *in vivo* was used (Moyal, 1998; Moyal and Fourtanier, 2001). The advantage of this approach was that it provided an in vivo measurement of the intact immune system, and it did not require prior immunization of the test subjects, making the logistics and expense associated with the protocol reasonable. We suggest that a similar protocol could be used to ask if applying jet fuel suppresses the immune reaction to the common recall antigens that most of us—and particularly military personnel—are immunized against (tetanus and diphtheria toxoids, *Streptococcus, Tuberculin, Candida, Trichophyton*, and Proteus antigens). Based on existing animal data indicating that JP-8 can suppress secondary immune reactions (Ramos et al., 2002), we suggest that these studies should be funded and performed to answer this important question and address this critical gap in our knowledge base.

ACKNOWLEDGMENTS

The work performed in our laboratory was supported by a grant from the United States Air Force Office of Scientific Research (F95550-05-1-0402). The animal, flow cytometry, and histology facilities at University of Texas MD Anderson Cancer Center are supported in part by a Cancer Center Support Grant from the National Cancer Institute (CA16672). GR was supported by a scholarship from the USAF Institute of Technology. The views and opinions expressed here are those of the authors and do not reflect the official policy or position of the United States Air Force, Department of Defense, or the U.S. Government.

REFERENCES

Abadin, H.G., C.H. Chou, and F.T. Llados. 2007. Health effects classification and its role in the derivation of minimal risk levels: Immunological effects. *Regul. Toxicol. Pharmacol.*, 47(3): 249–256.

Alappatt, C., C.A. Johnson, K.L. Clay, and J.B. Travers. 2000. Acute keratinocyte damage stimulates platelet-activating factor production. *Arch. Dermatol. Res.*, 292(5): 256–259.

Byrne, S.N., A.Y. Limón-Flores, and S.E. Ullrich. 2008. Mast cell migration from the skin to the draining lymph nodes upon ultraviolet irradiation represents a key step in the induction of immune suppression. *J. Immunol.*, 180(7): 4648–4655.

Boulares, A.H., F.J. Conteras, L.A. Espinoza, and M.E. Smulson. 2002. Roles of oxidative stress and glutathione depletion in JP-8 jet fuel-induced apoptosis in rat lung epithelial cells. *Toxicol. Appl. Pharmacol.*, 180(2): 92–99.

Chao, Y.C., L.L. Kupper, B. Serdar, P.P. Egeghy, S.M. Rappaport, and L.A. Nylander-French. 2006. Dermal exposure to jet fuel JP-8 significantly contributes to the production of urinary naphthols in fuel-cell maintenance workers. *Environ. Health Perspect.*, 114(2): 182–185.

Chou, C.C., J.E. Riviere, and N.A. Monteiro-Riviere. 2003. The cytotoxicity of jet fuel aromatic hydrocarbons and dose-related interleukin-8 release from human epidermal keratinocytes. *Arch. Toxicol.*, 77(7): 384–391.

Cooper, J.R., and D.R. Mattie. 1996. Developmental toxicity of JP-8 jet fuel in the rat. *J. Appl. Toxicol.*, 16(3): 197–200.

Dudley, A.C., M.M. Peden-Adams, J. EuDaly, R.S. Pollenz, and D.E. Keil. 2001. An aryl hydrocarbon receptor independent mechanism of JP-8 jet fuel immunotoxicity in Ah-responsive and Ah-nonresponsive mice. *Toxicol. Sci.*, 59(2): 251–259.

Harris, D.T., D. Sakiestewa, R.F. Robledo, and M. Witten. 1997a. Immuno-toxicological effects of JP-8 jet fuel exposure. *Toxicol. Indus. Health,* 13(1): 43–55.

Harris, D.T., D. Sakiestewa, R.F. Robledo, and M. Witten. 1997b. Short-term exposure to JP-8 jet fuel results in long-term immunotoxicity. *Toxicol. Indus. Health,* 13(5): 559–570.

Harris, D.T., D. Sakiestewa, R.F. Robledo, and M. Witten. 1997c. Protection from JP-8 jet fuel induced immunotoxicity by administration of aerosolized substance P. *Toxicol. Indus. Health,* 13(5): 571-588.

Harris, D.T., D. Sakiestewa, R.F. Robledo, R.S. Young, and M. Witten. 2000. Effects of short-term JP-8 jet fuel exposure on cell-mediated immunity. *Toxicol. Indus. Health,* 16(2): 78–84.

Harris, D.T., D. Sakiestewa, D. Titone, and M. Witten. 2007. JP-8 jet fuel exposure rapidly induces high levels of IL-10 and PGE_2 secretion and is correlated with loss of immune function. *Toxicol. Indus. Health,* 23(4): 223–230.

Hays, A.M., G. Parliman, J.K. Pfaff, R.C. Lantz, J. Tinajero, B. Tollinger, J.N. Hall, and M.L. Witten. 1995. Changes in lung permeability correlate with lung histology in a chronic exposure model. *Toxicol. Indus. Health,* 11(3): 325–336.

Ishii, S., and T. Shimizu. 2000. Platelet-activating factor (PAF) receptor and genetically engineered PAF receptor mutant mice. *Prog. Lipid Res.*, 39(1): 41–82.

Kanikkannan, N., T. Jackson, M. Sudhan Shaik, and M. Singh. 2000. Evaluation of skin sensitization potential of jet fuels by murine local lymph node assay. *Toxicol. Lett.*, 116(1-2): 165–170.

Keil, D.E., D.A. Warren, M.J. Jenny, J.G. EuDaly, J. Smythe, and M.M. Peden-Adams. 2003. Immunological function in mice exposed to JP-8 jet fuel in utero. *Toxicol. Sci.*, 76(2): 347–356.

Keil, D., A. Dudley, J. EuDaly, J. Dempsey, L. Butterworth, G. Gilkeson, and M. Peden-Adams. 2004. Immunological and hematological effects observed in B6C3F1 mice exposed to JP-8 jet fuel for 14 days. *J. Toxicol. Environ. Health A,* 67(14): 1109–1129.

Ko, H.M., K.H. Seo, S.J. Han, K.Y. Ahn, I.H. Choi, G.K. Koh, H.K. Lee, M.S. Ra, and S.Y. Im. 2002. Nuclear Factor kappaβ dependency of platelet activating factor-induced angiogenesis. *Cancer Res.*, 62(6): 1809–1814.

Kim, D., M.E. Andersen, and L.A. Nylander-French. 2006. Dermal absorption and penetration of jet fuel components in humans. *Toxicol. Lett.*, 165(1): 11–21.

Kinkead, E.R., S.A. Salins, and R.E. Wolfe. 1992. Acute irritation and sensitization potential of JP-8 jet fuel. *Acute Toxic. Data,* 11(6): 700.

Limón-Flores, A.Y., R. Chacón-Salinas, G. Ramos, and S.E. Ullrich. 2009. Mast cells mediate the immune suppression induced by dermal exposure to JP-8 jet fuel. *Toxicol., Sci.,* in press. (DOI:10.1093/toxsci/kfp181).

Mann, C.M., V.L. Peachee, G.W. Trimmer, J.E. Lee, L.E. Twerdok, and K.L. White, Jr. 2008. Immunotoxicity evaluation of jet a jet fuel in female rats after 28-day dermal exposure. *J. Toxicol. Environ. Health A,* 71(8): 495–504.

Mattie, D.R., G.B. Marit, C.D. Flemming, and J.R. Cooper. 1995. The effects of JP-8 jet fuel on male Sprague-Dawley rats after a 90-day exposure by oral gavage. *Toxicol. Indus. Health,* 11(4): 423–435.

McDougal, J.N., D.L. Pollard, W. Weisman, C.M. Garrett, and T.E. Miller. 2000. Assessment of skin absorption and penetration of JP-8 jet fuel and its components. *Toxicol. Sci.,* 55 (2): 247–255.

Moore, K.W., R. de Waal Malefyt, R.L. Coffman, and A. O'Garra. 2001. Interleukin-10 and the interleukin-10 receptor. *Annu. Rev. Immunol.,* 19: 683–765.

Moyal, D. 1998. Immunosuppression induced by chronic ultraviolet irradiation in humans and its prevention by sunscreens. *Eur. J. Dermatol.,* 8: 209–211.

Moyal, D.D., and A.M. Fourtanier. 2001. Broad-Spectrum sunscreens provide better protection from the suppression of the elicitation phase of delayed-type hypersensitivity response in humans. *J. Invest. Dermatol.,* 117: 1186–1192.

Muhammad, F., N.A. Monteiro-Riviere, R.E. Baynes, and J.E. Riviere. 2005. Effect of in vivo jet fuel exposure on subsequent in vitro dermal absorption of individual aromatic and aliphatic hydrocarbon fuel constituents. *J. Toxicol. Environ. Health A,* 68(9): 719–737.

National Research Council. 2003. *Toxicologic Assessment of Jet-Propulsion Fuel 8.* Washington, D.C.: National Academies Press.

Nessel, C.S., J.J. Freeman, R.C. Forgash, and R.H. McKee. 1999. The role of dermal irritation in the skin tumor promoting activity of petroleum middle distillates. *Toxicol. Sci.,* 49 (1): 48–55.

Pei, Y., L.A. Barber, R.C. Murphy, C.A. Johnson, S.W. Kelley, L.C. Dy, R.H. Fertel, T. M. Nguyen, D.A. Williams, and J.B. Travers. 1998. Activation of the epidermal platelet-activating factor receptor results in cytokine and cyclooxygenase-2 biosynthesis. *J. Immunol.,* 161(4): 1954–1961.

Pfaff, J., K. Parton, R.C. Lantz, H. Chen, A.M. Hays, and M.L. Witten. 1995. Inhalation exposure to JP-8 jet fuel alters pulmonary function and Substance P levels in Fischer 344 rats. *J. Appl. Toxicol.,* 15(4):249–256.

Ramos, G., D.X. Nghiem, J.P. Walterscheid, and S.E. Ullrich. 2002. Dermal application of jet fuel suppresses secondary immune reactions. *Toxicol. Appl. Pharmacol.,* 180(2): 136–144.

Ramos, G., N. Kazimi, D.X. Nghiem, J.P. Walterscheid, and S.E. Ullrich. 2004. Platelet activating factor receptor binding plays a critical role in jet fuel-induced immune suppression. *Toxicol. Appl. Pharmacol.,* 195(3): 331–338.

Ramos, G., A.Y. Limón-Flores, and S.E. Ullrich. 2007. Dermal exposure to jet fuel suppresses delayed-type hypersensitivity: a critical role for aromatic hydrocarbons. *Toxicol. Sci.,* 100(2):415–422.

Ramos, G., A.Y. Limón-Flores, and S.E. Ullrich. 2009. JP-8-induces immune suppression via a reactive Oxygen Species NF-kb-dependent mechanism. *Toxicol. Sci.,* 108(1): 100–109.

Rhodes, A.G., G.K. LeMasters, J.E. Lockey, J.W. Smith, J.H. Yiin, P. Egeghy, and R. Gibson. 2003. The effects of jet fuel on immune cells of fuel system maintenance workers. *J. Occup. Environ. Med.,* 45(1): 79–86.

Ritchie, G., K. Still, J. Rossi, 3rd, M. Bekkedal, A. Bobb, and D. Arfsten. 2003. Biological and health effects of exposure to kerosene-based jet fuels and performance additives. *J. Toxicol. Environ. Health B Crit. Rev.,* 6(4): 357–451.

Riviere, J.E., J.D. Brooks, N.A. Monteiro-Riviere, K. Budsaba, and C.E. Smith. 1999. Dermal absorption and distribution of topically dosed jet fuels Jet-A, JP-8, and JP-8(100). *Toxicol. Appl. Pharmacol.,* 160(1): 60–75.

Rogers, J.V., P.G. Gunasekar, C.M. Garrett, M.B. Kabbur, and J.N. McDougal. 2001. Detection of oxidative species and low-molecular-weight DNA in skin following dermal exposure with JP-8 jet fuel. *J. Appl. Toxicol.,* 21(6): 521–525.

Rogers, J.V., G.L. Siegel, D.L. Pollard, A.D. Rooney, and J.N. McDougal. 2004. The cytotoxicity of volatile JP-8 jet fuel components in keratinocytes. *Toxicology,* 197(2): 113–121.

Schmitt, D.A., J.P. Walterscheid, and S.E. Ullrich. 2000. Reversal of ultraviolet radiation-induced immune suppression by recombinant interleukin-12: suppression of cytokine production. *Immunology,* 101(1):90–96.

Seibert, K., Y. Zhang, K. Leahy, S. Hauser, J. Masferrer, W. Perkins, L. Lee, and P. Isakson. 1994. Pharmacological and biochemical demonstration of the role of cyclooxygenase 2 in inflammation and pain. *Proc. Natl. Acad. Sci. USA,* 91: 12013–12017.

Selgrade, M.K. 2007. Immunotoxicity: the risk is real. *Toxicol. Sci.,* 100(2): 328–332.

Shreedhar, V., T. Giese, V.W. Sung, and S.E. Ullrich. 1998. A cytokine cascade including prostaglandin E_2, IL-4, and IL-10 is responsible for UV-induced systemic immune suppression. *J. Immunol.,* 160(8): 3783–3789.

Smith, L.B., A. Bhattacharya, G. Lemasters, P. Succop, E. Puhala, 2nd, M. Medvedovic, and J. Joyce. 1997. Effect of chronic low-level exposure to jet fuel on postural balance of US Air Force personnel. *J. Occup. Environ. Med.,* 39(7): 623–632.

Ullrich, S.E. 1999. Dermal application of JP-8 jet fuel induces immune suppression. *Toxicol. Sci.,* 52(1): 61–67.

Ullrich, S.E., and H.J. Lyons. 2000. Mechanisms involved in the immunotoxicity induced by dermal application of JP-8 jet fuel. *Toxicol. Sci.,* 58(2): 290–298.

Walterscheid, J.P., S.E. Ullrich, and D.X. Nghiem. 2002. Platelet-activating factor, a molecular sensor for cellular damage, activates systemic immune suppression. *J. Exp. Med.,* 195(2): 171–179.

7 Absorption, Penetration, and Cutaneous Toxicity of Jet Fuels and Hydrocarbon Components

James E. Riviere, Alfred O. Inman, and Nancy Monteiro-Riviere

CONTENTS

INTRODUCTION

Jet propulsion (JP) fuels that power commercial and military aircraft consist of a complex kerosene mixture of aromatic and aliphatic hydrocarbons. The base fuel, commercial Jet A, consists of approximately 228 aromatic and aliphatic hydrocarbons with carbon chain lengths ranging from C_6 to C_{17}. In Jet A, 13 components at concentrations greater than 1% v/v (Table 7.1) make up 28.9% of the total formulation (Riviere et al., 1999; Ritchie et al., 2003). The primary military jet fuel, JP-8, consists of Jet A plus the required additives DC1-4A (corrosion inhibitor), Stadis 450 (antistatic compound), and diethylene glycol monomethylether (DiEGME; icing inhibitor) that make up less than 2% of the formulation. JP-8(100) fuel is JP-8 with antioxidant, chelator, detergent, and dispersant additives. These additives provide thermal stability, reduce fuel freezing, dissipate electrical charge, and improve overall engine dependability (White, 1999). Bulk fuel worldwide ending inventory of JP-8 was 28.4 million barrels in FY 2004 and 29.6 million barrels in FY 2005 (Defense Energy Support Center Fact Book, 2005). Approximately 5 billion gallons of JP-8 are used annually by United States and North Atlantic Treaty Organization forces primarily as fuel for aircraft, ground vehicles, tent heaters, and generators. JP-8 has also been

TABLE 7.1
Major Aliphatic and Aromatic Components of Jet A

Component	Class	Percent of Total	Number of Carbons
Dodecane	Aliphatic	4.7	C_{12}
Tridecane	Aliphatic	4.4	C_{13}
Undecane	Aliphatic	4.1	C_{11}
Tetradecane	Aliphatic	3.0	C_{14}
2,6-Dimethylundecane	Aliphatic	2.1	C_{13}
Pentadecane	Aliphatic	1.6	C_{15}
2-Methylundecane	Aliphatic	1.2	C_{12}
Heptylcyclohexane	Aliphatic	1.0	C_{13}
1-Methylnaphthalene	Aromatic	1.8	C_{11}
2-Methylnaphthalene	Aromatic	1.5	C_{11}
2,6-Dimethylnaphthalene	Aromatic	1.3	C_{12}
Naphthalene	Aromatic	1.1	C_{10}
1,2,3,4-Tetramethylbenzene	Aromatic	1.1	C_{10}

Source: From Riviere, J.E., Brooks, J.D., Monteiro-Riviere, N.A., Budsaba, K., and Smith, C.E. *Toxicol. Appl. Phamacol.,* 160, 60–75, 1999.

used as a vehicle for insecticides, herbicides, and pesticides, as well as a suppressant for dust. Therefore, the risk of occupational exposure to JP-8 is high for all military personnel in the field, and especially so for personnel who directly handle the fuel, work on flight lines, and maintain aircraft. Worldwide, more than 2 million military and civilian personnel are annually exposed to the military and civilian jet propulsion fuels (Ritchie et al., 2003). Air Force personnel who routinely worked with or were exposed to JP-8 were found to have elevated levels of the markers naphthalene (Chao et al., 2005) and urinary 1- and 2-naphthols (Chao et al., 2006).

Chronic topical exposure to jet fuels may reduce the integrity of the dermal barrier, the stratum corneum, and cause an increase in systemic exposure to jet fuel components as well as other occupational toxicants. The health effects to personnel exposed to jet fuels or hydrocarbon solvents in occupational settings have been assessed in a number of studies (Selden and Ahlborg, 1986, 1987, 1991; Bell et al., 2005; Chao et al., 2005, 2006; Cavallo et al., 2006; D'Este et al., 2008). As such, the health effects following dermal exposure to the jet fuel mixtures and component hydrocarbons have been studied in vitro (cultured human epidermal keratinocytes, flow-through diffusion cells with rodent or porcine skin, isolated perfused porcine skin) and in vivo (rodent, porcine).

IN VITRO STUDIES

Cultured normal human epidermal keratinocytes (HEK) were exposed to mixtures of Jet A, JP-8, and JP-8(100) in order to identify potential biomarkers of inflammation (Allen et al., 2000). The HEK were exposed to 0.1% Jet A, JP-8, and JP-8(100) for 24 hours and evaluated for release of the proinflammatory cytokines interleukin-8 (IL-8) and tumor necrosis factor-alpha (TNF-α). IL-8 concentration increased after 4 hours up to 24 hours relative to the controls following exposure to the three jet fuels. The IL-8 mRNA was also elevated by 4 hours post exposure. TNF-α release was greater than the controls, with the highest cytokine concentration noted at 4 hours, followed by a steady decrease through 24 hours. To determine the effect of Jet A, JP-8, and JP-8(100) on porcine epidermal keratinocytes (PEK), Allen et al. (2001a) exposed PEK and an immortalized PEK cell line (MSK3877) to the three fuel mixtures. PEK expression to the three fuels, unlike the MSK3877 cells, were similar to that of HEK, with IL-8 concentration increasing from 4 to 24 hours post treatment and TNF-α release peaking at 1 to 4 hours before a terminal decrease. The authors conclude that jet fuel is toxic to normal PEK, but urge caution when interpreting data from immortalized cells. To better understand which components of the jet fuel mixture are toxic, HEK were exposed to serial dilutions of the aliphatic hydrocarbons undecane (C_{11}), dodecane (C_{12}), tridecane (C_{13}), and hexadecane (C_{16}). IL-8 concentration peaked in the HEK exposed to a hydrocarbon concentration of 31.2 μM, with tridecane eliciting the greatest release of the biomarker, followed in decreasing order by hexadecane, dodecane, undecane, and untreated control (Allen et al., 2001b). Protein expression in HEK was determined following a 24-hour exposure to 0.1% JP-8 (Witzmann et al., 2005). JP-8 altered the expression of 33 proteins, which suggests that HEK inflammation (as indicated by IL-8 release) is associated with a limited group of functional changes in altered proteins. These include proteins that are involved in membrane trafficking and in inflammation and injury, as well as cytoskeletal and intermediate filaments that function in cell junction and motility.

To help rule out vehicle affects, HEK were exposed to neat jet fuel and constituent hydrocarbons. Chou et al. (2002) treated HEK with aliphatic hydrocarbons (C_6 to C_{16}) for 1, 5, and 15 minutes and determined IL-8 release after 0, 1, 4, 8, and 24 hours. HEK mortality, assessed 24 hours following exposure, increased with a decrease in the length of the aliphatic carbon chain. In addition, Jet A and JP-8 exhibited low toxicity by 24 hours, similar to the C_{14} to C_{16} hydrocarbons. The study also found an increase (3 to 10 times) of IL-8 following exposure to the mid-length hydrocarbon chains (C_9 to C_{13}). Higher toxicity by short-chain aliphatic hydrocarbons did not correlate to an increased stimulation of IL-8. While aliphatic hydrocarbons make up more than 75% of the jet fuels, the aromatic hydrocarbons were more instrumental in causing HEK mortality. Chou et al. (2003) found that the number and size of the side-chains attached to the aromatic ring correlated to an increase in HEK cytotoxicity. The ability to cause HEK mortality, however, did not directly correspond to IL-8 induction. Rogers et al. (2004) reported similar results, with HEK exposed for 4 hours in unsealed plates exhibiting more sensitivity to the aromatic hydrocarbons than the aliphatic hydrocarbons. Chou et al. (2003) also found that HEK exposed to aromatic hydrocarbons and the aliphatic hydrocarbons hexadecane or mineral oil (aliphatic mixture of C_{11} to C_{16}) concurrently had greatly reduced mortality than what was elicited by the aromatics alone. Yang et al. (2006) later demonstrated that the toxicity of individual hydrocarbons does not predict the toxicity of these hydrocarbons in mixtures, and suggested that aliphatic hydrocarbon mixtures play a more significant role in dermatotoxicity than the aromatic hydrocarbon mixtures.

Several compounds with the potential to inhibit the toxic effects of jet fuels and constituent hydrocarbons were investigated. Substance P (Sar[9], Met $(O_2)^{11}$), an antagonist of neurokinin receptor NK_1, was found to decrease IL-8 release from HEK following exposure to JP-8 (Monteiro-Riviere et al., 2004). This is consistent with earlier studies in the lung (Harris et al., 1997, 2000; Robledo and Witten, 1999) in which the pulmonary and immunotoxic effects of JP-8 or tetradecane following inhalation were paradoxically reduced with exposure to Substance P. Inman et al. (2008) studied the effectiveness of the inhibitors parthenolide, isohelenin, Substance P, SB 203580, and recombinant human IL-10 on reducing the cytotoxic effects of neat JP-8, S-8, and the constituent aliphatic hydrocarbons pentadecane, tetradecane, tridecane, and undecane in HEK. The synthetic fuel S-8, an alternative to JP-8, is a clean-burning fuel made up of a complex mixture of aliphatic hydrocarbons. IL-8 production was suppressed a minimum of 30% following concurrent treatment with each inhibitor. However parthenolide, a sesquiterpene lactone that is a potent inhibitor of NF-κB, was the most effective inhibitor of IL-8. Parthenolide caused a significant ($p < 0.05$) increase in viability in HEK exposed to JP-8 and S-8 (Figure 7.1), and a significant ($p < 0.05$) decrease in IL-8 release in the jet fuels and hydrocarbons (Figure 7.2). S-8 toxicity was similar to undecane but significantly ($p < 0.05$) less than JP-8. This study raises the possibility that inhibiting NF-κB in vitro may reduce hydrocarbon inflammation in vivo.

Dermal penetration of the jet fuels and components has been assessed in a number of in vitro studies. Percutaneous absorption and cutaneous deposition of Jet A,

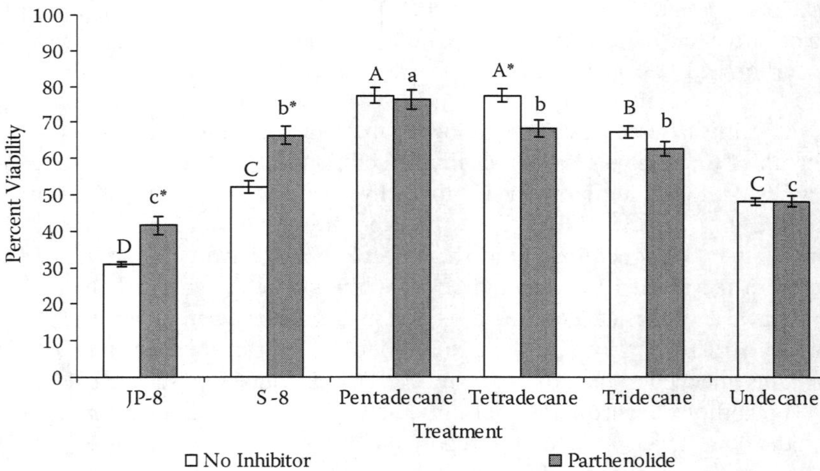

FIGURE 7.1 Viability of HEK with and without parthenolide and exposed to JP-8, S-8, and the aliphatic hydrocarbons pentadecane, tetradecane, tridecane, and undecane. Overall, viability decreased with a decrease in the length of the aliphatic carbon chain. Across No Inhibitor treatments: A,B,C,D are significantly different ($p < 0.05$). Across Parthenolide treatments: a,b,c are significantly different ($p < 0.05$). * significantly greater ($p < 0.05$) than paired treatment. (*Source:* From Inman, A.O., Monteiro-Riviere, N.A., and Riviere, J.E. *J. Appl. Toxicol.*, 28, 543–553, 2008.)

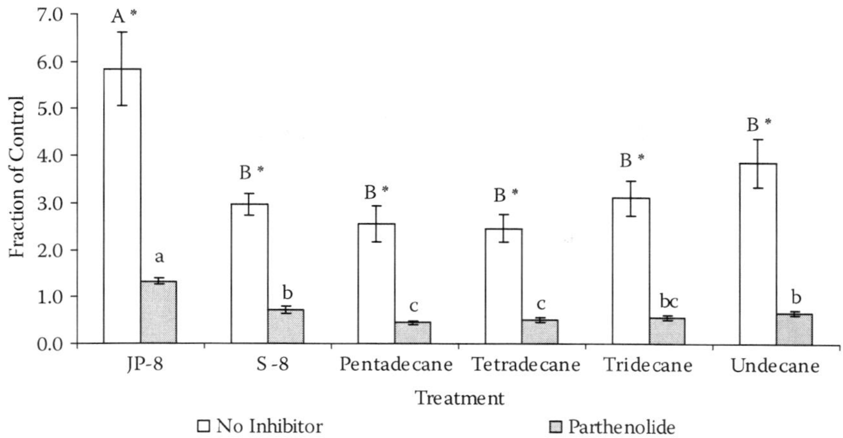

FIGURE 7.2 IL-8 Fraction of Control in HEK with and without parthenolide and exposed to JP-8, S-8, and the aliphatic hydrocarbons pentadecane, tetradecane, tridecane, and undecane. Across No Inhibitor treatments: A,B are significantly different ($p<0.05$). Across Parthenolide treatments: a,b,c are significantly different ($p<0.05$). * significantly greater ($p < 0.05$) than paired treatment. (*Source:* From Inman, A.O., Monteiro-Riviere, N.A., and Riviere, J.E. *J. Appl. Toxicol.*, 28, 543–553, 2008.)

JP-8, puddled (evaporated) JP-8, and JP-8(100) were determined by monitoring the absorptive flux of the marker components ^{14}C-naphthalene (C_{10}), ^{3}H-dodecane (C_{12}), and ^{14}C-hexadecane (C_{16}), applied simultaneously to the isolated perfused porcine skin flap (IPPSF) and perfused for 5 hours (Riviere et al., 1999). Naphthalene absorption into the perfusate peaked in less than 1 hour (~30 minutes), while the aliphatics dodecane and hexadecane plateaued between 1.5 and 2 hours with much lower absorption flux profiles (Figure 7.3). For JP-8, the rank order of absorption into the perfusate for all marker components was naphthalene > dodecane > hexadecane, while deposition within the upper layers of the skin was hexadecane > dodecane > naphthalene. Dodecane and hexadecane, long-chain hydrocarbons that are relatively nonvolatile, had the greatest surface concentrations but decreased perfusate fluxes and total penetration by the end of the skin flap experiments. The absolute mass of absorbed naphthalene was 2 times greater than dodecane even though the dodecane concentration was 5 times greater than the naphthalene. This is consistent with the findings of McDougal et al. (1998, 2000). Naphthalene absorption was similar across the four fuel types, with total penetration of naphthalene ranked as JP-8(100) > Jet A > JP-8 > puddled JP-8 (Figure 7.4a). Perfusate absorption for dodecane was JP-8 > JP-8(100) > Jet A > puddled JP-8 (Figure 7.4b). These studies indicate that while dermal absorption from the jet fuels may be consistent across fuels, the chemical characteristics of individual hydrocarbons (i.e., volatility, molecular weight, hydrophilicity, lipid solubility) and the effects of the performance additives may be more important in predicting deposition in skin, fat, and blood (Ritchie et al., 2003).

McDougal et al. (2000) measured the flux of JP-8 and major components across rat skin using static diffusion cells up to 3.5 hours. Thirteen individual components

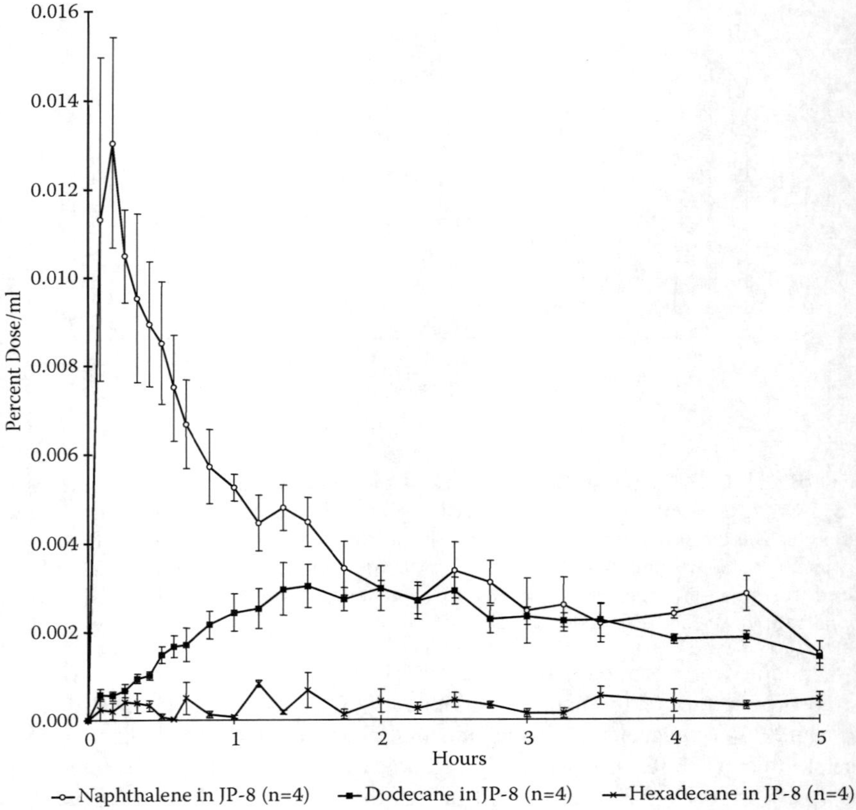

FIGURE 7.3 Comparison of naphthalene, dodecane, and hexadecane absorption into the IPPSF perfusate after dosing in JP-8 jet fuel. (*Source:* From Riviere, J.E., Brooks, J.D., Monteiro-Riviere, N.A., Budsaba, K., and Smith, C.E. *Toxicol. Appl. Phamacol.,* 160, 60–75, 1999. With permission.)

of JP-8 penetrated the rat skin into the receptor solution with a ranking DiEGME > decane > methylnaphthalene > trimethylbenzene > undecane > naphthalene > dimethylbenzene > dimethylnaphthalene > methylbenzene > dodecane > nonane > ethylbenzene > tridecane. DiEGME, which exhibited the highest flux (51.5 μg/cm^2-hr), made up only 0.08% w/w of JP-8 while tridecane was the lowest at 0.334 μg/cm^2-hr and made up 2.7% w/w. Six aliphatic components (undecane > decane > dodecane > tridecane > nonane > tetradecane) were identified in the skin, which correlates to their high octanol/water partition coefficients. The authors concluded that under normal occupational conditions, JP-8 was not expected to be absorbed well enough to be a systemic hazard, although the absorption of aliphatic hydrocarbons into the skin may be sufficient to induce dermal irritation.

Baynes et al. (2001b) conducted full factorial experiments to determine the effect of jet fuel additives on the dermal disposition of marker components ^{14}C-naphthalene and ^{14}C-dodecane applied to porcine skin in flow-through cells and to the IPPSF.

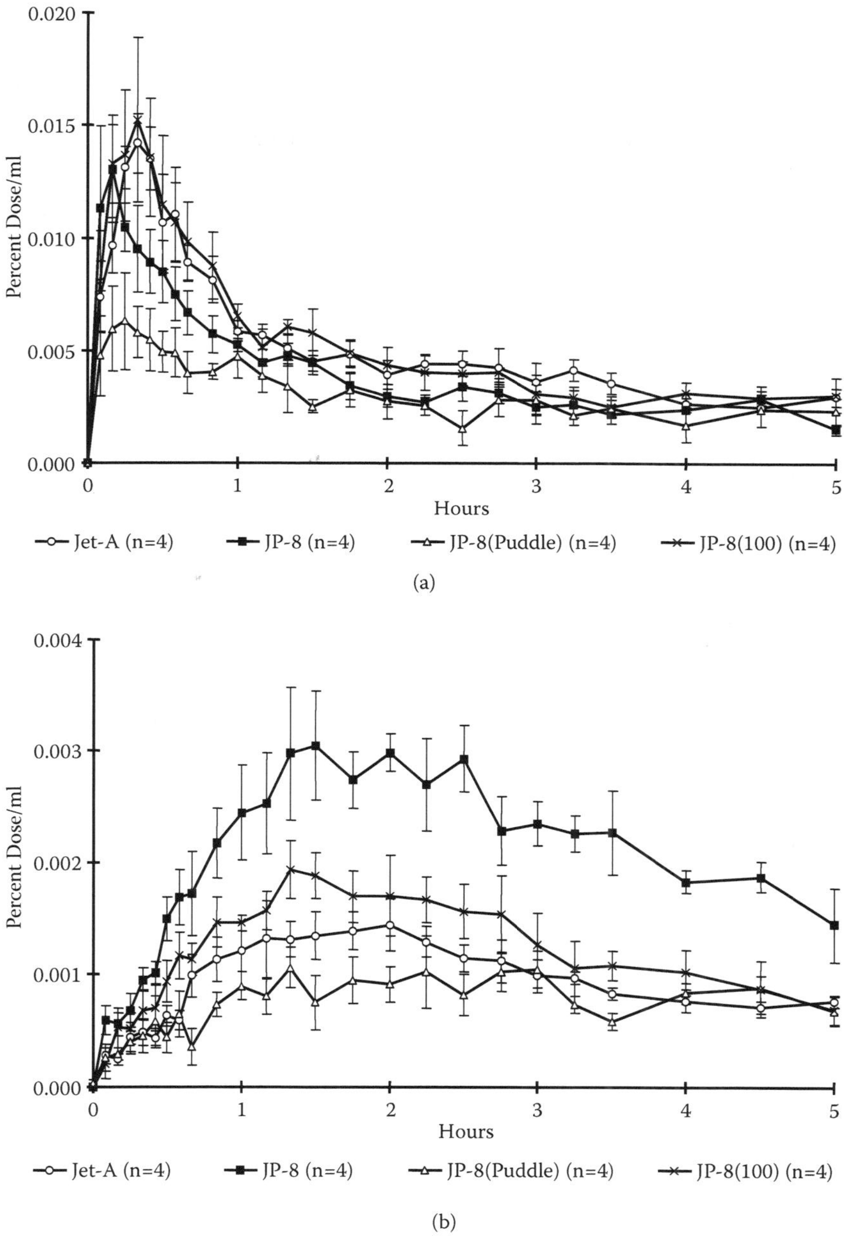

FIGURE 7.4 Hydrocarbon absorption after IPPSF dosing in Jet A, JP-8, JP-8 (puddled), and JP-8(100): (a) naphthalene, and (b) dodecane. (*Source:* From Riviere, J.E., Brooks, J.D., Monteiro-Riviere, N.A., Budsaba, K., and Smith, C.E. *Toxicol. Appl. Phamacol.*, 160, 60–75, 1999. With permission.)

Jet A, Jet A + DiEGME, Jet A + 8Q21, Jet A + Stadis 450, Jet A + DiEGME + 8Q21, Jet A + DiEGME + Stadis 450, Jet A + 8Q21 + Stadis 450, and JP-8 were topically applied and media perfused for 5 hours. Naphthalene absorption and peak fluxes were greater than that of dodecane, which is consistent with earlier studies with puddled JP-8 and JP-8(100) in porcine skin (Riviere et al., 1999; Baynes et al., 2001a) and with JP-8 in rodent skin (McDougal et al., 2000). In the flaps, synergistic interactions with 8Q21 + Stadis 450 enhanced systemic absorption of naphthalene and dodecane, while DiEGME + Stadis 450 increased naphthalene and dodecane penetration into the skin and fat. Retention of the aliphatic dodecane in the stratum corneum was greater than for naphthalene, with evaporative loss greater with the aromatic naphthalene. This study showed that significant interactions between two or three additives can have synergistic or antagonistic effects on dermal disposition of naphthalene or dodecane. Also, the products of two-factor interactions were not predictable from single-factor exposures and may prevent extrapolation to three-factor interactions.

Because a single membrane system may not extrapolate to multiple or complex membrane systems, Muhammad et al. (2004b) studied three membrane systems to determine mixture interactions. The marker hydrocarbons, ^{14}C-naphthalene and ^{14}C-dodecane, were added to specified jet fuel mixtures and applied topically to a synthetic silastic membrane or porcine skin in the flow-through system, or to the IPPSF. In the three systems, naphthalene absorption was more than 2× greater than dodecane, while the tissue retention of dodecane was more than 2× greater than naphthalene. Chemical-chemical interactions, as determined by the silastic membrane, were affected by the addition of a single fuel additive (MDA, BHT, 8Q405) to JP-8, which caused the naphthalene absorption to differ significantly from JP-8 alone or with the combination of additives. The porcine skin used to detect chemical-biological interactions found no significant interaction with any additive mixture. The data suggests that in the IPPSF, which closely mimics in vivo exposures, the addition of MDA to JP-8 significantly decreased naphthalene absorption, and the addition of BHT to JP-8 significantly enhanced naphthalene absorption (Figure 7.5a), but did not significantly affect dodecane absorption (Figure 7.5b).

Muhammad et al. (2004a) characterized the dermal absorption of marker aliphatic (undecane, dodecane, tridecane) and aromatic (naphthalene, dimethylnaphthalene) hydrocarbons in three dosing mixtures. There was a significant dose-related increase in the absorption of naphthalene and dimethylnaphthalene, consistent with Chou et al. (2003) in that aromatics cause a dose-related increase in IL-8 release from HEK. The authors concluded that increased human exposure to jet fuels might cause enhanced absorption of such hydrocarbons and cause local or systemic toxicity.

IN VIVO STUDIES

Low dose (7.96 μl/cm^2) or high dose (67 μl/cm^2) Jet A, JP-8, and JP-8(100) were exposed on the backs of pigs occluded (Hill Top® chamber or cotton fabric) or non-occluded for a 5-hour or 24-hour single exposure, or a 4-day repetitive exposure (Monteiro-Riviere et al., 2001). Transepidermal water loss increased after 5- and 24-hour fabric and Hill Top occluded treatments, and decreased in the 5-day

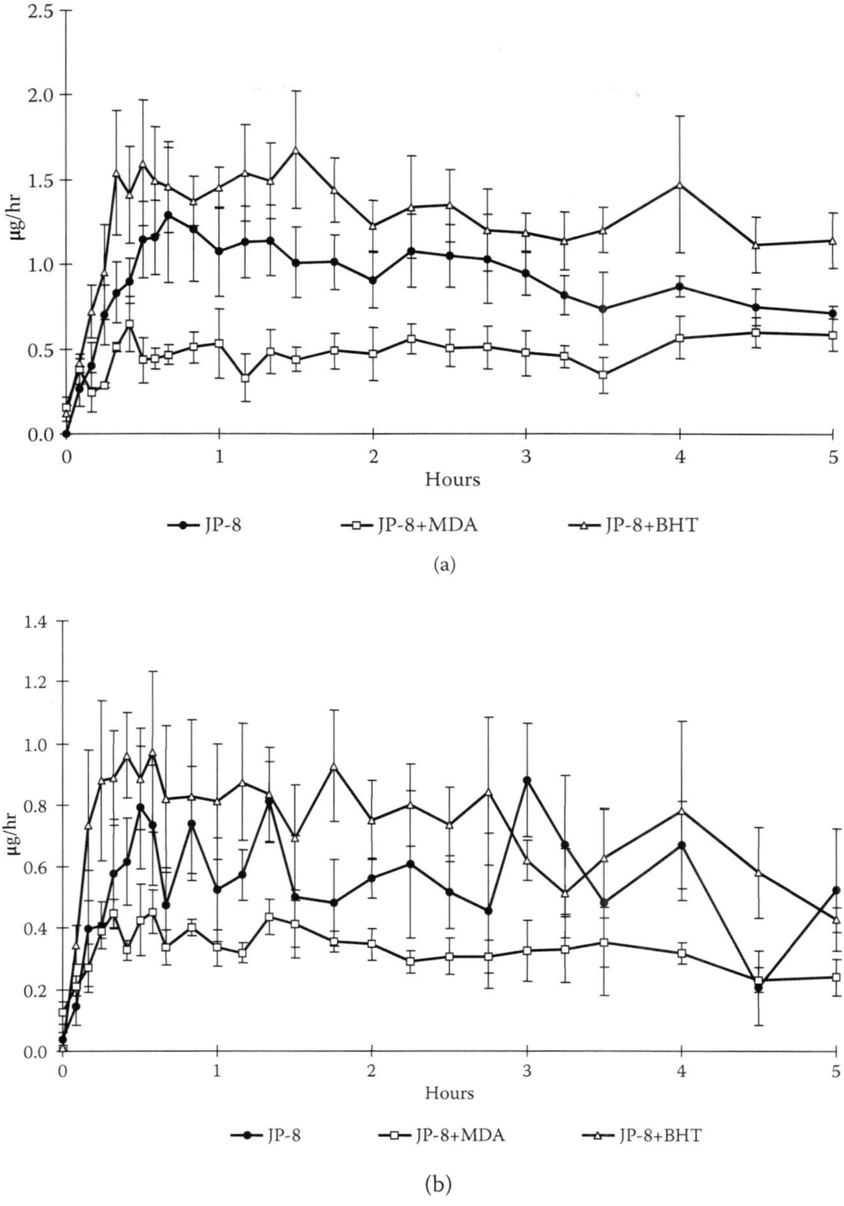

FIGURE 7.5 Perfusate absorption profiles following dosing of the IPPSF with JP-8, JP-8+MDA, and JP-8+BHT: (a) naphthalene and (b) dodecane. (*Source:* From Muhammad, F., Brooks, J.D., and Riviere, J.E. *Toxicol. Lett.*, 150, 351–365, 2004. With permission.)

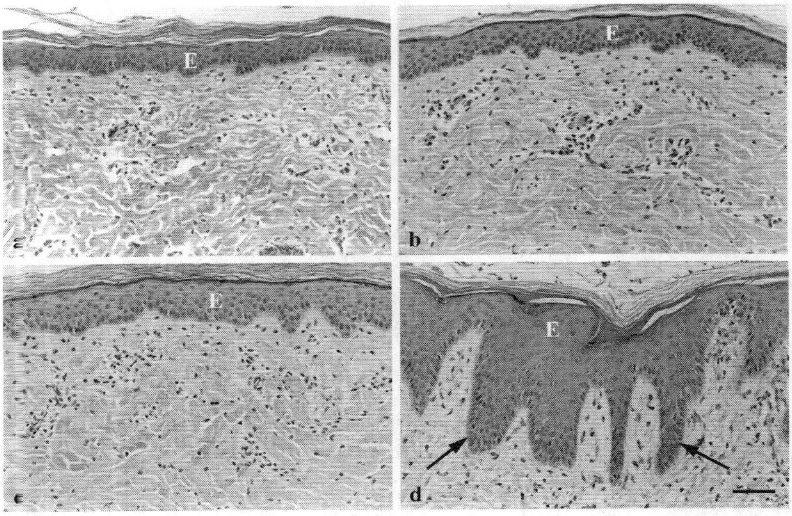

FIGURE 7.6 Light micrograph of pig skin exposed to fabric saturated with JP-8: (a) 5-day fabric only control, (b) 5-hour fabric, (c) 24-hour fabric, and (d) 5-day fabric. Note increase in epidermal (E) thickness and rete peg depth (arrows) with an increase in exposure time. H&E. Magnification bar equals 63 μm. (*Source:* From Monteiro-Riviere, N.A., Inman, A., and Riviere, J.E. *J. Appl. Toxicol.*, 21, 485–494, 2001.)

occluded sites. The high-dose fabric-occluded sites containing each of the three fuels exhibited epidermal thickening and an increase in rete peg (epidermal evagination) depth at 24 hours and 5 days (Figure 7.6), with an increase above control in the order JP-8(100) > JP-8 > Jet A (Table 7.2). In addition, the three fuels caused intracorneal microabscesses filled with inflammatory cells in the 24-hour and 5-day fabric sites (Figure 7.7). This study showed that an additive to JP-8(100) caused the formation of microabscesses and an increase in epidermal thickening. In a follow-up study, pigs were treated as above with cotton fabric containing 67 μl/cm^2 of Jet A, JP-8, and JP-8(100) and the epidermal ultrastructure evaluated (Monteiro-Riviere et al. 2004). At 5 hours and 24 hours, slight intercellular edema, an increase in the number of Langerhans cells, swollen mitochondria, and migrating leukocytes were observed in the epidermal layers. By 4 days, an increase in intercellular epidermal edema, epidermal inflammation, Langerhans cells, and dermal edema was present. Also, the stratum corneum/stratum granulosum interface was disorganized, and mitochondria became swollen to coalesce and form vacuoles. Staining of stratum corneum lipids with ruthenium tetroxide showed that the three jet fuels, especially JP-8(100), caused cleft formation or expansion within the lipid lamellar bilayers (Figure 7.8b) not present in normal pig skin (Figure 7.8a). These structural abnormalities greatly affect barrier function and may be a primary event in jet fuel toxicity. Muhammad et al. (2005b) studied the effect of cotton fabric soaked with JP-8, aliphatic hydrocarbons (n = 8; C_9 to C_{16}), or aromatic hydrocarbons (n = 6) on the back of the pig for a 1-day, single exposure or a 4-day repetitive exposure. Significant erythema, epidermal thickening, and increase in the number of cell layers was present following the 4-day

TABLE 7.2
Epidermal Thickness (±SEM), Rete Peg Depth (±SEM), and Number of Epidermal Cell Layers (±SEM) in the 5-hour, 24-hour, and 5-day Fabric-Occluded Sites Exposed to Jet A, JP-8, and JP-8(100)

	Epidermal Thickness (μm)	Rete Peg Depth (μm)	Number of Cell Layers
5-hr Control	48.9 ± 2.5 [A]	104.7 ± 10.3 [a, A]	4.0 ± 0.2
5-hr Jet A	54.0 ± 2.5 [D]	81.8 ± 3.1 [b, D]	4.3 ± 0.2 [B]
5-hr JP-8	48.4 ± 1.8 [F]	104.0 ± 8.9 [a, F]	3.9 ± 0.2 [E]
5-hr JP-8(100)	49.8 ± 1.9 [H]	79.5 ± 2.6 [b, I]	4.1 ± 0.2 [F]
24-hr Control	44.9 ± 2.5 [b, A]	88.0 ± 7.8	4.2 ± 0.2 [b]
24-hr Jet A	66.7 ± 6.0 [a]	125.1 ± 13.2 [C]	4.7 ± 0.3 [B]
24-hr JP-8	62.8 ± 4.1 [a, E]	138.5 ± 18.3	5.2 ± 0.2 [a, D]
24-hr JP-8(100)	71.7 ± 8.1 [a, G]	124.8 ± 14.7 [H]	4.6 ± 0.2 [G]
5-day Control	32.4 ± 1.4 [d, B]	73.9 ± 4.5 [c, B]	4.5 ± 0.2 [d]
5-day Jet A	70.2 ± 5.1 [c, f, C]	139.2 ± 10.6 [d, e, C]	6.8 ± 0.4 [c, f, A]
5-day JP-8	75.8 ± 6.7 [c, E]	169.1 ± 15.4 [d, E]	8.0 ± 0.5 [c, C]
5-day JP-8(100)	86.6 ± 6.1 [c, e, G]	181.4 ± 13.4 [d, f, G]	8.9 ± 0.5 [c, e, F]

Source: From Monteiro-Riviere, N.A., Inman, A., and Riviere, J.E. *J. Appl. Toxicol.*, 21, 485–494, 2001.

Mean Epidermal Thickness:

Within each time point: ([a] and [b]), ([c] and [d]), ([e] and [f]) are significantly different ($p < 0.05$). Across same treatment, different times: ([A] and [B]), ([C] and [D]), ([E] and [F]), ([G] and [H]) are significantly different ($p < 0.05$).

Mean Rete Peg Depth:

Within each time point: ([a] and [b]), ([c] and [d]), ([e] and [f]) are significantly different ($p < 0.05$). Across same treatment, different times: ([A] and [B]), ([C] and [D]), ([E] and [F]), ([G, H] and [I]) are significantly different ($p < 0.05$).

Mean Number of Cell Layers:

Within each time point: ([a] and [b]), ([c] and [d]), ([e] and [f]) are significantly different ($p < 0.05$). Across same treatment, different times: ([A] and [B]), ([C, D] and [E]), ([F] and [G]) are significantly different ($p < 0.05$).

exposure with JP-8, and the aliphatics tridecane, tetradecane, and pentadecane. JP-8 and the long-chain aliphatics caused the formation of subcorneal microabscesses containing inflammatory cells after 4 days. The short-chain aliphatic hydrocarbons (nonane, decane, undecane) and the aromatic hydrocarbons caused no damage to the skin. The morphological damage to the skin caused by JP-8 and the long-chain aliphatic, which increased with exposure time, supports cell culture studies that found that maximum IL-8 release in HEK was caused by tridecane (Allen et al., 2001b) and that IL-8 release peaked with the mid-chain length aliphatics (Chou et al., 2002; Inman et al., 2008). In addition, stratum corneum disorganization and extraction

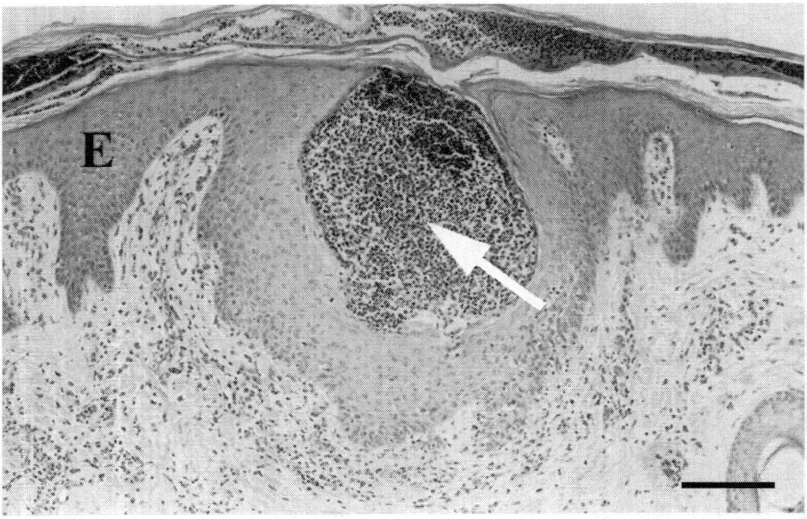

FIGURE 7.7 Light micrograph of subcorneal microabscess from a 5-day, fabric-occluded Jet A treatment site. Note inflammatory cells within the microabscess (arrow). E, epidermis. H&E. Magnification bar equals 63 μm. (*Source:* Monteiro-Riviere, N.A., Inman, A., and Riviere, J.E. *J. Appl. Toxicol.,* 21, 485–494, 2001.)

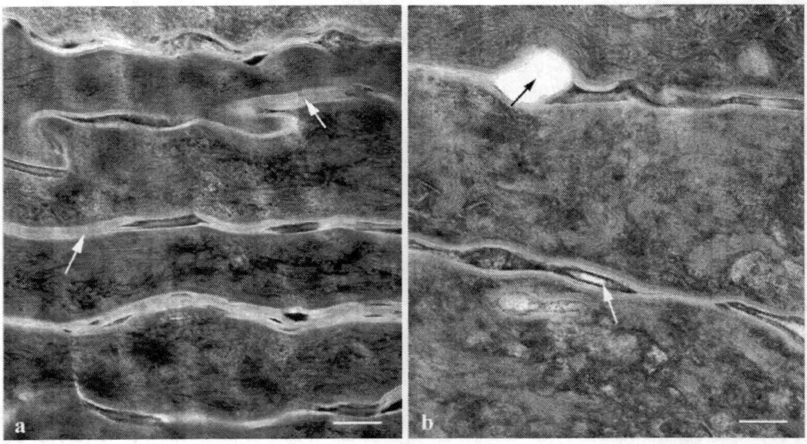

FIGURE 7.8 Transmission electron micrograph of porcine skin stained with ruthenium tetroxide. (a) Control skin exhibited intercellular lamellae (arrows) of electron-dense and electron-lucent bands. Magnification bar equals 290 nm. (b) Jet-fuel-treated skin showing desmosome separation from the central core (white arrow), creating a space. The expansion of the intercellular space (black arrow) creates an area where the lipid lamellae appear extracted. Magnification bar equals 270 nm. (*Source:* From Monteiro-Riviere, N.A., Inman, A., and Riviere, J.E. *Toxicol. Appl. Pharmacol.,* 195, 339–347, 2004. With permission.)

of lipids is similar to that found by Monteiro-Riviere et al. (2004). Based on these studies, the authors conclude that tridecane (C_{13}) and tetradecane (C_{14}) may be the primary hydrocarbons responsible for jet fuel-induced irritation to the skin.

To determine whether skin previously exposed to jet fuel increases percutaneous absorption of aliphatic and aromatic hydrocarbons, the back of the pig was treated with a 1-day single exposure or a 4-day repetitive exposure of JP-8 as above, the skin dermatomed and then loaded into flow through diffusion cells (Muhammad et al., 2005a). The hydrocarbons were used as a dosing mixture with 50% ethanol (in water), and dosed on the skin for 5 hours. Compared to the controls, there was a 2 to 3 times increase in the absorption of the aliphatics nonane and undecane through the 1-day exposed skin, and a 3 to 4 times increase of dodecane and tridecane through the 4-day exposed skin. The absorption of the aromatics naphthalene and dimethylnaphthalene was about 1.5 times greater in the JP-8-exposed skin compared to unexposed skin. The benzene-containing hydrocarbons (ethylbenzene, trimethylbenzene, *o*-xylene) showed a 2 times (1-day preexposed) and 4 times (4-day preexposed) increase in percutaneous absorption. Naphthalene, dimethylnaphthalene, and cyclohexylbenzene were significantly retained in the 4-day preexposed skin compared to the 1-day skin. Preexposure of skin to jet fuels may disrupt the stratum corneum barrier to enhance the absorption of hydrocarbons following subsequent exposure.

Dermal absorption and penetration of aliphatic and aromatic hydrocarbons following JP-8 exposure were also studied in humans (Kim et al., 2006). The forearms of ten healthy volunteers were exposed to JP-8 at a dose of 50 μl/cm^2 for 30 minutes. The skin was tape-stripped to determine absorption and blood sampled up to 3.5 hours following treatment to determine hydrocarbon penetration. The aliphatics decane, undecane, and dodecane and the aromatics naphthalene, 1-methylnaphthalene, and 2-methylnaphthalene were absorbed by the skin. The aromatic hydrocarbons in JP-8 penetrated the skin faster than the aliphatic hydrocarbons. These results are similar to Muhammad et al. (2004b), with comparable mean permeability constants for naphthalene and dodecane. In a recent study, Kim et al. (2008) used mathematical modeling of tape-stripped stratum corneum from the human volunteers to quantify levels of naphthalene, 1-methylnaphthalene, 2-methylnaphthalene, undecane, and dodecane in the stratum corneum layers. The values reported were a factor of 3.7 times less than those reported in rat skin by McDougal et al. (2000).

CONCLUSION

Jet propulsion fuels consist of approximately 228 aromatic and aliphatic hydrocarbons with carbon chain lengths ranging from C_6 to C_{17}. In vitro and in vivo studies show that aromatic hydrocarbons are absorbed and become systemic more readily than aliphatic hydrocarbons, while the less volatile aliphatics remain longer in the skin. Based on in vitro data, long-chain hydrocarbons are also cytotoxic. The contact of the jet fuels and long-chain aliphatic hydrocarbons (pentadecane, tetradecane, tridecane) with the skin leads to a time-dependent increase in epidermal damage and skin irritation (erythema). The difference between jet fuels appears to be related to the ability of additives to modulate long-chain hydrocarbon penetration into the skin. Exposure to jet fuel can also lead to disruption of the stratum corneum barrier to enhance absorption following a subsequent exposure. Synthetic fuels, such as S-8, composed entirely of aliphatic hydrocarbons appear to cause less damage to the skin and may not be absorbed as readily as the current jet fuels. The mechanism behind this phenomenon has not been studied, although it probably is secondary to reduced partitioning of toxic long-chain aliphatic hydrocarbons from a complete aliphatic fuel compared to the aliphatic/aromatic mixture typical of other jet fuels. This, and other fuel formulations, may ultimately reduce adverse health effects of the jet propulsion fuels.

REFERENCES

Allen, D.G., Riviere, J.E., and Monteiro-Riviere, N.A. 2000. Induction of early biomarkers of inflammation produced by keratinocytes exposed to jet fuels Jet-A, JP-8 and JP-8(100). *J. Biochem. Molec. Toxicol.,* 14, 231–237.

Allen, D.G., Riviere, J.E., and Monteiro-Riviere, N.A. 2001a. Cytokine induction as a measure of cutaneous toxicity in primary and immortalized porcine keratinocytes exposed to jet fuels and their relation to normal human keratinocytes. *Toxicol. Lett.,* 119, 209–217.

Allen, D.G., Riviere, J.E., and Monteiro-Riviere, N.A. 2001b. Analysis of interleukin-8 release from normal human epidermal keratinocytes exposed to aliphatic hydrocarbons: Delivery of hydrocarbons to cell cultures via complexation with α-cyclodextrin. *Toxicol. In Vitro,* 15, 663–669.

Baynes, R.E., Brooks, J.D., and Riviere, J.E. 2001a. Membrane transport of naphthalene and dodecane in jet fuel mixtures. *Toxicol. Indus. Health,* 16, 225–238.

Baynes, R.E., Brooks, J.D., Budsaba, K., Smith, C.E., and Riviere, J.E. 2001b. Mixture effects of JP-8 additives on the dermal disposition of jet fuel components. *Toxicol. Appl. Pharmacol.,* 175, 269–281.

Bell, I.R., Brooks, A.J., Baldwin, C.M., Fernandez, M., Figueredo, A.J., and Witten, M.L. 2005. JP-8 jet fuel exposure and divided attention test performance in 1991 Gulf War veterans. *Aviat. Space Environ. Med.*, 76, 1136–1144.

Cavallo, D., Ursini, C.L., Carelli, G., Iavicoli, I., Ciervo, A., Perniconi, B., Rondinone, B., Gismondi, M., and Iavicoli, S. 2006. Occupational exposure in airport personnel: characterization and evaluation of genotoxic and oxidative effects. *Toxicology,* 223, 26–35.

Chao, Y.C.E., Gibson, R.L., and Nylander-French, L.A. 2005. Dermal exposure to jet fuel (JP-8) in US Air Force personnel. *Ann. Occup. Hyg.,* 49, 639–645.

Chao, Y.-C.E., Kupper, L.L., Serdar, B., Egeghy, P.P., Rappaport, S.M., and Nylander-French, L.A. 2006. Dermal exposure to jet fuel JP-8 significantly contributes to the production of urinary naphthols in fuel-cell maintenance workers. *Environ. Health Perspect.,* 114, 182–185.

Chou, C.C., Riviere, J.E., and Monteiro-Riviere, N.A. 2002. Differential relationship between the carbon chain-length of jet fuel aliphatic hydrocarbons and their ability to induce cytotoxicity versus interleukin-8 release in human epidermal keratinocytes. *Toxicol. Sci.,* 69, 226–233.

Chou, C.C., Riviere, J.E., and Monteiro-Riviere, N.A. 2003. The cytotoxicity of jet fuel aromatic hydrocarbons and dose-related interleukin-8 release from human epidermal keratinocytes. *Arch. Toxicol.,* 77, 384–391.

D'Este, C., Attia, J.R., Brown, A.M., Gibson, R., Gibberd, R., Tavener, M., Guest, M., Horsley, K., Harrex, W., and Ross, J. 2008. Cancer incidence and mortality in aircraft maintenance workers. *Am. J. Indus. Med.,* 51, 16–23.

Defense Energy Support Center. 2005. Supporting Worldwide Missions with DESC Energy around the Globe. *Fact Book.* 28th edition.

Harris, D.T., Sakiestewa, D.F., Robledo, R., and Witten, M. 1997. Immunotoxicological effects of JP-8 jet fuel exposure. *Toxicol. Indus. Health,* 113, 43–55.

Harris, D.T., Sakiestewa, D.F., Titone, D., Robledo, R., Young, R.S., and Witten, M. 2000. Substance P as a prophylaxis for JP-8 jet fuel-induced immunotoxicity. *Toxicol. Indus. Health,* 16, 253–259.

Inman, A.O., Monteiro-Riviere, N.A., and Riviere, J.E. 2008. Inhibition of jet fuel aliphatic hydrocarbon induced toxicity in human epidermal keratinocytes. *J. Appl. Toxicol.,* 28, 543–553.

Kim, D., Andersen, M.E., and Nylander-French, L.A. 2006. Dermal absorption and penetration of jet fuel components in humans. *Toxicol. Lett.,* 165, 11–21.

Kim, D., Farthing, M.W., Miller, C.T., and Nylander-French, L.A. 2008. Mathematical description of the uptake of hydrocarbons in jet fuel into the stratum corneum of human volunteers. *Toxicol. Lett.,* 178, 146–151.

McDougal, J.N., and Miller, T.E. 1998. Dermal absorption of JP-8 and its components. *Proc. AFOSR JP-8 Fuel Toxicol. Workshop,* Tucson, AZ.

McDougal, J.N., Pollard, D.L., Weisman, W., Garrett, C.M., and Miller, T.E. 2000. Assessment of skin absorption and penetration of JP-8 fuel and its components. *Toxicol. Sci.,* 55, 247–255.

Monteiro-Riviere, N.A., Inman, A., and Riviere, J.E. 2001. Effects of short-term high-dose and low-dose dermal exposure to Jet A, JP-8 and JP-8+100 jet fuels. *J. Appl. Toxicol.,* 21, 485–494.

Monteiro-Riviere, N.A., Inman, A., and Riviere, J.E. 2004. Skin toxicity of jet fuels: Ultrastructural studies and the effects of Substance P. *Toxicol. Appl. Pharmacol.,* 195, 339–347.

Muhammad, F., Baynes, R.E., Monteiro-Riviere, N.A., Xia, X.R., and Riviere, J.E. 2004a. Dose related absorption of JP-8 jet fuel hydrocarbons through porcine skin with quantitative structure permeability relationship analysis. *Toxicol. Mech. Meth.,* 14, 159–166.

Muhammad, F., Brooks, J.D., and Riviere, J.E. 2004b. Comparative mixture effects of JP-8(100) additives on the dermal absorption and disposition of jet fuel hydrocarbons in different membrane model systems. *Toxicol. Lett.,* 150, 351–365.

Muhammad, F., Monteiro-Riviere, N.A., Baynes, R.E., and Riviere, J.E. 2005a. Effect of in vivo jet fuel exposure on subsequent in vitro dermal absorption of individual aromatic and aliphatic hydrocarbon fuel constituents. *J. Toxicol. Environ. Health, Part A,* 68, 719–737.

Muhammad, F., Monteiro-Riviere, N.A., and Riviere, J.E. 2005b. Comparative in vivo toxicity of topical JP-8 jet fuel and its individual hydrocarbon components: identification of tridecane and tetradecane as key constituents responsible for dermal irritation. *Toxicol. Pathol.,* 33, 258–266.

Ritchie, G.D., Still, K.R., Rossi, J., Bekkedal, M.Y.V., Bobb, A.J., and Arfsten, D.P. 2003. Biological and health effects of exposure to kerosene-based jet fuels and performance additives. *J. Toxicol. Environ. Health B,* 6, 357–451.

Riviere, J.E., Brooks, J.D., Monteiro-Riviere, N.A., Budsaba, K., and Smith, C.E. 1999. Dermal absorption and distribution of topically dosed Jet fuels jet-A, JP-8, and JP-8(100). *Toxicol. Appl. Phamacol.,* 160, 60–75.

Robledo, R.F., and Witten, M.L. 1999. NK1-receptor activation prevents hydrocarbon-induced lung injury in mice. *Am. J. Physiol.,* 276, L229–L238.

Rogers, J.V., Siegel, G.L., Pollard, D.L., Rooney, A.D., and McDougal, J.N. 2004. The cytotoxicity of volatile JP-8 jet fuel components in keratinocytes. *Toxicology,* 197, 113–121.

Selden, A., and Ahlborg, G., Jr. 1986. Causes of Death and Cancer Morbidity at Exposure to Aviation Fuels in the Swedish Armed Forces. ASF Project 84-0308. Department of Occupational Medicine, Orebro, Sweden.

Selden, A., and Ahlborg, G., Jr. 1987. Causes of Death and Cancer Morbidity at Exposure to Aviation Fuels in the Swedish Armed Forces. An update. Department of Occupational Medicine, Orebro, Sweden.

Selden, A., and Ahlborg, G., Jr. 1991. Mortality and cancer morbidity after exposure to military aircraft fuel. *Aviat. Space Environ. Med.,* 62, 789–794.

White, R.D. 1999. Refining and blending of aviation turbine fuels. *Drug Chem. Toxicol.,* 22, 143–153.

Witzmann, F.A., Monteiro-Riviere, N.A., Inman, A.O., Kimpel, M.A., Pedrick, N.M., Ringham, H.N., and Riviere, J.E. 2005. Effect of JP-8 jet fuel exposure on protein expression in human keratinocyte cells in culture. *Toxicol. Lett.,* 160, 8–21.

Yang, J.-H., Lee, C.-H., Monteiro-Riviere, N.A., Riviere, J., Tsang, C.-L., and Chou, C.-C. 2006. Toxicity of jet fuel aliphatic and aromatic hydrocarbon mixtures on human epidermal keratinocytes: evaluation based on in vitro cytotoxicity and interleukin-8 release. *Arch. Toxicol.,* 80, 508–523.

8 Methods of Assessing Skin Irritation and Sensitization of Jet Fuels

R. Jayachandra Babu, Ram Patlolla, and Mandip Singh

CONTENTS

INTRODUCTION

The skin is the main target tissue for many external insults such as chemicals, microbes, heat, and harmful toxins. Despite being considered a good barrier to the systemic absorption of chemicals, many chemical agents can breach the barrier and enter the skin. Jet fuels such as Jet A, JP-4, JP-8, JP-8+100, and S-8 are the major aviation fuels designed for use in aircraft and in the U.S. Air Force. According to U.S. Department of Energy statistics, jet fuels are consumed at 72 million gallons per day. They are complex mixtures of hundreds of hydrocarbons, categorized as aliphatic, aromatic, and naphthalene types, including their thousands of isomers. JP-8+100 contains JP-8 and an additive package consisting of an antioxidant (BHT), metal deactivator (MDA), and detergent and dispersant (8Q405).

Occupational exposures to jet fuel may occur during fuel transport, aircraft fueling and defueling, aircraft maintenance, cold aircraft engine starts, maintenance of equipment and machinery, use of tent heaters, and cleaning or degreasing with fuel (McDougal and Rogers, 2004). Direct liquid contact with jet fuels would come primarily from spills, splashes, or handling parts wet with fuel. The aromatic components of JP-8 were found to be penetrating more rapidly through the skin while aliphatic compounds predominantly retain in the skin layers (McDougal et al., 2000). The additives of JP-8 were found to modulate the percutaneous absorption and skin deposition of hydrocarbon components of JP-8 such as naphthalene and dodecane (Baynes et al., 2001). A mathematical model of the uptake of aromatic and aliphatic hydrocarbons into the stratum corneum (SC) of human skin in vivo was developed by Kim et al. (2008). This model predicts the spatiotemporal variation of naphthalene, 1- and 2-methylnaphthalene, undecane, and dodecane in the SC of humans. These results demonstrate the value of measuring dermal exposure using the tape-strip technique and the importance of quantification of dermal uptake of a hydrocarbon from the mixture. Kim et al. (2006) investigated the absorption and penetration of aromatic and aliphatic components of JP-8 in humans using the tape-stripping method and showed evidence of uptake into the skin for all the JP-8 components. The rank order of the apparent permeability coefficient (Kp) estimated from blood data was naphthalenes > decane > dodecane > undecane. Overall studies indicate that some of the hydrocarbons permeate in higher amounts than the others; aliphatic hydrocarbons retain in the skin while aromatic hydrocarbons permeate through the skin. Due to the complexity of jet fuels' chemical composition, it is difficult to assess the percutaneous absorption of individual fuel components. Furthermore, the components of jet fuels cause lipid and protein extraction as well as barrier perturbation in the skin, complicating the assessment of the percutaneous absorption of component hydrocarbons of a jet fuel (Singh et al., 2003). The skin irritation potential of individual hydrocarbons and their contribution to skin irritation effects of the jet fuel is not clearly established. This chapter provides an overview of the effect of jet fuels and the role of surrogate chemicals in jet fuels on skin irritation, and sensitization effects. Furthermore, various methods of measuring skin irritation and sensitization of jet fuels and their surrogate chemicals are reviewed.

BIOPHYSICAL METHODS OF ASSESSING SKIN IRRITATION OF JET FUELS

In general, chemically induced skin irritation can be divided into three basic subtypes: acute skin irritation, cumulative skin irritation, and delayed acute skin irritation. Acute skin irritation occurs rapidly after a single exposure to a powerful irritant. The cumulative type is the most common skin irritation and arises after repetitive exposures to mild irritants. It often occurs in humans who perform repetitive wet work and subsequently is often a cause of occupational skin disease (reviewed in Welss et al., 2004). A visual scoring method for erythema and edema has been followed for decades for the evaluation of skin irritation. However, this method is very useful only for the evaluation of visible changes on the skin due to severe irritation (Draize, 1944). In many cases it takes several days before visible changes can occur in the skin; and for mild skin irritants, there are no visual changes at all. Bioengineering techniques such as the measurement of transepidermal water loss (TEWL), skin moisture content, and skin blood flow have been used for evaluation of skin irritation during the early stages of irritation from mild irritants.

TEWL is a measure of the skin barrier function of normal as well as damaged skin and is one of the most important biophysical parameters for evaluating the efficiency of the skin-water barrier (Kanikkannan et al., 2002). Barrier damage cannot be measured directly and is detected only by its consequences: skin dryness with TEWL and inflammatory skin reactions (De Haan et al., 1996). The measurement of TEWL (i.e. the outward diffusion of water through the skin) has become increasingly important in studies evaluating the stratum corneum barrier function. Skin lipids have been postulated to be intimately involved in maintaining barrier function (Michaels et al., 1975). It is generally accepted that solvents remove intercellular lipid material, resulting in cutaneous barrier disruption (Scheuplein and Blank, 1971). In general, an increase in TEWL is observed when the skin structure is disrupted. We have done extensive work in our laboratory on the evaluation of skin irritation of jet fuels using various biophysical methods (Kanikkannan et al., 2001a, 2001b). Occlusive dermal exposure (24 hours) of JP-8 and JP-8+100 caused a substantial increase in the TEWL in minipigs, and the TEWL value remained high until 24 hours post removal of the patches. Exposure of minipig skin to JP-8 and JP-8+100 decreased the skin moisture content gradually up to 24 hours (Kanikkannan et al., 2001a, 2001b). Moisture content of the skin is one of the most important parameters used to evaluate the barrier function of the skin. An important parameter of the skin is the hydrolipidic film, which consists of sebum excreted by the sebaceous glands and moisture components excreted with the sweat. This emulsion protects the skin from drying out, keeps it supple, and due to the natural acid protection barrier, it prevents the penetration of harmful external substances (e.g., germs). The more balanced the hydrolipidic film, the higher the water content. Application of JP-8 and JP-8+100 might have disrupted the hydrolipidic film, resulting in a decrease in the skin moisture content, although it was not statistically significant. We further studied unocclusive dermal exposures of jet fuel aliphatic hydrocarbons on the TEWL and ethythema in hairless rats. The TEWL and erythema of the chemicals were in the order: tetradecane > dodecane > nonane. As shown in Figure 8.1, the affinity of the chemicals to the stratum corneum (SC) and their gradual accumulation in the skin is one of the reasons for the observed differences in the skin irritation profiles of different aliphatic chemicals (Babu et al., 2004a). Muhammad et al. (2005) also concluded that tetradecane was a key constituent of dermal irritation caused by jet fuels.

In general, the skin irritation potential of hydrocarbons decreases with an increase in the molecular weight or lipophilicity of the hydrocarbon, as this could make the chemical a poor permeant. However, repeated exposures to the poorly permeating hydrocarbons can gradually accumulate in the skin and cause severe cumulative irritation. The symptoms do not immediately follow after exposure (as is the case in acute irritant contact dermatitis), but repeat exposures induce cumulative irritation (Elsner and Maibach, 1993; Welss et al., 2004). Our earlier investigation with single occlusive exposure by tetradecane, which is less permeable than dodecane and nonane, did not show any skin irritation (Babu et al., 2004b), but the same chemical upon repeated exposures would have caused gradual skin accumulation and produced a severe increase in the TEWL and erythema (Babu et al., 2004a). In a different study, we investigated the effect of methyl substitution of benzene on the skin permeation kinetics and irritation in rats. As shown in Figure 8.2, the skin retention affinity of methylbenzenes were determined according to the degree of methyl substitution, in the order of tetramethylbenzene > xylene > benzene. Accordingly, the TEWL of the chemicals also increased with

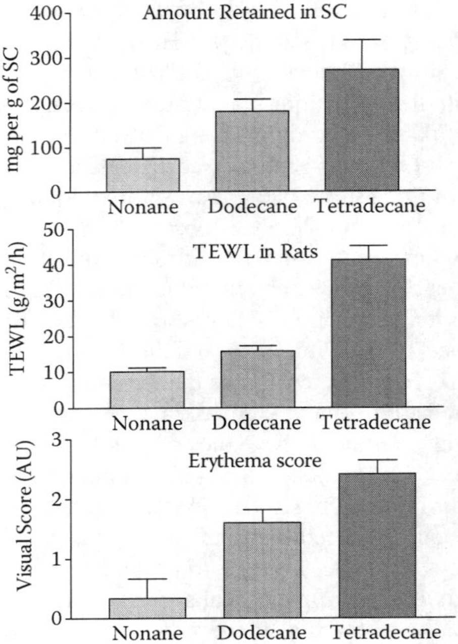

FIGURE 8.1 Relationship between SC (stratum corneum) concentration of different aliphatic hydrocarbons and TEWL or erythema in hairless rats.

an increase in the lipophilicity or molecular weight of the hydrocarbon (Ahaghotu et al., 2005).

Further, we investigated the effects of dermal exposure of jet fuels (Jet A, JP-8, and JP-8+100) on the magnetic resonance imaging (MRI) of skin in hairless rats. MRI was used to obtain proton images and water self-diffusion maps of hairless rat skin exposed to jet fuel. Exposure to JP-8 showed the largest difference from the control with regard to visual observations of the stratum corneum and hair follicles, while JP-8+100 appeared to affect the hair follicle region. The results demonstrated that MRI can be used as a tool to investigate the alterations in the skin morphology after exposure to jet fuels and toxic chemicals (Kanikkannan et al., 2002).

BIOCHEMICAL METHODS OF ASSESSING SKIN IRRITATION OF JET FUELS

Biomarker Measurements in Animals

Despite the high barrier effect of the stratum corneum, jet fuel component chemicals can breach the skin in significant quantities and cause skin irritation and inflammation (Babu et al., 2004a; Babu et al., 2004b; Ahaghotu et al., 2005). Experiments have shown that the concentration of the hydrocarbon and the time of exposure dependence of the response are strongly related to the barrier capacity of the stratum

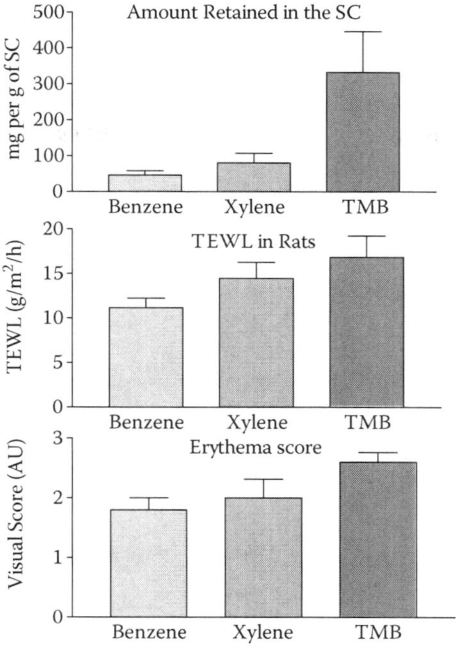

FIGURE 8.2 Relationship between SC (stratum corneum) concentration of different aromatic hydrocarbons and TEWL or erythema in hairless rats.

corneum (Babu et al., 2004a; Babu et al., 2004b). Application of jet fuels under occluded dosing dramatically reduced the skin barrier function (Kanikkannan et al., 2002), which is consistent with lipid and protein extraction from porcine skin observed after jet fuel exposure (Singh et al., 2003). The skin barrier damage leads to enhanced TEWL and, more seriously, to increased penetration of further irritants to deeper epidermal layers of living keratinocytes. The effect of irritant chemicals on the keratinocytes to induce inflammation is widely reported in the literature (Corsini and Galli, 1998; Smith et al., 2002; Welss et al., 2004). Short-term (1 hour) dermal exposure of JP-8 in rats induced IL-1α, and inducible nitric oxide synthase (iNOS) protein in the skin and produced local inflammation as measured by histological changes and granulocyte infiltration (Kabbur et al., 2001). Increased levels of oxidative species and low-molecular-weight DNA were also present after brief exposures to JP-8 (Rogers et al., 2001).

Occlusive dermal exposure (1 hour) of jet fuel aliphatic hydrocarbon components in hairless rats induced the expression of IL-1α, TNFα, MCP-1, and the release of these cytokines was further supported by the activation of NF$\kappa\beta$ p65 and corresponding degradation of I$\kappa\beta$ (Babu et al., 2004). Repeated low-level exposures of jet fuel aromatic hydrocarbon components (methylbenzenes) induced the expression of IL-1α and TNFα in proportion to the degree of methyl substitution of the benzene ring (Ahaghotu et al., 2005). It has also been reported that repeated JP-8 exposure induces proinflammatory cytokines (IL-1β, IL-6) and chemokines (CXCL 1, 2, 3, and 10, as well as CCL 3, 5, 11, and 20) in skin that tends to favor neutrophil accumulation. The increased expression of these cytokines and chemokines may lead to increased inflammatory infiltrate and injury of the skin, resulting in JP-8-induced irritant dermatitis (Galluchi et al., 2004; Galluchi and Mickle, 2006).

In an attempt to detect whether aliphatic or aromatic components could mimic the JP-8-induced gene expression response, McDougal and Garrett (2007) exposed rats to JP-8, undecane tetradecane, trimethylbenzene, and dimethylnaphthalene for 1 hour and examined the epidermis to characterize the gene expression response using the gene microarray technique. Undecane and trimethylbenzene exposures caused the greatest number of changes in transcript levels compared to dimethylnaphthalene and tetradecane. There was no single component that mimicked the gene expression resulting from the JP-8 exposure but undecane had the most similar responses. They found consistent twofold increases in gene expression of 27 transcripts at 1, 4, and 8 hours after 1 hour of chemical exposure. The JP-8-induced epidermal stress response activates gene expression in the signaling pathways and results in the inflammatory, apoptotic, and growth responses.

Biomarker Measurements in the Cultured Skin Tissue

The skin irritation potentials of new chemicals, drugs, and formulations are still evaluated by application to animals, and the investigation of visible changes such as erythema and edema (Testing for skin irritation in animals involves pain and discomfort, and the results generally are not predictive for those found in humans) (York et al., 1996). Due to this reason, the use of in vitro reconstructed organotypic skin equivalents is mostly favored because of their close resemblance to human skin. In vitro human skin equivalents (Mattek Corp., Ashland, Massachusetts; Skin Ethic, Nice, France) are used for irritancy testing of chemicals and cosmetics. To date, two different kinds of reconstituted skin equivalents are available: (1) epidermal equivalents consisting of multilayering, differentiating human keratinocyte cultures grown on synthetic matrices, and (2) full skin equivalents with multilayering, differentiating human keratinocyte cultures grown on fibroblast-containing collagen matrices. At the air-medium interface, the keratinocytes of both systems develop an organized stratum corneum, which resembles a functional barrier. All models consist of primary keratinocytes (neonatal or adult origin), seeded on matrices of either dermal components or nonbiological origins. Using sophisticated protocols, the keratinocytes fully differentiate and form a reconstituted epidermis (Welss et al., 2004). Keratinocytes play an important role in the initiation, modulation, and regulation of inflammation (Coquette et al., 1999). The use of keratinocyte cultures has been gaining acceptance as an alternative to animal studies for skin irritation research (Roguet, 1998; Bernstein and Vaughan, 1999). Normal human epidermal keratinocytes (NHEK) exposed to three jet fuels (Jet A, JP-8, and JP-8+100) in a culture medium demonstrated that these chemicals induce production and release of proinflammatory cytokines as TNF-α and IL-8 (Allen et al., 2000). Similar results were obtained with porcine keratinocytes (PKC) exposed to jet fuels; both TNF-α, IL-8 were up-regulated (Allen et al., 2001). Jet fuel aliphatic hydrocarbons (C_6 to C_{16}) were dosed on NHEK to evaluate their effect on cytotoxicity and IL-8 expression in the tissue. Short-chain hydrocarbons (C_6 to C_{11}) were more cytotoxic, while C_9 to C_{13} hydrocarbons were more effective in inducing proinflammatory cytokine IL-8 in the cultures (Chou et al., 2002). The exposure of the aromatic hydrocarbon components of jet fuels (e.g., cyclohexylbenzene, trimethylbenzene, xylene, dimethylnaphthalene, ethylbenzene, toluene, and benzene) to NHEK resulted in a dose-related differential response in IL-8 release (Chou et al., 2003). Espinoza et al. (2004) studied the use of cDNA microarray to identify the gene expression profile in NHEK exposed to JP-8. The pattern of expressions in response to JP-8 provides evidence that detoxificant-related and cell growth regulator genes with the most variability in the level of expression may be useful genetic markers in adverse health effects of personnel exposed to JP-8. Furthermore, protein expression in NHEK exposed to 0.1% JP-8 in culture medium was studied (Witzmann et al., 2005). JP-8 exposure resulted in significant expression differences in 35 of the 929 proteins. In another study, the gene expression changes in NHEK after 1-min JP-8 exposure was studied (Chou et al., 2006). A total of 151 JP-8 responsive genes were identified; the genes involved in basal transcription and translations were up-regulated, whereas genes related to DNA repair, metabolism, and keratin were mostly down-regulated. These results indicated that the human keratinocyte responds to a single dose of JP-8 insult and revealed several cellular processes previously not associated with jet fuel exposure.

All the above studies utilized monolayer keratinocytes growing submerged in a culture medium. In contrast to cells in monolayer culture, engineered skin equivalents mimic human epidermis in terms of tissue architecture and barrier function (Andreadis et al., 2001). After a week of culture at the air-liquid interface, they form a well-stratified epidermis with basal, spinous, granular, and cornified layers. Biochemical and ultrastructural studies have shown that this epidermal culture was similar to human skin (Asbill et al., 2000; Hayden et al., 2003). Therefore, three-dimensional engineered tissues may provide realistic models to study the molecular mechanism of skin irritation. We have used the three-dimensional EpiDerm™ culture (EPI-200) as a human skin equivalent to study the effect of JP-8 and JP-8+100 on cytotoxicity, and IL-1α and TNF-α expression. We found that the in vitro cytokine response data paralleled in vivo cytokine expression in hairless rats. The similarities between the in vivo and in vitro studies suggested that in vitro cultures may be a good, reliable model to study skin irritation of jet fuels and various other chemicals (Chatterjee et al., 2006). In another investigation, we have used three-dimensional Epiderm Full Thickness (EFT-300) skin culture as a human skin equivalent to study the structural activity relationship (SAR) of saturated hydrocarbons (C_9 to C_{16}) on the cytotoxicity, skin morphology, and cytokine (IL-1α, IL-6, and IL-8) expression in the EFT-300. The dermal exposures of all hydrocarbons significantly increased the expression of IL-1α, IL-6, and IL-8 in the skin as well as culture medium in proportion to the hydrocarbon chain length. The in vivo skin irritation data also showed that both TEWL and erythema scores were increased with increases in hydrocarbon chain length (C_9 to C_{16}). Therefore, EFT-300 was found to be an excellent in vitro model to understand the structural activity relationship of hydrocarbons (Mallampati et al., 2009).

Dermal Microdialysis of Neuropeptides and Cytokines

The microdialysis (MD) technique involves the insertion of a semipermeable membrane underneath the epidermis or in the dermis using a guiding cannula and continuous sampling of biomarkers from the extracellular fluids by pumping a physiological fluid through the semipermeable membrane at a constant rate. The extracellular substances diffuse through the membrane against a concentration gradient. This is a well-established technique for the continuous sampling of biomarkers of disease within the extracellular fluid space in vivo. It has several advantages over other sampling techniques in that it can be used to follow temporal variations in the generation and release of a biomarker at a discrete location within the tissue space. Furthermore, this technique can be directly adapted to humans to measure molecular responses in the skin, without having to collect biopsy samples. There are some excellent review articles in the literature describing skin MD technique (e.g., Kreilgaard, 2002; Clough, 2005; Ao and Stenken, 2006). Microdialysis of large molecular compounds is often difficult because of poor recovery of the macromolecules in the dialysate samples. Several different factors such as dilution effect inherent to microdialysis, low concentrations of inflammatory mediators, and relatively low sensitivity of ELISA techniques have been attributed to the limited use of the microdialysis technique in inflamed skin. By using a larger surface area of membranes (longer probes), large-molecular-weight cutoff probes, and the addition of bovine serum albumin (or dextran and surfactants), higher biomarkers recoveries can be achieved. Few studies have used MD to assess the release of neuropeptides and inflammatory cytokines in dermal neurogenic inflammation (Schmelz et al., 1997). We studied the effect of jet fuel (JP-8) and xylene (a representative aromatic component of JP-8) on skin irritation in rats by measuring the expression of inflammatory biomarkers, Substance P (SP) and prostaglandin E_2 (PGE_2), in the skin using the MD technique (Fulzele et al., 2007). Occlusive exposure (2 hours) of hairless rats to JP-8 induced significantly higher release of SP and PGE_2 as compared to control values. The JP-8 skin irritation effect was prevented by pretreatment with SR-140333 and celecoxib (Fulzele et al., 2007). The MD technique was also used to assess the skin irritation effect of JP-8 aliphatic hydrocarbons such as nonane, dodecane, and tetradecane by measuring SP, PGE_2 α-MSH, and IL-6 in hairless rats (Patlolla et al., 2009). As shown in Table 8.1, nonane increased all the biomarker levels in significant amounts within 2 hours of chemical exposure compared to dodecane and tetradecane. In this study we demonstrated the potential of the MD technique to quantify the biomarkers in the skin as function of time.

SKIN SENSITIZATION AND IMMUNE SUPPRESSION OF JET FUELS

Contact hypersensitivity (CHS) is a dendritic cell (DC)-dependent, T cell-mediated, cutaneous inflammatory reaction elicited by epicutaneous exposure to reactive chemicals, known as haptens, through environmental and occupational exposures. The best-studied haptens are low-molecular-weight chemicals that bind discrete amino acid residues on self- or exogenous proteins/peptides in the skin and become immunogenic. Clinically, CHS typically occurs as a delayed type of allergic contact

TABLE 8.1
Assessment of Skin Irritation of Aliphatic Hydrocarbons (Nonane, Dodecane, and Tetradecane) by Measuring the SP, PGE2, and IL-6 (pg/ml) Levels in the Rat Skin using the MD Technique

Chemical	EQ	1 hr	2 hr	3 hr	4 hr	5 hr	6 hr	7 hr
				(I) Nonane				
SP	20.7	68.5[c]	70.4[c]	67.7[c]	58.9[c]	52.5[c]	51.9[c]	49.4[c]
	(± 6.4)	(± 6.4)	(± 4.7)	(± 5.4)	(± 2.6)	(± 4.1)	(± 4.4)	(± 4.2)
PGE_2	174.5	419.0[c]	396.8[c]	344.8[c]	307.1[c]	242.6[c]	15.7	ND
	(± 19.4)	(± 8.4)	(± 7.3)	(±10.2)	(±13.5)	(± 12.7)	(± 5.7)	
IL-6	86.4	109.1	122.3	144.1	178.1	292.9[a]	274.5[a]	431.7[c]
	(± 2.1)	(± 13.9)	(±13.3)	(± 14.3)	(± 23.6)	(± 83.4)	(± 62.5)	(± 31.5)
				(II) Dodecane				
SP	20.7	43.7[c]	49.7[c]	47.5[c]	37.4[c]	32.3[c]	30.3[a]	28.8
	(± 6.4)	(± 3.1)	(± 6.7)	(± 4.7)	(± 2.6)	(± 1.7)	(± 1.2)	(± 0.7)
PGE2	174.5	219.7	162.8	129.8	122.9	98.8	57.7	ND
	(± 19.4)	(± 26.7)	(± 21.8)	(± 18.1)	(15.2)	(± 22.8)	(± 35.8)	
IL-6	116.7	180.8	197.5	197.5	209.2	270.0	301.6[a]	420.4[c]
	(± 5.0)	(± 33.2)	(± 36.5)	(± 36.5)	(± 11.7)	(± 12.7)	(± 69.6)	(± 53.9)
				(III) Tetradecane				
SP	20.73	16.4	17.5	20.5	14.5	12.1	9.2	8.8
	(± 6.4)	(± 1.0)	(± 2.3)	(± 2.7)	(± 0.5)	(± 1.0)	(± 0.6)	(± 0.2)
PGE2	186.2	193.1	129.3	86.8	105.0	85.9	14.1	ND
	(± 21.2)	(± 14.5)	(± 18.3)	(± 4.2)	(± 32.9)	(± 45.5)	(± 4.6)	
IL-6	80.6	116.6	110.7	100.3	104.1	143.6	109.2	121.0
	(± 3.0)	(± 20.5)	(± 13.5)	(± 18.3)	(± 14.1)	(± 41.3)	(± 15.9)	(± 21.2)

Note: The perfusion fluid used for SP and PGE_2 recovery was Krebs ringer solution at a flow rate of 2 μg/ml and IL-6 recovered using 0.1 % w/v bovine serum albumin (BSA) in Krebs ringer solution, pumped at a flow rate of 0.5 μg/ml and sample vials were diluted with 30 μl of 0.1 % w/v BSA solution to stabilize the protein. The samples were collected every 60 min and the values represent the mean with standard error. EQ: Equilibration period before applying the chemicals, ND: Not detected; [a] $P < 0.05$; [b] $P < 0.01$, and [c] $P < 0.001$ versus control.

dermatitis. Haptens penetrate the skin and bind to self-proteins to form complete antigens, which are taken by antigen presenting cells to start a cascade of actions resulting in a delayed hypersensitivity reaction (reviewed in Elsaie et al., 2008). JP-8 is classified as a weak skin sensitizer based on studies where guinea pigs were treated with JP-8 four times in a 10-day period before the challenge (Kinkead et al., 1992). We studied the skin sensitization potential of three jet fuels (JP-8, JP-8 + 100, and Jet A) in mice exposed to 25 μl of jet fuels for 3 days. The mean disintegrations per minute (dpm) and stimulation index (SI) of various jet fuels and controls are presented in Figure 8.3. All three jet fuels caused a proliferative activity significantly greater than the control. The SI of JP-8 was 3.2, indicating it to be a weak skin sensitizer. The SI

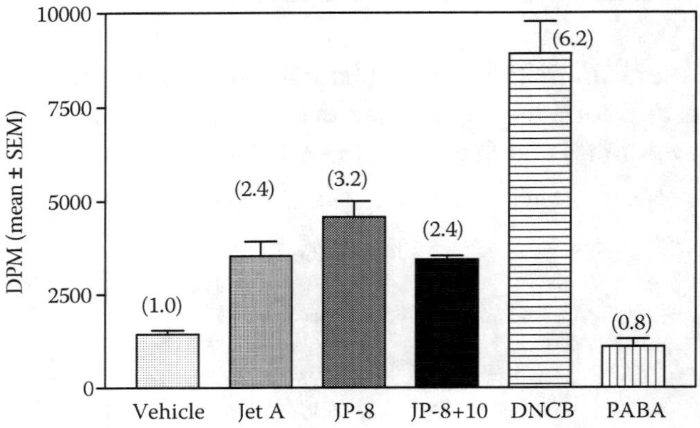

FIGURE 8.3 The local lymph node proliferative activity following topical treatment of mice with jet fuels (Jet A, JP-8, and JP-8+100), positive control (DNCB, 0.25%), negative control (PABA, 2.5%), and vehicle control. The values in parentheses indicate stimulation index (SI), which was determined by dividing the dpm of the treatment group by that of the vehicle control.

of both Jet A and JP-8+100 were 2.4. According to the principles of the LLNA, Jet A and JP-8+100 are not skin sensitizers though their SI values are close to the required SI of 3 to be classified as a skin sensitizer. Overall, our studies demonstrate that JP-8 is a weak skin sensitizer (Kanikkannan et al., 2000).

Dermal exposure of JP-8 in mice either multiple small exposures (50 μl for 5 days) or a single large dose (300 μl) induced immune suppression (Ullrich, 1999). Cell-mediated immune reactions are particularly sensitive to the effects of JP-8. It was also found that delayed type hypersensitivity (DTH), contact hypersensitivity, and T cell proliferation were also suppressed by JP-8 dermal exposures (Ullrich, 1999; Ullrich and Lyons, 2000). Both primary and secondary immune reactions were suppressed by jet fuel application at either multiple small doses (50 μl over 4 days) or a single large dose (approximately 300 μl) of JP-8 and/or Jet-A in mice (Ramos et al., 2002). The activation of cytokine production by JP-8 treatment appears to play an important role in the activation of immune suppressive pathways. By blocking prostaglandin E_2 (PGE_2) production with a selective cyclooxygenase-2 (COX-2) inhibitor, or neutralizing IL-10 activities with a monoclonal antibody, blocked JP-8–induced immune suppression (Ullrich and Lyons, 2000). The synthetic jet fuel (S-8), which is refined from natural gas, and is devoid of aromatic hydrocarbons, was found to have no immune suppressive properties. Ramos and co-workers report that applying S-8 to the skin of mice neither up-regulates the expression of epidermal cyclooxygenase-2 nor does it induce immune suppression. Adding back a cocktail of seven of the most prevalent aromatic hydrocarbons found in jet fuel (benzene, toluene, ethylbenzene, xylene, 1,2,4-trimethlybenzene, cyclohexylbenzene, and dimethylnaphthalene) to S-8 up-regulated epidermal COX-2 expression and suppressed a delayed type hypersensitivity (DTH) reaction. This study indicates that aromatic hydrocarbons present in jet fuel are responsible for immune suppression (Ramos et al., 2007). A 28-day chronic dermal exposure study in female rats in Jet A showed no immune responses (Mann et al., 2008). Exposure of mice to JP-8 for 1 hour/day resulted in immediate secretion of two immunosuppressive agents, namely, IL-10 and PGE_2 (Harris et al., 2007). Short-term, low-concentration JP-8 jet fuel exposures have significant suppressive effects on the immune system that can result in increased severity of viral infections, as determined with influenza virus in mice (Harris et al., 2008). Recently, Ramos and co-workers (2009) found that JP-8 induces immune suppression via a reactive oxygen species NF-κβ-dependent mechanism: When JP-8-treated mice, or JP-8-treated keratinocytes, were treated with a selective NF-κβ inhibitor (parthenolide), COX-2 expression and immune suppression were abrogated. Similarly, when JP-8-treated keratinocytes were treated with small interfering RNA specific for the p65 subunit of NF-κβ, COX-2 up-regulation was blocked. These data indicate that reactive oxygen species (ROS) and NF-κβ are activated by JP-8, and these pathways are involved in COX-2 expression and the induction of immune suppression by jet fuel (Ramos et al., 2009).

CONCLUSIONS

Upon skin contact, these fuels can cause non-immunogenic skin irritation (irritant contact dermatitis) and immunogenic skin irritation (allergic contact dermatitis). Several biophysical and biochemical methods are used to determine skin irritation caused by jet fuels. It is generally difficult to distinguish if the skin irritation is of the immunogenic or non-immunogenic type because the type and release of the biomarker pattern are the same in both cases. More research is needed in identifying the key compounds of jet fuels that are responsible for skin irritation response in human subjects.

REFERENCES

Ahaghotu, E., Babu, R.J., Chatterjee, A., and Singh, M., Effect of methyl substitution of benzene on the percutaneous absorption and skin irritation in hairless rats, *Toxicol. Lett.*, 159, 261, 2005.

Allen, D.G., Riviere, J.E., and Monteiro-Riviere, N.A., Cytokine induction as a measure of cutaneous toxicity in primary and immortalized porcine keratinocytes exposed to jet fuels, and their relationship to normal human epidermal keratinocytes, *Toxicol. Lett.*, 119, 209, 2001.

Allen, D.G., Riviere, J.E., and Monteiro-Riviere, N.A., Identification of early biomarkers of inflammation produced by keratinocytes exposed to jet fuels jet A, JP-8, and JP-8(100), *J. Biochem. Mol. Toxicol.*, 14, 231, 2000.

Andreadis, S.T., Hamoen, K.E., Yarmush, M.L., and Morgan, J.R., Keratinocyte growth factor induces hyperproliferation and delays differentiation in a skin equivalent model system, *FASEB. J.*, 15, 898, 2001.

Ao, X., and Stenken, J.A., Microdialysis sampling of cytokines, *Methods*, 38, 331, 2006.

Asbill, C.S., El-Kattan, A.F., and Michniak, B., Enhancement of transdermal drug delivery: chemical and physical approaches, *Crit. Rev. Ther. Drug Carrier Syst.*, 17, 621, 2000.

Babu, R.J., Chatterjee, A., Ahaghotu, E., and Singh, M., Percutaneous absorption and skin irritation upon low-level prolonged dermal exposure to nonane, dodecane and tetradecane in hairless rats, *Toxicol. Ind. Health*, 20, 109, 2004a.

Babu, R.J., Chatterjee, A., Ahaghotu, E., and Singh, M., Percutaneous absorption and skin irritation upon low-level prolonged dermal exposure to nonane, dodecane and tetradecane in hairless rats, *Toxicol. Indus. Health,* 20, 109, 2004b.

Baynes, R.E., Brooks, J.D., Budsaba, K., Smith, C.E., and Riviere, J.E., Mixture effects of JP-8 additives on the dermal disposition of jet fuel components, *Toxicol. Appl. Pharmacol.,* 175, 269, 2001.

Bernstein, I.A., and Vaughan, F.L., Cultured keratinocytes in in vitro dermatotoxicological investigation: a review, *J. Toxicol. Environ. Health B: Crit. Rev.,* 2, 1, 1999.

Chatterjee, .A, Babu, R.J., Klausner, M., and Singh, M., In vitro and in vivo comparison of dermal irritancy of jet fuel exposure using EpiDerm (EPI-200) cultured human skin and hairless rats. *Toxicol. Lett.,* 167, 85, 2006.

Chou, C.C., Riviere, J.E., and Monteiro-Riviere, N.A., Differential relationship between the carbon chain length of jet fuel aliphatic hydrocarbons and their ability to induce cytotoxicity vs. interleukin-8 release in human epidermal keratinocytes, *Toxicol. Sci.,* 69, 226, 2002.

Chou, C.C., Riviere, J.E., and Monteiro-Riviere, N.A., The cytotoxicity of jet fuel aromatic hydrocarbons and dose-related interleukin-8 release from human epidermal keratinocytes, *Arch. Toxicol.,* 77, 384, 2003.

Chou, C.C., Yang, J.H., Chen, S.D., Monteiro-Riviere, N.A., Li, H.N., and Chen, J.J., Expression profiling of human epidermal keratinocyte response following 1-minute JP-8 exposure, *Cutan. Ocul. Toxicol.,* 25, 141, 2006.

Clough, G.F., Microdialysis of large molecules, *AAPS J.*, 26, E686, 2005.

Coquette, A., Berna, N., Vandenbosch, A., Rosdy, M., and Poumay, Y., Differential expression and release of cytokines by an in vitro reconstructed human epidermis following exposure to skin irritant and sensitizing chemicals, *Toxicol. In Vitro,* 13, 867, 1999.

Corsini, E., and Galli, C.L., Cytokines and irritant contact dermatitis, *Toxicol. Lett.,* 102, 277, 1998.

De Haan, P., Meester, H.H.M., and Bruynzeel, D.P., Irritancy of alcohols. In P. Van der Valk and H.I. Maibach, Editors, *The Irritant Contact Dermatitis Syndrome*, CRC Press, New York, NY, 1996, pp. 65–70.

Draize, J.H., Woodard, G., and Calvery, H.O., Methods for the study of irritation and toxicity of substances applied topically to the skin and mucous membranes, *J. Pharmacol. Exp. Ther.*, 82, 377, 1944.

Elsaie, M.L., Olasz, E., and Jacob, S.E., Cytokines and Langerhans cells in allergic contact dermatitis, *G. Ital. Dermatol. Venereol.,* 143, 195, 2008.

Elsner, P., and Maibach, H.I., Irritant and allergic contact dermatitis. In P. Elsner and J. Martius, Editors, *Vulvovaginitis,* Marcel Dekker Inc., New York, NY, 1993.

Espinoza, L.A., Li, P., Lee, R.Y., Wang, Y., Boulares, A.H., Clarke, R., and Smulson, M.E., Evaluation of gene expression profile of keratinocytes in response to JP-8 jet fuel, *Toxicol. Appl. Pharmacol.,* 200, 93, 2004.

Fulzele, S.V., Babu, R.J., Ahaghotu, E., and Singh, M., Estimation of proinflammatory biomarkers of skin irritation by dermal microdialysis following exposure with irritant chemicals, *Toxicology,* 31, 77, 2007.

Gallucci, R.M., and Mickle, B.M., Inflammatory cytokine expression patterns in rat skin exposed to JP-8 jet fuel, *Am. J. Pharmacol. Toxicol.,* 1, 48, 2006.

Gallucci, R.M., O'Dell, S.K., Rabe, D., and Fechter, L.D., JP-8 jet fuel exposure induces inflammatory cytokines in rat skin, *Int. Immunopharmacol.,* 4, 1159, 2004.

Harris, D.T., Sakiestewa, D., He, X., Titone, D., and Witten, M., Effects of in utero JP-8 jet fuel exposure on the immune systems of pregnant and newborn mice, *Toxicol. Indus. Health,* 23, 545, 2007.

Harris, D.T., Sakiestewa, D., Titone, D., He, X., Hyde, J., and Witten, M., JP-8 jet fuel exposure suppresses the immune response to viral infections, *Toxicol. Indus. Health,* 24, 209, 2008.

Hayden, P.J., Ayehunie, S., Jackson, G.R., Kupfer-Lamore, S., Last, T.J., Klausner, M., and Kubilus, J., In vitro skin equivalent models for toxicity testing. In H. Salem and S.A. Katz, Editors, *Alternative Toxicological Methods*, CRC Press LLC, Boca Raton, FL, 2003, pp. 229–247.

Kabbur, M.B., Rogers, J.V., Gunasekar, P.G., Garrett, C.M., Geiss, K.T., Brinkley, W.W., and McDougal, J.N., Effect of JP-8 jet fuel on molecular and histological parameters related to acute skin irritation, *Toxicol. Appl. Pharmacol.*, 83, 175, 2001.

Kanikkannan, N., Burton, S., Patel, R., Jackson, T., and Shaik, M.S., and Singh, M., Percutaneous permeation and skin irritation of JP-8+100 jet fuel in a porcine model, *Toxicol. Lett.,* 119, 133, 2001.

Kanikkannan, N., Jackson, T., Shaik, S. M., and Singh, M., Evaluation of skin sensitization potential of jet fuels by murine local lymph node assay, *Toxicol. Lett.,* 116, 165, 2000.

Kanikkannan, N., Locke, B.R., and Singh, M., Effect of jet fuels on the skin morphology and irritation in hairless rats, *Toxicology,* 175, 35, 2003.

Kanikkannan, N., Patel, R., Jackson, T., Shaik, M.S., and Singh, M., Percutaneous absorption and skin irritation of JP-8 (jet fuel), *Toxicology*, 161, 1, 2001.

Kim, D., Andersen, M.E., and Nylander-French, L.A., Dermal absorption and penetration of jet fuel components in humans, *Toxicol. Lett.,* 165, 11, 2006.

Kim, D., Farthing, M.W., Miller, C.T., and Nylander-French, L.A., Mathematical description of the uptake of hydrocarbons in jet fuel into the stratum corneum of human volunteers, *Toxicol. Lett.,* 178, 146, 2008.

Kinkead, E.R., Salins, S.A., and Wolfe, R.E., Acute irritation and sensitization potential of JP-8 jet fuel, *J. Am. Coll. Toxicol.,* 11, 700, 1992.

Kreilgaard, M., Assessment of cutaneous drug delivery using microdialysis, *Adv. Drug Deliv. Rev*., 54(Suppl. 1) S99, 2002.

Mallampati, R., Patlolla, R.R., Babu, R.J., Klausner, M., Hayden, P., and Sachdeva, M.S., Dermal irritancy of aliphatic hydrocarbons (C9-C14) using the in vitro EpiDerm full thickness (EFT-300) skin model, Presented at *48th Annu. Meeting Soc. Toxicol.*, Baltimore, MD, 2009.

Mann, C.M., Peachee, V.L., Trimmer, G.W., Lee, J.E., Twerdok, L.E., and White, K.L. Jr., Immunotoxicity evaluation of jet a jet fuel in female rats after 28-day dermal exposure, *J. Toxicol. Environ. Health A,* 71, 495, 2008.

McDougal, J.N., and Garrett, C.M., Gene expression and target tissue dose in the rat epidermis after brief JP-8 and JP-8 aromatic and aliphatic component exposures, *Toxicol. Sci.,* 97, 569, 2007.

McDougal, J.N., Garrett, C.M., Amato, C.M., and Berberich, S.J., Effects of brief cutaneous JP-8 jet fuel exposures on time course of gene expression in the epidermis, *Toxicol. Sci.,* 95, 495, 2007.

McDougal, J.N., and Rogers, J.V., Local and systemic toxicity of JP-8 from cutaneous exposures, *Toxicol. Lett.,* 149, 301, 2004.

McDougal, J.N., Pollard, D.L., Wiesman, W., Garrett, C.M., and Miller, T.E., Assessment of skin absorption and penetration of JP-8 jet fuel and its components, *Toxicol. Sci.,* 55, 247, 2000.

Michaels, A.S., Chandrasekaran, S.K., and Shaw, J.E., Permeation through human skin: theory and in vitro experimental measurement, *Am. Inst. Chem. Eng. J.,* 21, 985, 1975.

Muhammad, F., Monteiro-Riviere, N.A., and Riviere, J.E., Comparative in vivo toxicity of topical JP-8 jet fuel and its individual hydrocarbon components: Identification of tridecane and tetradecane as key constituents responsible for dermal irritation, *Toxicol. Pathol.,* 33, 258, 2005.

Patlolla, R.R., Mallampati, R., Fulzele, S.V., Babu, R.J., and Singh, M., Dermal microdialysis of inflammatory markers induced by aliphatic hydrocarbons in rats, *Toxicol. Lett.,* 185, 168, 2009.

Ramos, G., Limon-Flores, A.Y., and Ullrich, S.E., Dermal exposure to jet fuel suppresses delayed-type hypersensitivity: a critical role for aromatic hydrocarbons, *Toxicol. Sci.,* 100, 415, 2007.

Ramos, G., Limon-Flores, A.Y., and Ullrich, S.E., JP-8 induces immune suppression via a reactive oxygen species NF-kappabeta-dependent mechanism, *Toxicol. Sci.,* 108, 100, 2009.

Ramos, G., Nghiem, D.X., Walterscheid, J.P., Ullrich, S.E., Dermal application of jet fuel suppresses secondary immune reactions, *Toxicol. Appl. Pharmacol.*, 136, 180, 2002.

Riviere, J.E., Brooks, J.D., Monteiro-Riviere, N.A., Budsaba, K., and Smith, C.E., Dermal absorption and distribution of topically dosed jet fuels Jet-A, JP-8, and JP-8 (100), *Toxicol. Appl. Pharmacol.,* 160, 60, 1999.

Rogers, J.V., Gunasekar, P.G., Garrett, C.M., Kabbur, M.B., McDougal, J.N., Detection of oxidative species and low-molecular-weight DNA in skin following dermal exposure with JP-8 jet fuel. *J. Appl. Toxicol.,* 21, 521, 2001.

Roguet, R., Use of skin cell culture for in vitro assessment of corrosion and cutaneous irritancy, *Cell Biol. Toxicol.,* 15, 63, 1998.

Scheuplein, R.J., and Blank, I.J., Permeability of the skin, *Physiol. Rev.,* 51, 702, 1971.

Schmelz, M., Luz, O., Averbeck, B., and Bickel, A., Plasma extravasation and neuropeptide release in human skin as measured by intradermal microdialysis, *Neurosci. Lett.,* 230, 117, 1997.

Singh, S., Zhao, K., and Singh, J., In vivo percutaneous absorption, skin barrier perturbation, and irritation from JP-8 jet fuel components, *Drug Chem. Toxicol.*, 26, 135, 2003.

Smith, H.R., Basketter, D.A., and McFadden, J.P., Irritant dermatitis, irritancy and its role in allergic contact dermatitis, *Clin. Exp. Dermatol.*, 27, 138, 2002.

Ullrich, S.E., and Lyons, H.J., Mechanisms involved in the immunotoxicity induced by dermal application of JP-8 jet fuel, *Toxicol. Sci.,* 58, 290, 2000.

Ullrich, S.E., Dermal application of JP-8 jet fuel induces immune suppression, *Toxicol. Sci.,* 52, 61, 1999.

Welss, T., Basketter, D.A., Schröder, K.R., In vitro skin irritation: Facts and future. State of the art review of mechanisms and models, *Toxicol. In Vitro*, 18, 231, 2004.

Witzmann, F.A., Monteiro-Riviere, N.A., Inman, A.O., Kimpel, M.A., Pedrick, N.M., Ringham, H.N., and Riviere, J.E., Effect of JP-8 jet fuel exposure on protein expression in human keratinocyte cells in culture, *Toxicol. Lett.,* 160, 8, 2005.

York, M., Griffiths, H.A., Whittle, E., and Basketter, D.A., Evaluation of a human patch test for the identification and classification of skin irritation potential, *Contact Dermatitis,* 34, 204, 1996.

9 Understanding Systemic and Local Toxicity of JP-8 after Cutaneous Exposures

James N. McDougal, James V. Rogers, and Richard Simman

CONTENTS

INTRODUCTION

JP-8 (Jet Propellant-8) is similar to hydrodesulfurized kerosene, but is specially formulated with additives to improve performance in the military aviation environment. This hydrocarbon fuel is the primary fuel used by the U.S. and North Atlantic Treaty Organization militaries and is similar to commercial aviation fuels (Jet A and Jet A-1). The transition to JP-8 from the previous jet fuel (JP-4) was a 20-year process that began in the 1980s and was intended to improve safety and improve distribution problems. JP-8 has a higher flash point than JP-4 and is therefore less likely to accidentally ignite. This multipurpose fuel is also used in ground vehicles, generators, heaters, and stoves (Makris, 1994). Estimates of worldwide use reach approximately

5 billion gallons per year (Henz, 1998; Zeiger and Smith, 1998). Because of the large volumes produced and the multipurpose nature of the fuel, there is potential for a variety of cutaneous exposures to liquid, aerosol, or vapor forms of JP-8.

Occupational exposures to JP-8 may occur during aircraft fueling and defueling, aircraft maintenance, cold aircraft engine starts, maintenance of equipment and machinery, use of tent heaters, and cleaning or degreasing with fuel (CDC, 1999; Committee on Toxicology, 2003). Skin may be the most important route of exposure because of the potential for liquid and aerosol contact with fuel, and the low volatility of JP-8 reduces the inhalation hazard. Skin contact with JP-8 vapors would not be expected to be a concern because research with whole-body vapor penetration of volatile organic chemicals in rodents (McDougal et al., 1985) and humans (Corley et al., 1997; Riihimaki and Pfaffli, 1978) shows that these vapors exhibit very low penetration rates. These publications suggest that unless an individual was wearing a respirator, inhalation of vapors would overwhelm the small amount of vapor that would penetrate through the skin. Liquid contact with jet fuels would come primarily from spills and splashes or handling parts wet with fuel. Some aircraft fuel maintenance workers may be cutaneously exposed to liquid fuel for more than 10 minutes at a time (Committee on Toxicology, 2003). Also, in cold climates, a visible JP-8 aerosol may be formed when a cold jet engine is started because of the delay in fuel ignition at low ambient temperatures (Committee on Toxicology, 2003). This plume coming from the jet engine, for a short period of time on startup, appears as a white cloud that contains both fuel aerosol droplets and ice crystals. An individual standing in this plume might get enough JP-8 aerosol on clothing and skin to cause concern about skin penetration. JP-8 aerosol would be expected to have a different composition from liquid JP-8 over time, due to different rates of loss of the volatile components to the air.

JP-8 is a complex mixture that may contain aliphatic and aromatic hydrocarbons ranging from about 9 to 17 hydrocarbons, including thousands of isomers and three to six performance additives (Committee on Toxicology, 2003). Because JP-8 is a performance-based fuel, its composition can vary from batch to batch and refinery to refinery. The average composition is 33 to 61% n-alkanes and iso-alkanes, 10% to 45% naphthalenes, 12% to 22% aromatics, and 0.5% to 5% olefins (Vere, 2003). Because of this variability in composition, researchers funded by or associated with the U.S. Air Force have used fuel from a common source. Our analysis of this fuel is shown in Table 9.1 (McDougal et al., 2000). The only components that we identified with a proportion greater than 1.5% (on a weight-to-weight basis) were aliphatic hydrocarbons containing 10 to 14 carbons. Undecane was the aliphatic hydrocarbon present at the largest percentage (6%), and the most prominent aromatic components were methylnaphthalene (just over 1.0%), trimethylbenzene (1.0%), and dimethylnaphthalenes (0.8%). The only additive detected was the ice inhibitor diethylene glycol monomethylether, which was present at less than 0.1%.

SYSTEMIC TOXICITY

In general, the toxicity of petroleum hydrocarbons has been fairly well studied in laboratory animals and extensively reviewed (Committee on Toxicology, 1996;

TABLE 9.1
Composition of USAF-Provided JP-8 Sample as Analyzed by Gas Chromatography

Component	Percent (w/w)
Undecane	6.0
Dodecane	4.5
Decane	3.8
Tridecane	2.7
Tetradecane	1.8
Methylnaphthalenes	1.2
Nonane	1.1
Trimethylbenzene	1.0
Pentadecane	1.0
Dimethylnaphthalenes	0.78
Dimethylbenzene (xylene)	0.59
Naphthalene	0.26
Ethylbenzene	0.15
Diethylene glycol monomethylether	0.08
Methylbenzene	0.06

Source: From McDougal, J.N., Pollard, D.L., Weisman, W., Garrett, C.M., and Miller, T.E. *Toxicol. Sci.*, 55(2), 247–255, 2000.

ATSDR, 1998; Koschier 1999; Nessel 1999; Committee on Toxicology 2003; Ritchie et al., 2003). The primary toxic effects are hepatic, renal, neurologic, immunologic, and pulmonary (Gaworski et al., 1984; Committee on Toxicology, 1996; ATSDR, 1998). JP-8 and related jet fuels would be classified (Neely, 1994) as relatively harmless orally, based on the acute oral LD_{50} of much more than 5 g/kg (ATSDR, 1998). Epidemiological studies suggest that the primary effects in humans may be related to the central nervous system (Knave et al., 1978; Knave et al., 1979; Committee on Toxicology, 2003). Effects from very high cutaneous exposures in laboratory animals cause systemic effects similar to the other routes (ATSDR, 1998). Therefore, the focus of our initial studies was to predict whether realistic cutaneous exposures in humans might cause systemic toxic effects.

In Vitro Skin Penetration Studies

Measurements of chemical flux through excised skin in the laboratory is frequently used to predict human absorption using Fick's law of diffusion (Bronaugh et al., 1982). Skin penetration studies in the laboratory are most frequently accomplished in diffusion cells (Figure 9.1). Excised pieces of animal or human skin are placed between a donor chamber and a receptor chamber, and the appearance of chemical

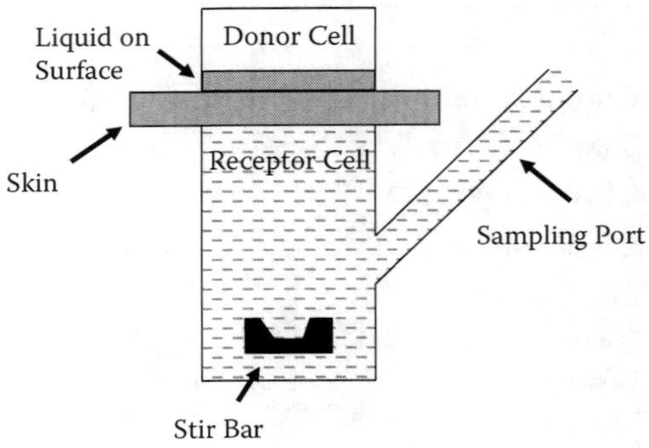

FIGURE 9.1 Static diffusion cell: the apparatus for measuring flux through skin.

in the receptor chamber over time is measured. There are many variations in these methods, including vehicles, skin preparation, receptor solutions, and type of diffusion cell (flow-through or static) (Franz 1978; Bronaugh and Stewart, 1985; Bronaugh, 1995; Vere 2003). Additionally, the way in which penetration data are expressed (e.g., flux, permeability, or percent absorbed) can vary among the experimental approaches (see Poet and McDougal, 2002, for advantages and disadvantages of each).

Because JP-8 is a complex mixture, difficulties arise with studies designed to measure penetration of JP-8 and its components through the skin. Measurements of JP-8 component penetration through the skin can be used to estimate the internal dose of each component that would be absorbed from a JP-8 exposure based on a projected skin exposure area and exposure time. As mentioned above, no individual hydrocarbon component of JP-8 makes up more than 6% of the mixture (Table 9.1); therefore, an analytical detection limit low enough to measure penetration of all of the hundreds of hydrocarbon components through the skin is not possible. There are two approaches described in the literature for measuring the penetration of JP-8 components.

The most popular approach to measure JP-8 penetration through the skin is to add one or more radioactive constituent hydrocarbons as markers to the fuel mixture and detect the penetration of radioactivity through the skin (Riviere et al., 1999; Baynes et al., 2001; Kanikkannan et al., 2001a, 2001b). Generally, these studies show that the aromatics (toluene and naphthalene) penetrate the skin faster than the aliphatics (decane, dodecane, tridecane, and hexadecane). The flux of each component is proportional to the concentration of the component in JP-8 (McDougal et al., 2000; Kanikkannan et al., 2001b). For example, if the concentration of a JP-8 component is doubled, then the flux should theoretically double.

Another way to measure JP-8 penetration is to apply the complete mixture and measure the penetration of all individual hydrocarbons that penetrate the skin in sufficient quantities to be reliably detectable (McDougal et al., 2000). Table 9.2 shows 12 components of JP-8 that penetrated rat skin at concentrations detected and resolved individually, and Figure 9.2 shows the plot of cumulative chemical absorbed, which is used to estimate the flux from the slope of the cumulative

TABLE 9.2
Fluxes (± Srd Deviation) and Permeability Coefficients for JP-8 Components from 4-Hour Diffusion Cell Experiments

Component	Flux ($\mu g/cm^2$-hr)	Permeability Coefficient (cm/hr)
Methylbenzene (toluene)	0.54 ± 0.09	1.1×10^{-3}
Naphthalene	1.04 ± 0.38	5.1×10^{-4}
Ethylbenzene	0.38 ± 0.15	3.1×10^{-4}
Dimethylbenzene (xylene)	0.80 ± 0.24	1.7×10^{-4}
Methylnaphthalenes	1.55 ± 0.52	1.6×10^{-4}
Trimethylbenzene	1.25 ± 0.50	1.3×10^{-3}
Dimethylnaphthalenes	0.59 ± 0.17	9.3×10^{-5}
Decane	1.65 ± 0.68	5.5×10^{-5}
Nonane	0.38 ± 0.24	4.2×10^{-5}
Undecane	1.22 ± 0.81	2.5×10^{-5}
Tridecane	0.33 ± 0.19	1.5×10^{-5}
Dodecane	0.51 ± 0.36	1.4×10^{-5}

Source: From McDougal, J.N., Pollard, D.L., Weisman, W., Garrett, C.M., and Miller, T.E. *Toxicol. Sci.*, 55(2), 247–255, 2000.

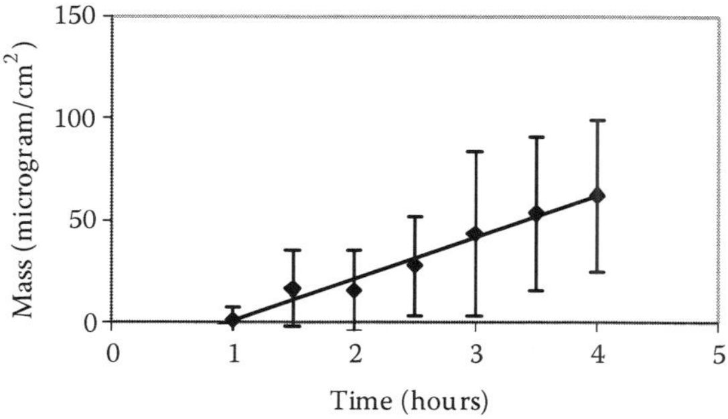

FIGURE 9.2 Diffusion cell time course of the sum of detected hydrocarbon peaks from JP-8 penetration through rat skin dermatomed to 560 μm. (*Source:* From McDougal, J.N., Pollard, D.L., Weisman, W., Garrett, C.M., and Miller, T.E. *Toxicol. Sci.*, 55(2), 247–255, 2000.)

absorbed curve. Flux, which depends on the proportion of the hydrocarbon component in the fuel, varies by about a factor of 3; but permeability coefficients, which are normalized for concentration, differ by almost two orders of magnitude.

A common correlation approach to estimate penetration rates—the Potts/Guy correlation (Guy and Potts, 1993) is not accurate for JP-8 or its components. This correlation approach is used to predict rates of penetration from aqueous solutions for chemicals where there is no permeability data (Bunge and McDougal, 1998). The correlation is based on the molecular weight and octanol/water partition coefficient, and a set of experimental results compiled by Flynn (1990). This equation can be used to estimate the permeability coefficient of a chemical from aqueous solutions within about an order of magnitude (McDougal and Robinson, 2002). The problem is that the correlation is based on penetration from aqueous solutions, and nonaqueous vehicles have interactions with organic chemicals that are in the opposite direction expected according to the correlation (Fasano and McDougal, 2008). Figure 9.3 shows that the correlation of molecular weight with penetration from a lipophilic vehicle (isopropyl myristate) predicts the opposite dependence on molecular weight for a set of industrially important chemicals. This comparison shows that penetration rates from the lipophilic vehicle actually decrease with increasing molecular weight, which is the opposite to what is predicted by the Potts/Guy model. This result points out the importance of measuring the rates of penetration of the JP-8 components from neat-applied JP-8 (essentially a lipophilic vehicle) rather than trying to predict from other studies.

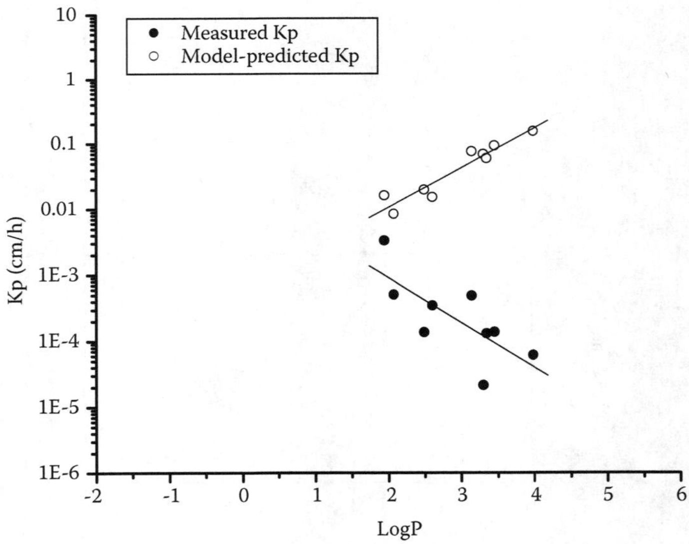

FIGURE 9.3 Comparison of predicted penetration rate (Kp) using the Potts and Guy model for aqueous solutions with the measured penetration rate from isopropyl myristate. (*Source:* From Fasano, W.J., and McDougal, J.N. *Regul. Toxicol. Pharmacol.*, 51(2), 181–194, 2008.)

TABLE 9.3
Determination of a Composite Oral Reference Dose for the Fractions of JP-8 that Come through the Skin

Fraction	RfD (mg/kg/day)	Percent of Flux*	Weighted RfD (mg/kg/day)
C_7–C_8 aromatic	0.2	16.7	.033
C_8–C_{16} aliphatic	0.1	39.8	.040
C_8–C_{16} aromatic	0.04	43.5	.017
Total			.091

Source: From McDougal, J.N., Pollard, D.L., Weisman, W., Garrett, C.M., and Miller, T.E. *Toxicol. Sci.*, 55(2), 247–255, 2000.

* Percent of flux based on the proportion of the total measured fluxes for the hydrocarbon components. For example, dimethylbenzene, methylbenzene, and ethylbenzene would be categorized as C_7–C_8 aromatic and they makeup 16.7% of all the fluxes.

Route-to-Route Extrapolation

The total amount of JP-8 absorption has been estimated for a realistic exposure scenario of exposure of two hands (McDougal et al., 2000). Based on a steady-state JP-8 flux of 20.3 μg/cm^2-hr from rat skin studies, about 17 μg would penetrate the skin of both hands in 1 hour. This internal dose can be compared with oral toxicity studies using a route-to-route extrapolation to determine if it might be toxic. Oral reference doses (RfD) are defined as an estimate of the daily exposure dose that is likely to be without appreciable health effect even if it occurs over a lifetime. Using RfD (mg/kg-day) for hydrocarbon components (aromatic and aliphatic) adjusted for differential flux (Table 9.3), a route-to-route extrapolation suggests that a 70-kg person could absorb 6.37 mg/day, 7 days/week for a lifetime without appreciable risk. This would equate to about 22 minutes/day exposure to both hands (McDougal et al., 2000) and suggests that exposure to JP-8 in most realistic scenarios would not cause systemic toxicities that have been identified in the U.S. EPA Reference Doses. Based on these data, the focus of our subsequent studies was on the local toxicity (skin irritation) of JP-8 rather than the systemic toxicity.

Local Toxicity

Skin toxicity is most frequently a direct result of chemical interaction with cellular and molecular constituents within the skin tissue. Fortunately, the skin functions as a barrier to the movement of foreign objects, such as viruses, bacteria, and small particulates, across the skin. It is also well suited to maintaining our well-hydrated state in dry environments and during immersion in water. The primary barrier to penetration of water and hydrophilic compounds is the outermost layer, the stratum corneum, which is a good barrier to most chemical substances. The stratum corneum is a dead protein-lipid matrix that protects the underlying viable epidermis and dermis. If a

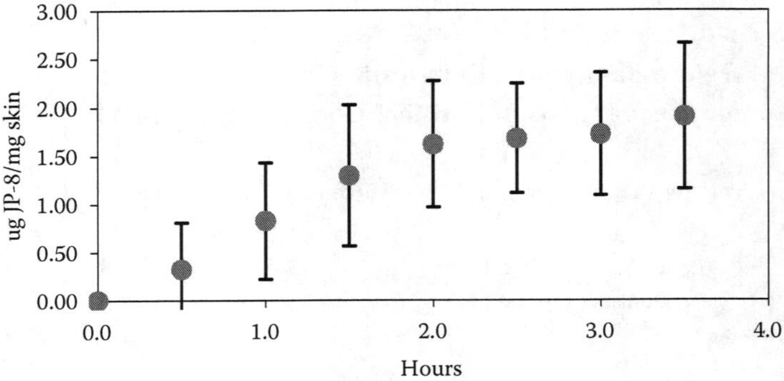

FIGURE 9.4 Concentration of JP-8 (total peaks) in rat skin during a 3.5-hour experiment with neat JP-8 in a static diffusion cell.

substance penetrates the stratum corneum barrier, it can cause local toxicity (e.g., sensitization, skin cancer, or skin irritation) that should be related to the amount of chemical in the skin; however, there are only a few studies that actually measure the concentration of chemicals in the skin tissue after cutaneous exposures. Figure 9.4 shows that with continued exposure in a diffusion cell, the appearance of JP-8 in the skin slowly increases over a period of hours; however, the appearance of some chemical in the skin starts nearly immediately on exposure (McDougal and Robinson, 2002). In another experiment (Figure 9.5), we measured the skin concentration due to a 20-minute exposure to dibromomethane in aqueous solution measured at 4-minute intervals in static diffusion cells (McDougal and Jurgens-Whitehead, 2001). A 1-hour in vivo JP-8 cutaneous exposure to rats (Figure 9.6) shows that the components are absorbed to different extents and the skin concentration decreases fairly rapidly for 5 hours after exposure (McDougal, unpublished data). The skin concentration of the components (normalized for the component concentration in JP-8) during and after the exposure shows that the aliphatics with longer carbon chains were present

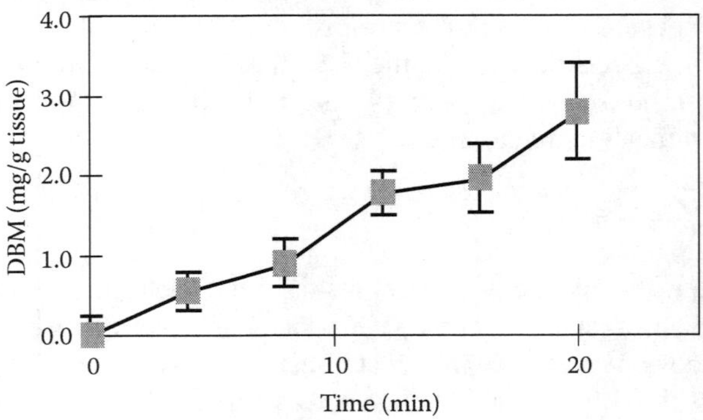

FIGURE 9.5 Rat skin concentration of dibromomethane during a 20-minute experiment in a static diffusion cell. (*Source:* From McDougal, J.N., and Jurgens-Whitehead, J.L. *Risk Anal.,* 21(4), 719–726, 2001.)

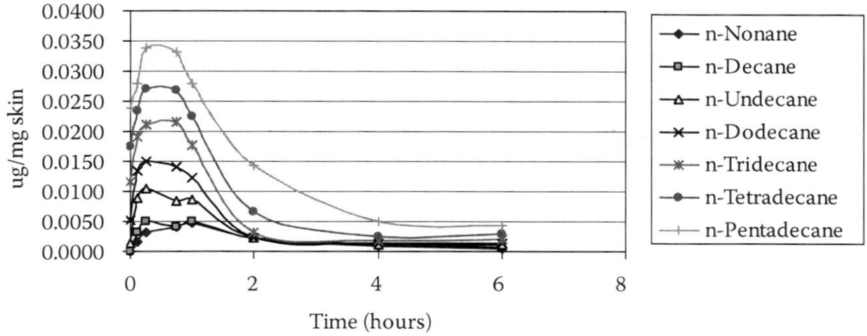

FIGURE 9.6 Concentration of seven JP-8 components in rat skin after a 1-hour in vivo exposure to JP-8. The concentrations were normalized for the concentration of each component in JP-8.

in highest concentrations. Other chemicals, including aromatics, were not present at detectable concentrations even though care was taken to flash-freeze the skin samples to minimize evaporative loss. The results of these studies suggest that the effect of a chemical on the skin could occur as soon as the threshold for damage is reached.

Skin Sensitization

The clinical name for skin sensitization is allergic dermatitis, which is a delayed (type IV) hypersensitivity reaction that involves cell-mediated immune responses, including T cells and the thymus (Marzulli and Maibach, 1996). JP-8 was classified as a weak skin sensitizer (Kinkead et al., 1992) based on studies that revealed that guinea pigs cutaneously treated with 0.1 ml JP-8 four times in a 10-day period became sensitized. The three jet fuels tested (JP-8, JP-8+100, and Jet A) were active in the murine local lymph node assay, indicating that they are weak sensitizers (Kanikkannan et al., 2000). JP-8 may affect the sensitization to other antigens, because after single or repeated JP-8 treatments to the ears of mice, the induction of contact hypersensitivity to dinitrofluorobenzene (Ullrich, 1999) and *Candida albicans* (Ramos et al., 2002) was impaired in a dose-dependent manner. Similar studies (Ullrich, 1999; Ullrich and Lyons, 2000) found that T cells isolated from JP-8-treated mice had reduced proliferation potential in vitro, but antibody production was identical in treated mice compared to controls. The proposed mechanism of interference with an immune response was through the release of immune biological-response modifiers such as prostaglandin E_2 (PGE_2) and interleukin-10 (IL-10) (Ullrich and Lyons, 2000). Much more research must be done to understand if the immunological changes seen with cutaneous dosing of JP-8 in laboratory animals would relate to any human health consequences other than skin sensitization.

Skin Cancer

Two-year carcinogenicity studies (skin painting) in mice with Jet A produced marked skin irritation and skin tumors in 44% of the mice (Freeman et al., 1993). In similar studies using an intermittent schedule to reduce skin irritation, the number of tumors decreased to 2% (Freeman et al., 1993). Freeman (1993) suggests that irritation is necessary but not sufficient for tumor formation. Jet A applied to the skin of mice three times a week for 105 weeks promoted squamous cell carcinoma and fibrosarcoma tumors (Clark et al., 1988). However, hydro-treated fuel (which has less sulfur) was less tumorigenic (Clark et al., 1988; Freeman et al., 1990; Freeman et al., 1993). Skin tumors caused by occupational exposures to JP-8 are not likely for two reasons:

1. The chronic repeated application of jet fuel to the skin is required to cause severe irritation and tumors in rodent studies. This is not a realistic scenario for human exposures, because workers would undoubtedly limit exposures to avoid the repeated irritation.
2. The production specification that limits the sulfur content to less than 0.3% would reduce the tumorigenicity of JP-8 if prolonged exposure occurred.

Skin Irritation

JP-8 and other kerosene-based fuels have been shown to cause skin irritation with repeated or prolonged contact. Skin irritation is a non-immune-related response characterized by direct action of a compound on skin tissues (Weltfriend et al., 1996). Irritant dermatitis can be caused by a wide variety of compounds such as surfactants, solvents, oils, and hydrocarbons (Wahlberg, 1993; Wigger-Alberti et al., 2000), but the underlying mechanisms are not completely known.

Methods to assess irritation include measurements of skin condition, release of inflammatory cytokines, and in vitro studies of the primary epidermal cell (i.e., the keratinocyte). JP-8 exposure to the skin causes erythema and edema, as indicated visually and by changes in histology at periods of 1 hour to 4 weeks in hairless rats, rats, and pigs (Baker et al., 1999a, 1999b; Kabbur et al., 2001; Kanikkannan et al., 2001b; Monteiro-Riviere et al., 2001; Kanikkannan et al., 2002). Transepidermal water loss, which increases with skin damage, increases with JP-8 exposure (Kanikkannan et al., 2001b; Monteiro-Riviere et al. 2001; Kanikkannan et al., 2002). Kabbur and co-workers (2001) showed that a 1-hour cutaneous JP-8 exposure to rats caused increases in inducible nitric oxide synthase (iNOS) and interleukin-1α (IL-1α) protein. Increased levels of oxidative species and low-molecular-weight DNA were also present after brief exposures to JP-8 (Rogers et al., 2001a).

Cultured skin cells are often used to assess the irritation of chemicals to the skin. In vitro assessment of the cytotoxic or irritating potential of volatile organic compounds (VOCs) is problematic due to their physical properties. VOCs are practically insoluble in water and can evaporate rapidly from the exposure medium over time, leading to a nonuniform or unknown chemical dose throughout the exposure period that may ultimately affect the interpretation of dose-response relationships (Rogers and McDougal, 2002). Exposure of cultured cells to jet fuel (including JP-8 and JP-8 components) has led to the measurement of observed biological effects, including cytotoxicity, cytokine release, DNA damage, and oxidative stress, indicating the toxic and proinflammatory ability of JP-8 and JP-8 components (Boulares et al.,

2002; Chou et al., 2002; Jackman et al., 2002; Wang et al., 2002; Chou et al., 2003). The components of JP-8 possess varying degrees of water solubility and volatility, which can make it difficult to relate the experimental exposure dose with the dose added to the cells (Rogers and McDougal, 2002; Coleman et al., 2003; Rogers et al., 2004). Therefore, an important step toward utilizing in vitro assessment of JP-8 irritation potential would be to measure or estimate the chemical dose at the target tissue or cellular site that is promoting the biological responses.

Chou et al. (2003) demonstrated differences in the cytotoxicity and interleukin-8 (IL-8) release in human epidermal keratinocytes (HEK) exposed to JP-8 components while cultured in unsealed plates. It appears that in HEK cells, the aromatic hydrocarbons cause greater direct cytotoxicity than aliphatic hydrocarbons, while the aliphatic compounds induce a higher proinflammatory response as measured by IL-8 release. In murine keratinocytes exposed in an unsealed culture system, we have also seen a similar trend in which aromatic JP-8 components (*m*-xylene and 1-methylnaphthalene) appear more cytotoxic than the aliphatic n-nonane when comparing EC_{50} values (Rogers et al., 2004). However, in murine keratinocytes exposed in sealed containers where the chemical concentration is stable throughout exposure, this relationship of the cytotoxic potency (based on EC_{50} values) of aromatic and aliphatic hydrocarbons is different. It appears that *m*-xylene is the most potent, followed by n-nonane, and 1-methylnaphthalene as the least potent. This difference could be attributed to chemical partitioning in the absence of evaporation, which seems to play an important role in the cytotoxicity of VOCs (Rogers and McDougal, 2002; Rogers et al., 2004). In conclusion, there is no doubt that JP-8, like most of the kerosene-based fuels, causes skin irritation with prolonged or repeated contact to the skin.

GENE EXPRESSION

Gene expression studies have become a useful approach to enhance our understanding of the mechanisms of toxicity (Battershill, 2005; Currie et al., 2005; Coe et al., 2006). After repeated exposures, JP-8 has been shown to change gene expression in brain and lung, and it has various effects on gene expression in cultured cells. Studies of gene expression in the rat brain after JP-8 inhalation for 6 hours per day for 91 consecutive days resulted in neurotransmitter signaling and stress response transcript changes that increased with increasing dose (Lin et al., 2001, 2004). An inhalation study monitoring gene changes in lung tissue after 1-hour JP-8 exposures for 7 consecutive days revealed that genes associated with antioxidant and detoxification mechanisms were increased with increasing dose (Espinoza et al., 2005). Exposures of Jurkat cells (a human T lymphocyte line) to JP-8 for 4 hours resulted in changes in gene expression patterns associated with the cell cycle, transcriptional, apoptotic, stress, and metabolic genes (Espinoza and Smulson, 2003). Human epidermal keratinocytes treated with JP-8 for 1 to 7 days responded with changes in genes related to cytoskeleton, enzymes, and signaling (Espinoza et al., 2004). JP-8 applied to the skin of rats once a day for 7 days induced the expression of numerous proinflammatory cytokine and chemokine mRNAs (Gallucci et al., 2004). Taken together, these studies demonstrate that a variety of inflammatory-, stress-, and damage-related genes

TABLE 9.4
Mass of all Detectible JP-8 Components in Skin at the end of a 3.5-Hour Static Diffusion Cell Experiment

Component	Mass in Skin (mg/g) ± S.D.	Concentration in JP-8 (mg/ml)	Ratio (× 10^{-3})
Nonane	0.077 ± 0.018	9.2	8.4
Decane	0.196 ± 0.047	30.2	6.4
Undecane	0.266 ± 0.070	48.3	5.5
Dodecane	0.143 ± 0.041	36.1	4.0
Tridecane	0.092 ± 0.035	21.9	4.2
Tetradecane	0.055 ± 0.022	14.6	3.8

Source: From McDougal, J.N., and Robinson, P.J. *Sci. Total Environ.*, 288(1-2), 23–30, 2002.

Note: The ratio of mass to concentration in JP-8 approximately correlates with the carbon chain length.

respond to prolonged or repeated JP-8 exposures. Therefore, we were interested in trying to identify the "trigger" or initiating signal that starts the variety of inflammatory responses occurring with JP-8.

Basal Gene Expression in the Rat

Skin is a dynamic and intricate organ that is involved in temperature regulation (sweating and regulation of blood flow); metabolism (carbohydrates, collagen, keratin, lipids, melanin, vitamin D, and xenobiotics); mechanical support and protection (transport barrier, response to oxidative stress, and immune responses); as well as sensory functions (Monteiro-Riviere, N.E., 1996). The level of mRNA transcripts in full-thickness normal skin should give some indication of the relative proportion and importance of each of these functions. Figure 9.7 shows that transcripts involved in metabolism, oxidative and cellular stress, and signal transduction are the most abundant of the gene transcripts that have been identified—see Rogers et al. (2003) for details. This proportion of transcripts constitutively present may reflect the portion of the genome (based on the Affymetrix Rat Toxicology U34 array) involved in each function and represents the background of normal gene expression that could be changed by chemical treatments to the skin.

Brief Exposure Studies with Rats

Gene expression changes induced by brief, 1-hour cutaneous exposures to rats with *m*-xylene, a minor component of JP-8, was compared with the same exposure to a surfactant (sodium lauryl sulfate, SLS) and a "green" solvent (d-limonene) (Rogers et al., 2003). We analyzed gene expression at the end of the 1-hour exposure and 4 hours after the beginning of the exposure, and the results suggest that these chemicals

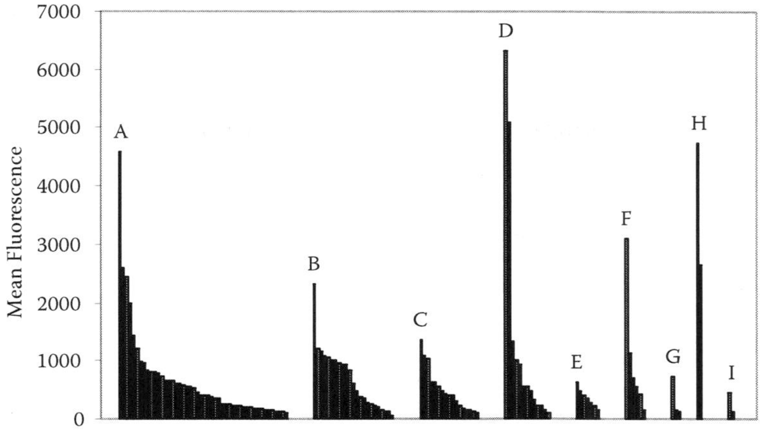

FIGURE 9.7 Mean fluorescence of genes detected as present or marginally present in normal skin. Categories sorted by number of genes. Labels indicate metabolism (A), oxidative/cellular stress (B), signal transduction (C), miscellaneous (D), differentiation/cell division (E), cell structure (F), transporter/ligands (G), extracellular matrix (H), and cytokines/growth factors/receptors (I). (*Source:* From Rogers, J.V., Garrett, C.M., and McDougal, J.N. *J. Biochem. Mol. Toxicol.,* 17(3), 123–137, 2003.)

acted by different mechanisms. Four hours after the exposure started, SLS showed no histological changes, but showed slightly more changes in gene expression than the other two chemicals. Histologically, skin exposed to the *m*-xylene or d-limonene solvents showed minor epidermal separation from the basement membrane at the end of the exposure, which became more apparent at the later skin sampling time. There was accumulation of eosinophilic material between the epidermis and basement membrane in the solvents in the later sampling time. In *m*-xylene-exposed skin, we previously reported significant oxidative species formation as early as 2 hours following the beginning of a 1-hour exposure (Rogers et al., 2001b). In a subsequent study, we observed changes in genes associated with oxidative/cellular stress such as the induction of metallothionein-1 and –2, and the down-regulation of catalase at both 1 and 4 hours (Rogers et al., 2003). The elevation of metallothionein suggests an early wound-healing response, as the induction of metallothionein has been observed in models of wound healing (Iwata et al., 1999). The transcript for major acute phase α_1-protein was up-regulated at both 1 and 4 hours in skin exposed to *m*-xylene, but only at 4 hours in skin exposed to SLS. The major acute phase α_1-protein is an important member of the kininogen family and plays a role in the inflammatory response. During tissue inflammation, kininogens can induce vasodilatation and increased capillary permeability. The levels of major acute phase α_1-protein mRNA and protein have been shown to change during the acute phase response of inflammation (Cole et al., 1985). Therefore, the increases in the transcript for major acute phase α_1-protein that we observed in skin exposed to *m*-xylene and SLS suggest the early induction of inflammatory mediators in skin in response to irritating chemicals. This study suggested that the aliphatic and aromatic components of JP-8 may

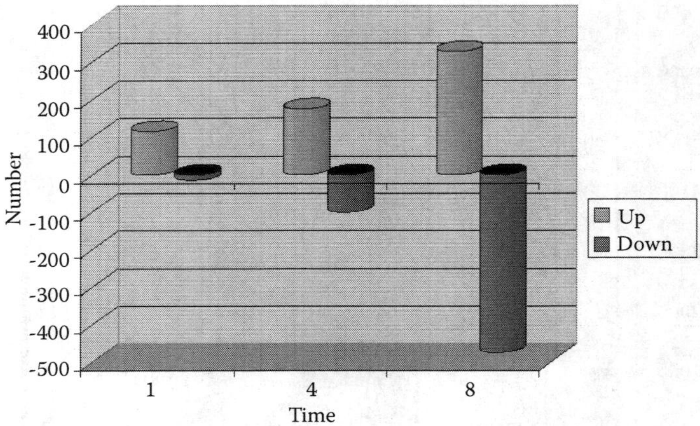

FIGURE 9.8 Plot of the number of genes changed twofold with time in the epidermis after the beginning of the 1-hour JP-8 exposure to the rat. (*Source:* From McDougal, J.N., Garrett, C.M., Amato, C.M., and Berberich, S.J. *Toxicol. Sci.,* 95(2), 495–510, 2007.)

elicit different responses in the skin, and if the "bad actors" could be identified, perhaps they could be eliminated from the fuel mixtures.

We have extensively investigated the time course of gene expression changed due to a 1-hour JP-8 exposure in the epidermis of the rat (McDougal et al., 2007; McDougal and Garrett, 2007). Although there were never any visible signs of irritation after the 1-hour exposure, Figure 9.8 shows that the number of genes changed twofold continues to increase temporally after the beginning of exposure, and that by 8 hours, the number of genes decreased was larger than the number that was increased. The significant decrease in transcript level compared to control may reflect cellular damage, especially since the largest category of genes decreased by 8 hours was related to metabolic and physiological processes (McDougal et al., 2007). According to gene ontology (generic GO slim), the nucleus was most affected in the cellular component category at the end of the exposure. More specifically, these were changes in the molecular binding of nucleic acid and DNA as well as transcription and translation regulator activity. Several signaling pathways related to inflammation, apoptosis, cell growth, and proliferation were rapidly activated by JP-8 exposure. The very rapid response in gene expression is consistent with a "physical" stimulus due to interaction of the JP-8 with critical epidermal structures. JP-8 contains a mixture of aromatic and aliphatic hydrocarbons with log octanol/water partition coefficients between 2.7 (toluene) and 7.6 (tridecane) that have been shown to penetrate into the skin (McDougal et al., 2000) and rapidly dissipate (Figure 9.6). Once in the epidermis, the JP-8 could move into the extracellular space between keratinocytes and partition into cellular, mitochondrial, and nuclear membranes where they could initiate the stress responses that trigger signaling events. The interaction of JP-8 with membranes could be similar to the classic physical effect of ethanol on membranes (Goldstein, 1984). From these results, we hypothesize that the "trigger" for the inflammatory process is "physical" stress involving JP-8 disruption of membrane

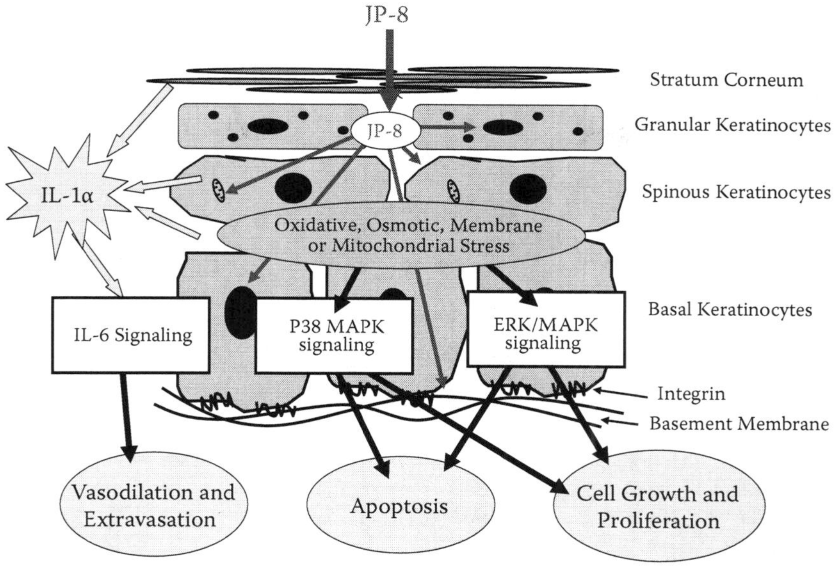

FIGURE 9.9 (See color insert following page 272) Schematic of the proposed mechanism of "physical" damage in the epidermis caused by JP-8 exposure. The stress response involves increases in signaling pathways that result in inflammation, apoptosis, growth, and proliferation. (Source: From McDougal, J.N., Garrett, C.M., Amato, C.M., and Berberich, S.J. *Toxicol. Sci.*, 95(2), 495–510, 2007.)

integrity and the oxidative or osmotic balance that ultimately activates signaling pathways resulting in the recognized effects of JP-8 on the skin (Figure 9.9).

Comparisons of JP-8 with Other Chemicals

Based on the suggested differences in irritant characteristics that we found with xylene, SLS, and d-limonene (Rogers et al., 2003), we compared the epidermal gene expression responses to two aromatic (dimethylnaphthalene [DMN] and trimethylbenzene [TMB]) and two aliphatic (undecane [UND] and tetradecane [TET]) chemicals with JP-8 responses. We wanted to determine if one of these components would mimic the gene expression response of JP-8 at the end of similar 1-hour exposures. A secondary purpose was to compare the potency of these JP-8 components as neat chemicals in an attempt to determine if aliphatics or aromatics were responsible for the irritation induced by JP-8. Table 9.1 shows that the proportions of the selected aliphatics in the JP-8 range from 6% (for UND) to 1.8% (for TET), and the proportion of the aromatics range from 1% (for TMB) to 0.8% (for DMN). Figure 9.10 confirms that the aliphatic chemicals have higher concentrations in the skin than the aromatic chemicals (not detected) at the end of a JP-8 exposure (see Figure 9.6). The epidermal concentrations of UND, TET, TMB, and DMN in the skin when applied as a neat chemical showed no statistically significant differences. After analysis of gene expression, the number of genes changed twofold and was statistically significant for

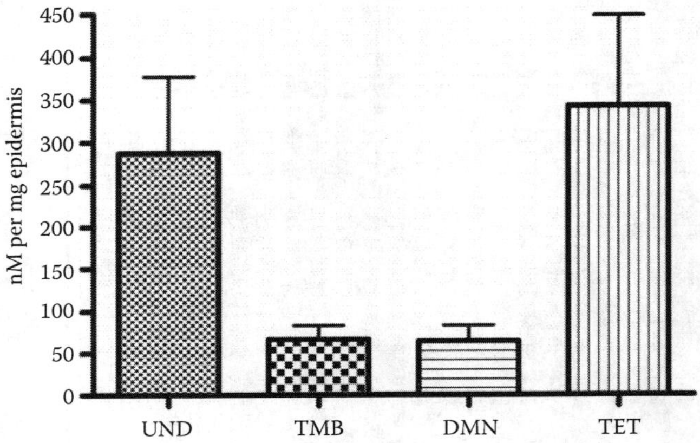

FIGURE 9.10 The concentration in the epidermis of four components measured by gas chromatography at the end of a 1-hour JP-8 exposure. $P \le 0.001$ with ANOVA, N = 5. UND = undecane, TMB = trimethylbenzene, DMN = dimethylnaphthalene, and TET = tetradecane. (*Source:* From McDougal, J.N., and Garrett, C.M. *Toxicol. Sci.*, 97(2), 569–581, 2007.)

each treatment compared to a sham treatment, as shown in Table 9.5. JP-8, UND, and TMB caused changes in the largest number of genes, followed by DMN and TET. One-way ANOVA Post Hoc analysis (Student-Neuman-Keuls) showed that the UND and TMB responses were similar, but JP-8, DMN, and TET were all different. With each treatment, there were more genes decreased than increased, ranging from 57% (for UND) to 78% (for TMB) of the total genes changed. UND and TMB caused the greatest number of gene changes, more than twice as many as DMN, and about tenfold more than TET. According to Ingenuity Pathways Analysis (IPA), the top functions affected by JP-8 were cell death, cellular growth and proliferation, and cellular movement. Table 9.6 shows that UND and TMB most closely mimicked the number of genes related to cell damage, growth and proliferation, and cellular movement changed by JP-8. Figure 9.11 demonstrates that none of the component exposures

TABLE 9.5
Number of Transcripts Changed Twofold and Statistically Significant ($P \le 0.05$) in Each Cutaneous Treatment Compared to Sham

	JP-8	TMB	UND	DMN	TET
Increased	476	289	683	207	44
Decreased	635	1,018	523	263	110
Total	1,100	1,307	1,206	470	154

Source: From McDougal, J.N., and Garrett, C.M. *Toxicol. Sci.*, 97(2), 569–581, 2007.

TABLE 9.6
Number of Transcripts Changed by Each of the Treatments that Were Related to the Top Three Functions Changed by JP-8

	JP-8	UND	TMB	DMN	TET
Cell Death:	212	210	197	88	40
Apoptosis	169	170	153	71	35
Survival of eukaryotic cells	62	65	59	29	NS
Cellular Growth and Proliferation:	228	238	232	98	45
Growth	153	157	156	66	39
Proliferation	160	92	150	65	31
Proliferation of keratinocytes	8	7	6	6	NS
Cellular Movement:	128	141	105	55	34
Chemotaxis	46	47	16	19	17
Rolling of eukaryotic cells	NS	7	NS	NS	NS

Source: From McDougal, J.N., and Garrett, C.M. *Toxicol. Sci.*, 97(2), 569–581, 2007.

Note: Indented categories are subcategories of the one above and their numbers are included in the category above. NS = none significantly changed.

exactly mimicked the biological processes changed by JP-8. Chemical-induced changes in gene expression with the four components showed consistent differences in magnitude, whether total gene expression or functional pathway-specific changes were investigated. When only the genes related to specific pathways or functions changed by JP-8 were considered, we found that these pathways were nearly all activated by the components, but to different extents. Analysis of chemical concentrations in the skin after the component exposures indicated that the nearly tenfold differences in gene expression response (Table 9.5) were not due to different target tissue (epidermis) concentrations (Figure 9.12). UND and TMB appear to be more potent inducers of gene expression in the epidermis than DMN and TET, but both aliphatic and aromatic compounds cause responses that may result in irritation. As far as we know, this study was the first attempt to relate in vivo changes in gene expression to tissue concentrations. We concluded that no one component will mimic the response of the complex JP-8 mixture, but in our study UND had the most similar responses. This study (McDougal and Garrett, 2007) does not allow us to state that either aromatics or aliphatics are responsible for JP-8-induced skin irritation.

The U.S. Air Force (USAF) is considering replacing some JP-8 with a synthetic fuel (S-8) that does not contain the aromatic hydrocarbons and is made from coal, natural gas, or biomass using the Fischer-Tropsch process. A couple of studies have compared the effects of S-8 and JP-8 in USAF-supported model systems. Immune suppression from S-8 was compared with JP-8 (Ramos et al., 2007), with the result that S-8 applied to the skin of mice did not up-regulate COX-2 or reduce footpad

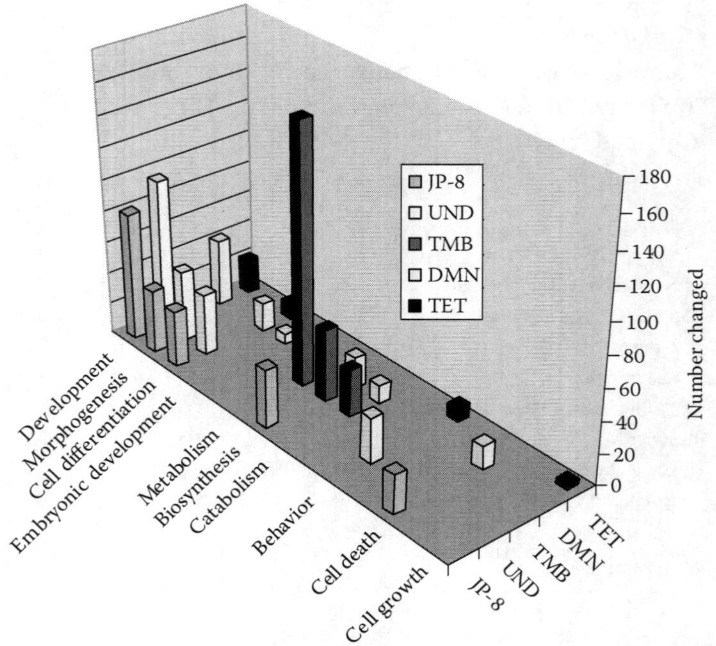

FIGURE 9.11 Graph of Gene Ontology Biological processes with numbers of transcripts changed for each of the treatments. UND = undecane, TMB = trimethylbenzene, DMN = dimethylnaphthalene, and TET = tetradecane. (*Source:* From McDougal, J.N., and Garrett, C.M. *Toxicol. Sci.*, 97(2), 569–581, 2007.)

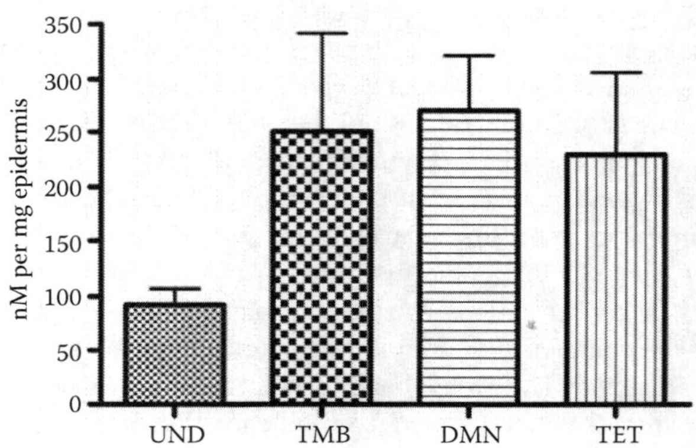

FIGURE 9.12 The epidermal concentrations of the four components applied as pure chemicals as measured by gas chromatography at the end of 1-hour exposures. UND = undecane, TMB = trimethylbenzene, DMN = dimethylnaphthalene, and TET = tetradecane. $P \leq 0.25$ with ANOVA; N = 5. (*Source:* From McDougal, J.N., and Garrett, C.M. *Toxicol. Sci.*, 97(2), 569–581, 2007.)

swelling to the extent that JP-8 did. When several aromatic hydrocarbons were added to the S-8 and the tests were repeated, they found that the COX-2 up-regulation and immune suppression returned, supporting their hypothesis that the aromatic hydrocarbons were responsible for these effects in JP-8. In human epidermal keratinocytes exposed for 5 minutes, viability and IL-8 release were compared for S-8, JP-8, and some of the aliphatic components (Inman et al., 2008). They reported that S-8 had better viability and less IL-8 release compared to JP-8 for up to 24 hours. Epithelial responses to JP-8 and S-8 in mice exposed by inhalation at 1 hour per day for 7 days showed that S-8 did not have the increase in inspiratory lung resistance that was seen with JP-8; however, both fuels increased expiratory resistance (Wong et al., 2008). Morphometric analysis of lung tissue showed that the epithelial cells in the bronchioles were targeted with S-8 and the epithelial cells in the alveoli/terminal bronchioles were targeted with JP-8, suggesting different deposition patterns with inhalation exposures.

We have investigated the changes in epidermal gene expression after brief (1 hour) exposures to S-8 or JP-8 using our standard exposure model and were unable to find dramatic changes in epidermal mRNA levels at the end of the exposure (McDougal, unpublished results). Of the 29,740 gene transcripts that were identified in the skin, 1147 of the S-8 samples had expression levels that were significantly different from the JP-8 samples with four skin samples in each treatment group. At the significance level of $p = 0.05$, 1487 differences would be expected by chance alone, suggesting that the differences were not real. Almost none of these genes were recognizable as genes we have previously seen altered in epidermis with chemical treatments. When the Benjamini-Hochberg multiple testing correction was used, there were no statistically significant differences at the $p = 0.05$ level. Figure 9.13 shows that of all the genes that were marked as changed due to S-8 treatment *without* the Benjamini-Hochberg multiple testing correction, only three transcripts had differences that were greater than twofold increased with S-8 compared to JP-8. These transcripts were the cytokine CCL2, heat shock protein 1b (HSPA1b), and pleckstrin (PLEK). The cytokine responds strongly to irritating chemicals on the skin (McDougal et al., 2007), and a twofold difference is probably not a significant event. This heat shock protein and pleckstrin are not commonly affected in irritation. We conclude that with our brief exposures model to rat skin, there are no dramatic differences due to S-8 and JP-8 treatments.

In an attempt to compare the gene expression effects of repeated JP-8 exposures with two other irritants in our in vivo rat model, we investigated 10-minute JP-8 exposures, once a day for 6 days, with identical exposures to octane (a minor component of JP-8) and cumene (an irritant) (McDougal et al., in preparation). Table 9.7 shows that the magnitude of gene expression changes due to JP-8 was about sixfold less than that of octane and cumene. Visual observations and histological analysis of skin sections indicate that the superficial and microscopic damage from JP-8 was less than the other chemicals (data not shown). Figure 9.14 shows that most of the genes changed by repeated JP-8 exposures (84%) were also changed by the other irritants. According to IPA 6.5, four of the top five biological functions changed by these treatments were similar—cellular movement, cell death, cell-to-cell signaling and interaction, cellular growth and proliferation. The only unique biological functions were post-translational modification for JP-8, cell signaling for octane, and lipid

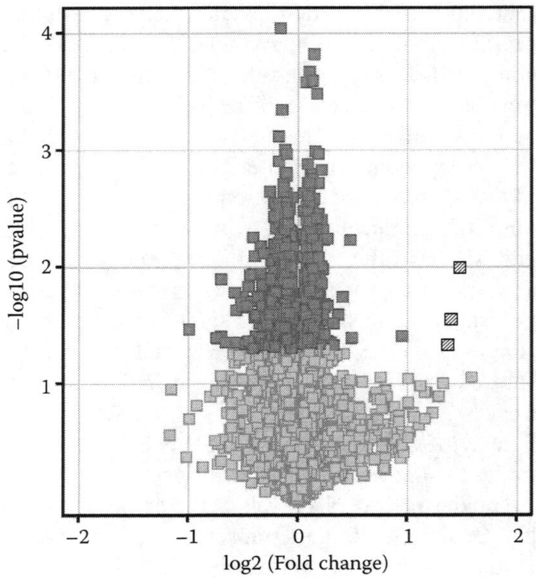

FIGURE 9.13 Volcano plot of the comparison of expression of individual genes between S-8 and JP-8 treatments. The dark transcripts have a P value of 0.05 or greater (uncorrected for multiple tests), and the three striped ones have both a P value of 0.05 or greater and twofold differences.

TABLE 9.7
Biologically (Twofold) and Statistically Significant ($p \leq 0.05$) Changes in Gene Expression after Six 10-Minute Cutaneous Exposures Compared to Control

	JP-8	Octane	Cumene
Up-regulated	79	424	439
Down-regulated	47	325	324
Total	126	749	763

metabolism for cumene. In addition, Table 9.8 shows that the relative magnitudes (fold-change) of changes for representative genes related to irritation for JP-8 were less than those modulated by exposure to octane and cumene. These results suggest that JP-8 and the other irritants cause quantitative differences in responses of the skin, but qualitatively the changes are very similar. These quantitative differences relate both to the number of transcripts significantly changed more than the twofold threshold and to the magnitude of the changes when we compare individual genes across the treatments.

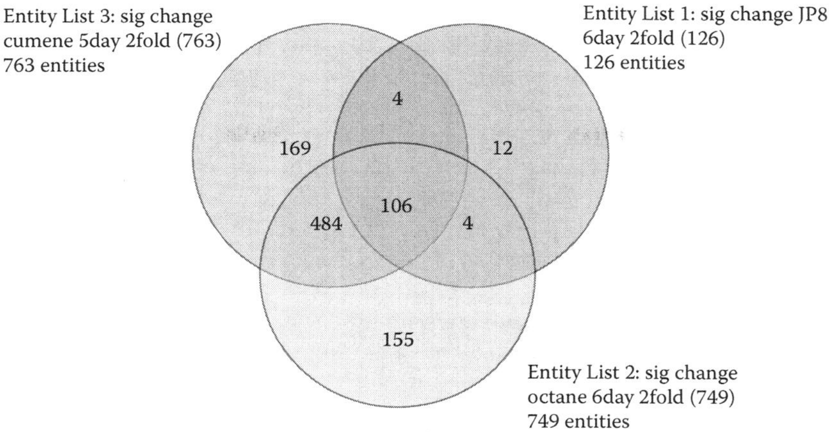

FIGURE 9.14 Venn diagram of the changes in gene expression with the three chemical treatments, showing numbers of common and unique transcript changes.

TABLE 9.8
Magnitude of Fold Changes for Some Irritant-Related Genes Up-Regulated by the JP-8, Octane, and Cumene Treatments for 6 Days

	JP-8	Octane	Cumene
CXCL1	35.6	97.1	115.8
MMP13	18.3	29.3	30.8
CXCL3	9.3	26.6	42.2
S100A9	7.8	21.9	16.3
PTSG2	7.4	18.7	21.3
IL-6	4.9	16.9	23.7

Brief Exposure Studies with Human Volunteers

We studied changes in gene expression due to brief (1 hour) JP-8 cutaneous exposures in humans (McDougal et al., in preparation) over the same time course (up to 8 hours) as previously studied in rats (McDougal et al., 2007). After approval by the appropriate institutional review boards, 12 adult male volunteers (ages 18 to 57 years) were divided into three groups and exposed to 0.5 ml JP-8 in a Hilltop chamber on the surface of one buttock. The other buttock was sham-treated with an empty chamber. The chambers were held in place with surgical tape, and the individuals were allowed to move around normally during the exposure. All treatment groups had the chambers removed after 1 hour, the site wiped with ethanol on a gauze pad, and one 8-mm biopsy punch was taken from each site under local anesthesia at 1, 4, and 8 hours after the beginning of the exposure. The epidermis was separated from the rest of the skin with a cryotome and total mRNA was isolated. Gene expression was measured with standard Affymetrix microarray (HG-U133 Plus 2) techniques and changes from sham treatment were analyzed with GeneSpring GX 9.0 Expression Analysis and Ingenuity Pathway Analysis 5.5.1. After appropriate quality control, probe sets with expression levels below 20 percentile were removed from the analysis, leaving 49,027 of the 54,645 available probe sets on the microarray chip.

The type of transcripts that are constitutively expressed in human skin can provide information about the cutaneous functions and the ability of the skin to respond to insults. When the sham-treated skin samples from all 12 subjects were analyzed using IPA, there were 38,401 transcripts present in the epidermis and 37,922 transcripts present in the dermis. Of these, 10,036 and 9,993 were mapped to the IPA knowledge base for epidermis and dermis, respectively. In both the epidermis and dermis, the top five functional categories that these transcripts belonged to were cell death, cell growth and proliferation, gene expression, cell cycle and post-translational modification. This is in contrast to the results we found in rat skin 5 years earlier (Rogers et al., 2003) in which metabolism, oxidative stress, and signal transduction were the most prevalent genes with the Affymetrix Rat Toxicology U34 array; however, this array only contained 850 genes. Since then, Affymetrix arrays (rat and human) have covered the complete genome and give better representations of the normal functions involved in the skin.

In these human exposure samples, individual variability was at least as great as the responses to JP-8 treatment, because the two-way ANOVA revealed that there were no differences between the treatments (sham and JP-8), or between the post-exposure times (1, 4, or 8 hours) and no interactions between treatment and post-exposure time at $p = 0.05$. Unlike the recent rat studies (McDougal et al., 2007) where the same treatments and times could be grouped together and averaged, the human variability in gene expression responses was so high that each individual had to serve as his own control, and the gene expression data was reanalyzed and uploaded to IPA. Figure 9.15 shows the number of genes for each individual that were changed twofold by JP-8 treatment compared to sham. Three individuals (two sampled at 4 hours and one sampled at 8 hours) showed changes in gene expression that were much greater in magnitude than the rest of the individuals, thereby providing evidence for the popular notion of "sensitive skin" in some individuals. Because of variable sensitivity and testing of each individual at only one time point, we cannot conclude that 4 hours after the beginning of the exposures is the time of maximum response. Figure 9.16 shows that the responses of the sensitive volunteers had 900 genes in common, which were similar to many of these genes, and the functions and pathways they affect were also changed in the rat studies. Four of the top five functions consistently altered in the rat studies (McDougal et al., in preparation) were also consistently altered in the sensitive volunteers (Table 9.3). The exceptions were transcripts related to the gene expression function, which was not in the top five in the rat and cell cycle function transcripts, which were not in the top five in the humans.

Many studies have investigated interleukin-8 (IL-8) release from human keratinocyte cultures in response to JP-8 dosing (Allen et al., 2000; Allen et al., 2001a,

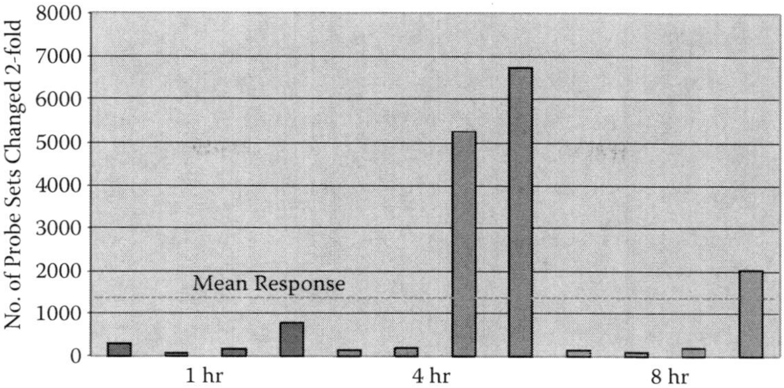

FIGURE 9.15 Number of probe sets altered twofold in the individual subjects. Mean response is shown as a dotted line.

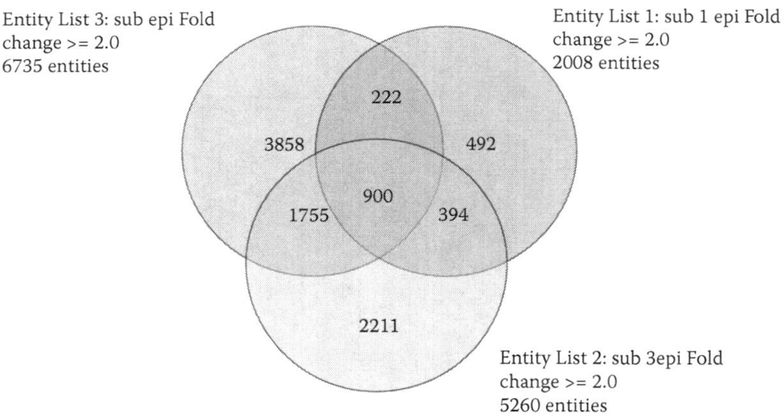

FIGURE 9.16 (See color insert following page 272) Venn diagram of the changes in gene expression for each responsive volunteer showing numbers of common and unique transcript changes.

2001b; Monteiro-Riviere et al., 2004; Yang et al., 2006; Inman et al., 2008). IL-8 is a proinflammatory cytokine, found in humans and mice, but not rats, that has been shown to be one of the major mediators of irritant responses. Figure 9.17 shows some genes related to IL-8 that were changed in each responsive individual. IL-8 mRNA was up-regulated by about 17- to 27-fold in the three responders; none of the non-responders had increases in IL-8 transcripts over the control levels. IL-8 is a CXC chemokine that is one of the major mediators of the inflammatory response, where it functions as a chemoattractant (Sticherling et al., 1991) and a potent angiogenic factor (Rosenkilde and Schwartz, 2004). Vascular endothelial growth factor (VEGFA), which was also up-regulated (4.2- to 6.0-fold), has been shown to increase the expression of IL-8 mRNA in endothelial cells (Lee et al.,

TABLE 9.9
Cellular and Molecular Functions of the Genes Changed in Each Responsive Individual

Function	p-Value Range	No. of Genes (mean)
Cellular growth and proliferation	1.8×10^{-19}–3.3×10^{-4}	554
Cell death	8.2×10^{-19}–4.7×10^{-4}	487
Cell cycle	1.4×10^{-13}–4.4×10^{-4}	242
Cellular development	8.1×10^{-13}–4.2×10^{-4}	393
Cellular movement	1.1×10^{-9}–4.7×10^{-4}	294

Note: Mean number of genes changed by the JP-8 treatment is listed.

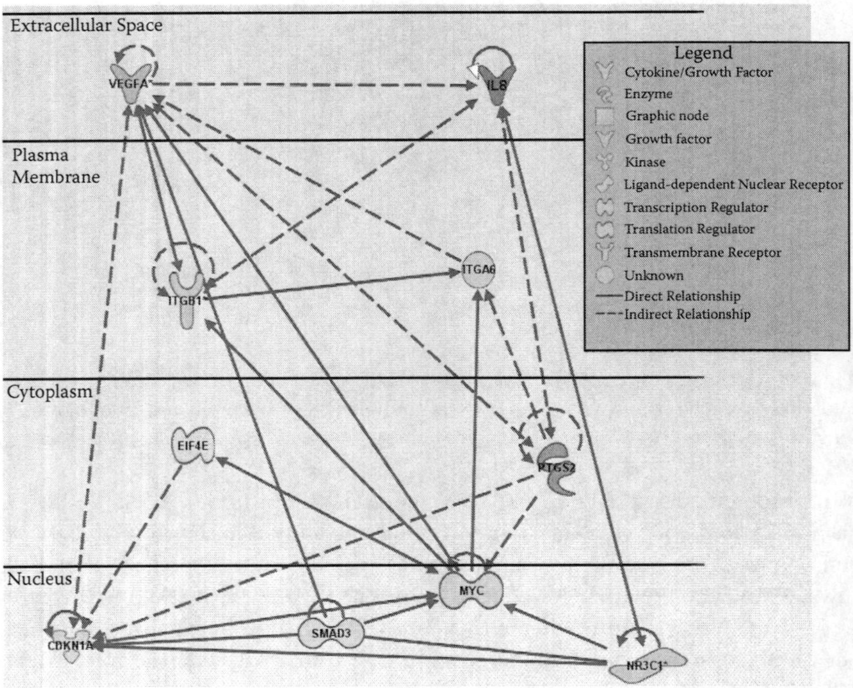

FIGURE 9.17 (See color insert following page 272) Genes related to proliferation of epithelial cells that were changed in each responsive individual. Genes colored red were up-regulated and genes down-regulated were colored green. The intensity of the color is related to the level of up- and down-regulation of subject 4. See text for abbreviations.

2002). Both beta-1 integrin (ITGB1) and alpha-6 integrin (ITGA6) proteins found in plasma membranes were also up-regulated and as a complex have been shown to be involved in VEGF expression (Chung et al., 2004). The mRNA for prostaglandin-endoperoxide synthase 2 (PTGS2), which is involved in cell growth (Murakami et al., 2000), was increased about sevenfold. Transcript levels were increased in several signaling molecules, including eukaryotic translation initiation factor 4E (EIF4E), cyclin-dependent kinase inhibitor 1A or p21 (CDKN1A), SMAD family member 3 (SMAD3), and v-myc myleocytomatosis viral oncogene homolog (MYC). Transcripts for the nuclear receptor (NR3C1) were decreased more than threefold, which might be expected to decrease cellular proliferation, but there is conflicting evidence in knockout mice that reduction of this protein can increase proliferation (Wintermantel et al., 2005). These changes are consistent with the stimulation of epithelial cell growth by brief JP-8 exposure as part of the homeostatic response. Our studies in human volunteers confirm the importance of IL-8 in the inflammatory responses to JP-8, which have only previously been shown in human keratinocyte cultures. These in vivo studies suggest that IL-8 release is a very good measure of the potential for an irritant response. In general, it appears that the brief exposure in vivo rat skin model may be a good surrogate for the gene expression responses of "sensitive" individuals. These studies confirm that many of the functional responses and signaling pathways are similarly affected by irritation of the rat and human epidermis.

CONCLUSIONS

JP-8 is widely used, and like all hydrocarbon fuels, there is potential for toxicity from inhalation, dermal, and oral exposures. Most toxicity studies have been accomplished with inhalation or oral exposures and, as a result, assessing systemic toxicity from cutaneous exposures requires estimating the rate of penetration of the fuel (and its components) through the skin and relating skin exposures to the other routes. We have estimated the flux of pure JP-8 through the skin to be 20.3 $\mu g/cm^2$-hr in diffusion cells (McDougal et al., 2000), which is a moderate flux based on a range of pure chemical penetration recently published (Fasano and McDougal, 2008). A route-to-route extrapolation based on known systemic toxicity suggests that small and brief exposures to areas of skin the size of both hands should not have toxic consequences systemically. Repeated exposures are much more likely to result in local toxicity such as skin sensitization, skin cancer, or skin irritation. We have chosen to investigate skin irritation using the fairly new technique of measuring changes in global gene expression in the skin from cutaneous exposures. Changes in gene expression levels are expected to provide information about the adaptive and toxic responses to any exogenous hydrocarbons that enter the skin, because the mRNA levels generally precede production of new proteins in the skin, particularly those related to metabolism, oxidative and cellular stress, and signal transduction. Microarray studies provide the ability to survey tens of thousands of gene products in one experiment and are very useful for elucidating patterns of response and generating hypothesis that can assist in understanding the mechanism of any deleterious effects. Our studies with brief cutaneous exposures in rats suggest that the "trigger" of the irritant response may be a "physical" response, which disrupts cellular or mitochondrial membranes. Changes in gene expression in response to JP-8 suggest the activation of signaling pathways that result in the inflammatory response as well as both apoptosis and cell growth and proliferation. We have found that the response of the four components of JP-8 that were compared with the JP-8 gene expression response in the epidermis were different than that of the complete mixture, and that no single component (or aromatic or aliphatic classification) mimicked the JP-8 effects, although undecane was most similar. From our research, the differences in gene expression between S-8 (the Fischer-Tropsch prepared fuel) were not dramatic. We did find that repeated exposures to JP-8 were much less irritating than similar exposures to cumene or octane. When we repeated JP-8 skin exposure studies that we accomplished in rats in human volunteers, we found that there was tremendous variability in the human epidermal response. We confirmed, in our in vivo model, that in vitro assays that measure interleukin-8 release as a measure of irritation are valid. When the responses of the most affected volunteers were compared with the rat responses, we observed that the rats were good models for the gene expression responses of the "sensitive" humans.

REFERENCES

Allen, D.G., Riviere, J.E., and Monteiro-Riviere, N.A. (2000). Identification of early biomarkers of inflammation produced by keratinocytes exposed to jet fuels jet A, JP-8, and JP-8(100). *J. Biochem. Mol. Toxicol.,* 14(5), 231–237.

Allen, D.G., Riviere, J.E., and Monteiro-Riviere, N.A. (2001a). Analysis of interleukin-8 release from normal human epidermal keratinocytes exposed to aliphatic hydrocarbons: Delivery of hydrocarbons to cell cultures via complexation with alpha-cyclodextrin. *Toxicol. In Vitro,* 15(6), 663–669.

Allen, D.G., Riviere, J.E., and Monteiro-Riviere, N.A. (2001b). Cytokine induction as a measure of cutaneous toxicity in primary and immortalized porcine keratinocytes exposed to jet fuels, and their relationship to normal human epidermal keratinocytes. *Toxicol Lett.,* 119(3), 209–217.

ATSDR. Toxicological Profile for Jet Fuels (JP-5 and JP-8). 11-11-1998. Atlanta GA, U.S. Department of Health and Human Services, Public Health Service, Agency for Toxic Substances and Disease Registry.

Baker, W., Dodd, D., McDougal, J.N., and Miller, T.E. Repeated Dose Skin Irritation Study on Jet Fuels—A Histopathological Study. AFRL-HE-WP-TR-1999-0022. (1999a). Wright Patterson AFB, Air Force Research Laboratory.

Baker, W., Dodd, D., McDougal, J.N., and Miller, T.E. (1999b). Repeated dose skin irritation study on jet fuels - preliminary dose range finding study. AFRL-HE-WP-TR-1999-0008. Wright Patterson AFB, Air Force Research Laboratory.

Battershill, J.M. (2005). Toxicogenomics: Regulatory perspective on current position. *Hum. Exp. Toxicol.,* 24(1), 35–40.

Baynes, R.E., Brooks, J.D., Budsaba, K., Smith, C.E., and Riviere, J.E. (2001). Mixture effects of JP-8 additives on the dermal disposition of jet fuel components. *Toxicol. Appl. Pharmacol.,* 175(3), 269–281.

Boulares, A.H., Contreras, F.J., Espinoza, L.A., and Smulson, M.E. (2002). Roles of oxidative stress and glutathione depletion in JP-8 jet fuel-induced apoptosis in rat lung epithelial cells. *Toxicol. Appl. Pharmacol.,* 180(2), 92–99.

Bronaugh, R.L. (1995). In vitro methodology for percutaneous absorption studies. 20, 307–312.. Washington, D.C., Division of Toxicology: Food and Drug Administration.

Bronaugh, R.L., and Stewart, R.F. (1985). Methods for in vitro percutaneous absorption studies. IV. The flow through diffusion cell. *J. Pharm. Sci.,* 74(1), 64–67.

Bronaugh, R.L., Stewart, R.F., Congdon, E.R., and Giles, A.L., Jr. (1982). Methods for in vitro percutaneous absorption studies. I. Comparison with in vivo results. *Toxicol. Appl. Pharmacol.,* 62(3), 474–480.

Bunge, A.L., and McDougal, J.N. (1998). Dermal Uptake. In *Exposure to Contaminants in Drinking Water* (S.S. Olin, Ed.), pp. 137–181. ILSI Press, Washington, D.C.

CDC (Centers for Disease Control). (1999).Background document on Gulf War-related research for the health impact of chemical exposures during the Gulf War: A research planning document. Centers for Disease Control.

Chou, C.C., Riviere, J.E., and Monteiro-Riviere, N.A. (2002). Differential relationship between the carbon chain length of jet fuel aliphatic hydrocarbons and their ability to induce cytotoxicity vs. interleukin-8 release in human epidermal keratinocytes. *Toxicol. Sci.,* 69(1), 226–233.

Chou, C.C., Riviere, J.E., and Monteiro-Riviere, N.A. (2003). The cytotoxicity of jet fuel aromatic hydrocarbons and dose-related interleukin-8 release from human epidermal keratinocytes. *Arch. Toxicol.,* 77(7), 384–391.

Chung, J., Yoon, S., Datta, K., Bachelder, R.E., and Mercurio, A.M. (2004). Hypoxia-induced vascular endothelial growth factor transcription and protection from apoptosis are dependent on alpha6beta1 integrin in breast carcinoma cells. *Cancer Res.,* 64(14), 4711–4716.

Clark, C.R., Walter, M.K., Ferguson, P.W., and Katchen, M. (1988). Comparative dermal carcinogenesis of shale and petroleum-derived distillates. *Toxicol Indus. Health,* 4(1), 11–22.

Coe, K.J., Nelson, S.D., Ulrich, R.G., He, Y.D., Dai, X., Cheng, O., Caguyong, M., Roberts, C.J., and Slatter, J.G. (2006). Profiling the hepatic effects of flutamide in rats: A microarray comparison with classical AhR ligands and atypical CYP1A inducers. *Drug Metab. Dispos,* 34(7), 1266–1275.

Cole, T., Inglis, A.S., Roxburgh, C.M., Howlett, G.J., and Schreiber, G. (1985). Major acute phase alpha 1-protein of the rat is homologous to bovine kininogen and contains the sequence for bradykinin: Its synthesis is regulated at the mRNA level. *FEBS Lett.,* 182(1), 57–61.

Coleman, C.A., Hull, B.E., McDougal, J.N., and Rogers, J.V. (2003). The effect of *m*-xylene on cytotoxicity and cellular antioxidant status in rat dermal equivalents. *Toxicol. Lett.* 142(1-2), 133–142.

Committee on Toxicology, N.R.C. (1996). Permissible Exposure Levels for Selected Military Fuel Vapors. The National Academies Press, Washington, D.C.

Committee on Toxicology, N.R.C. (2003) *Toxicologic Assessment of Jet-Propulsion Fuel 8.* National Academies Press, Washington, D.C.

Corley, R.A., Markham, D.A., Banks, C., Delorme, P., Masterman, A., and Houle, J.M. (1997). Physiologically based pharmacokinetics and the dermal absorption of 2-butoxyethanol vapor by humans. *Fundamen. Appl. Toxicol.,* 39, 120–130.

Currie, R.A., Bombail, V., Oliver, J.D., Moore, D.J., Lim, F.L., Gwilliam, V., Kimber, I., Chipman, K., Moggs, J.G., and Orphanides, G. (2005). Gene ontology mapping as an unbiased method for identifying molecular pathways and processes affected by toxicant exposure: application to acute effects caused by the rodent non-genotoxic carcinogen diethylhexylphthalate. *Toxicol. Sci.,* 86(2), 453–469.

Espinoza, L.A., Li, P., Lee, R.Y., Wang, Y., Boulares, A.H., Clarke, R., and Smulson, M.E. (2004). Evaluation of gene expression profile of keratinocytes in response to JP-8 jet fuel. *Toxicol. Appl. Pharmacol.,* 200(2), 93–102.

Espinoza, L.A., and Smulson, M.E. (2003). Macroarray analysis of the effects of JP-8 jet fuel on gene expression in Jurkat cells. *Toxicology,* 189(3), 181–190.

Espinoza, L.A., Valikhani, M., Cossio, M.J., Carr, T., Jung, M., Hyde, J., Witten, M.L., and Smulson, M.E. (2005). Altered expression of gamma-synuclein and detoxification-related genes in lungs of rats exposed to JP-8. *Am. J. Respir. Cell Mol. Biol.,* 32(3), 192–200.

Fasano, W.J., and McDougal, J.N. (2008). In vitro dermal absorption rate testing of certain chemicals of interest to the Occupational Safety and Health Administration: Summary and evaluation of USEPA's mandated testing. *Regul. Toxicol. Pharmacol.,* 51(2), 181–194.

Flynn, G.L. (1990). Physicochemical determinants of skin absorption. In *Principles of Route-to-Route Extrapolation for Risk Assessment* (T.R. Gerrity and C.J. Henry, Eds.), pp. 93–127. Elsevier Science Publishing Co., Inc., New York.

Franz, T.J. (1978). The finite dose technique as a valid in vitro model for the study of percutaneous absorption. *Curr. Probl. Dermatol.,* 7, 55–68.

Freeman, J.J., Federici, T.M., and McKee, R.H. (1993). Evaluation of the contribution of chronic skin irritation and selected compositional parameters to the tumorigenicity of petroleum middle distillates in mouse skin. *Toxicology,* 81(2), 103–112.

Freeman, J.J., McKee, R.H., Phillips, R.D., Plutnick, R.T., Scala, R.A., and Ackerman, L.J. (1990). A 90-day toxicity study of the effects of petroleum middle distillates on the skin of C3H mice. *Toxicol. Indus. Health,* 6(3-4), 475–491.

Gallucci, R.M., O'Dell, S.K., Rabe, D., and Fechter, L.D. (2004). JP-8 jet fuel exposure induces inflammatory cytokines in rat skin. *Int. Immunopharmacol.,* 4(9), 1159–1169.

Gaworski, C.L., MacEwen, J.D., Vernot, E.H., Bruner, R.H., and Cowan Jr., M.J. (1984). Comparison of the subchronic inhalation toxicity of petroleum and oil shale JP-5 fuels. In *Advances in Modern Environmental Toxicology: Volume VI Applied Toxicology of Petroleum Hydrocarbons* (H.N. MacFarland, C.E. Holdsworth, J.A. MacGregor, R.W. Call, and M.L. Lane, Eds.), pp. 33–47. Princeton Scientific Publishers, Princeton, NJ.

Goldstein, D.B. (1984). The effects of drugs on membrane fluidity. *Annu. Rev. Pharmacol. Toxicol.,* 24, 43–64.

Guy, R.H., and Potts, R.O. (1993). Penetration of industrial chemicals across the skin: a predictive model. *Am. J. Ind. Med.,* 23(5), 711–719.

Henz, K. (1998). Survey of Jet Fuels Procured by the Defense Energy Support Center, 1990–1996. Defense Logistics Agencies, Ft. Belvoir, VA.

Inman, A.O., Monteiro-Riviere, N.A., and Riviere, J.E. (2008). Inhibition of jet fuel aliphatic hydrocarbon induced toxicity in human epidermal keratinocytes. *J. Appl. Toxicol.,* 28(4), 543–553.

Iwata, M., Takebayashi, T., Ohta, H., Alcalde, R.E., Itano, Y., and Matsumura, T. (1999). Zinc accumulation and metallothionein gene expression in the proliferating epidermis during wound healing in mouse skin. *Histochem. Cell Biol.,* 112(4), 283–290.

Jackman, S.M., Grant, G.M., Kolanko, C.J., Stenger, D.A., and Nath, J. (2002). DNA damage assessment by comet assay of human lymphocytes exposed to jet propulsion fuels. *Environ. Mol. Mutagen.,* 40(1), 18–23.

Kabbur, M.B., Rogers, J.V., Gunasekar, P.G., Garrett, C.M., Geiss, K.T., Brinkley, W.W., and McDougal, J.N. (2001). Effect of JP-8 jet fuel on molecular and histological parameters related to acute skin irritation. *Toxicol. Appl. Pharmacol.,* 175(1), 83–88.

Kanikkannan, N., Burton, S., Patel, R., Jackson, T., Shaik, M.S., and Singh, M. (2001a). Percutaneous permeation and skin irritation of JP-8+100 jet fuel in a porcine model. *Toxicol. Lett.,* 119(2), 133–142.

Kanikkannan, N., Jackson, T., Sudhan, S.M., and Singh, M. (2000). Evaluation of skin sensitization potential of jet fuels by murine local lymph node assay. *Toxicol Lett.,* 116(1-2), 165–170.

Kanikkannan, N., Locke, B.R., and Singh, M. (2002). Effect of jet fuels on the skin morphology and irritation in hairless rats. *Toxicology,* 175(1-3), 35–47.

Kanikkannan, N., Patel, R., Jackson, T., Shaik, M.S., and Singh, M. (2001b). Percutaneous absorption and skin irritation of JP-8 (jet fuel). *Toxicology,* 161(1-2), 1–11.

Kinkead, E.R., Salins, S.A., and Wolfe, R.E. (1992). Acute irritation and sensitization potential of JP-8 jet fuel. *J. Am. Coll. Toxicol.,* 11, 700.

Knave, B., Gamberale, F., Bergstrom, S., Birke, E., Iregren, A., Kolmodin-Hedman, B., and Wennberg, A. (1979). Long-term exposure to electric fields. A cross-sectional epidemiologic investigation of occupationally exposed workers in high-voltage substations. *Scand. J. Work Environ. Health,* 5(2), 115–125.

Knave, B., Olson, B.A., Elofsson, S., Gamberale, F., Isaksson, A., Mindus, P., Persson, H.E., Struwe, G., Wennberg, A., and Westerholm, P. (1978). Long-term exposure to jet fuel. II. A cross-sectional epidemiologic investigation on occupationally exposed industrial workers with special reference to the nervous system. *Scand. J. Work Environ. Health,* 4(1), 19–45.

Koschier, F.J. (1999). Toxicity of middle distillates from dermal exposure. *Drug Chem. Toxicol.,* 22(1), 155–164.

Lee, T. H., Avraham, H., Lee, S.H., and Avraham, S. (2002). Vascular endothelial growth factor modulates neutrophil transendothelial migration via up-regulation of interleukin-8 in human brain microvascular endothelial cells. *J Biol. Chem.,* 277(12), 10445–10451.

Lin, B., Ritchie, G.D., Rossi, J., III, and Pancrazio, J.J. (2001). Identification of target genes responsive to JP-8 exposure in the rat central nervous system. *Toxicol. Indus. Health,* 17(5-10), 262–269.

Lin, B., Ritchie, G.D., Rossi, J., III, and Pancrazio, J.J. (2004). Gene expression profiles in the rat central nervous system induced by JP-8 jet fuel vapor exposure. *Neurosci. Lett.,* 363(3), 233–238.

Makris, N.J. (1994). JP-8 a conversion update. *Flying Safety,* 50(10), 12–13.

Marzulli, F.N., and Maibach, H.I. (1996). Allergic contact dermatitis. In *Dermatotoxicology.* Taylor & Francis, Washington, D.C., pp. 143–160.

McDougal, J.N., and Garrett, C.M. (2007). Gene expression and target tissue dose in the rat epidermis after brief JP-8 and JP-8 aromatic and aliphatic component exposures. *Toxicol. Sci.,* 97(2), 569–581.

McDougal, J.N., Garrett, C.M., Amato, C.M., and Berberich, S.J. (2007). Effects of brief cutaneous JP-8 jet fuel exposures on time course of gene expression in the epidermis. *Toxicol. Sci.,* 95(2), 495–510.

McDougal, J.N., Garrett, C.M., Kannanayakal, T.J., Schliemann, S., and Elsner, P. (2009). Gene expression after repeated cumene, octane and JP-8 exposures in rats, manuscript in preparation.

McDougal, J.N., Garrett, C.M., and Simman, R. (2009). Effects of brief cutaneous JP-8 jet fuel exposures on gene expression in human volunteers, manuscript in preparation.

McDougal, J.N., Jepson, G.W., Clewell III, H.J., and Andersen, M.E. (1985). Dermal absorption of dihalomethane vapors. *Toxicol. Appl. Pharmacol.,* 79(1), 150–158.

McDougal, J.N., and Jurgens-Whitehead, J.L. (2001). Short-term dermal absorption and penetration of chemicals from aqueous solutions: Theory and experiment. *Risk Anal.,* 21(4), 719–726.

McDougal, J.N., Pollard, D.L., Weisman, W., Garrett, C.M., and Miller, T.E. (2000). Assessment of skin absorption and penetration of JP-8 jet fuel and its components. *Toxicol. Sci.,* 55(2), 247–255.

McDougal, J.N., and Robinson, P.J. (2002). Assessment of dermal absorption and penetration of components of a fuel mixture (JP-8). *Sci. Total Environ.,* 288(1-2), 23–30.

Monteiro-Riviere N.E. (1996). Anatomical factors affecting barrier function. In *Dermatotoxicology* (Marzulli F.N. and Maibach H.I., Eds.), pp. 3–17. Taylor & Francis, Washington, D.C.

Monteiro-Riviere, N., Inman, A., and Riviere, J. (2001). Effects of short-term high-dose and low-dose dermal exposure to Jet A, JP-8 and JP-8 + 100 jet fuels. *J. Appl. Toxicol.,* 21(6), 485–494.

Monteiro-Riviere, N.A., Inman, A.O., and Riviere, J.E. (2004). Skin toxicity of jet fuels: ultrastructural studies and the effects of substance P. *Toxicol. Appl. Pharmacol.,* 195(3), 339–347.

Murakami, M., Naraba, H., Tanioka, T., Semmyo, N., Nakatani, Y., Kojima, F., Ikeda, T., Fueki, M., Ueno, A., Oh, S., and Kudo, I. (2000). Regulation of prostaglandin E2 biosynthesis by inducible membrane-associated prostaglandin E2 synthase that acts in concert with cyclooxygenase-2. *J Biol. Chem.,* 275(42), 32783–32792.

Neely, W.B. (1994). *Introduction to Chemical Exposure and Risk Assessment.* CRC Press, Boca Raton, FL, p. 32.

Nessel, C.S. (1999). A comprehensive evaluation of the carcinogenic potential of middle distillate fuels. *Drug Chem. Toxicol.,* 22(1), 165–180.

Poet, T.S., and McDougal, J.N. (2002). Skin absorption and human risk assessment. *Chem. Biol. Interact.,* 140(1), 19–34.

Ramos, G., Limon-Flores, A.Y., and Ullrich, S.E. (2007). Dermal exposure to jet fuel suppresses delayed-type hypersensitivity: a critical role for aromatic hydrocarbons. *Toxicol. Sci.,* 100(2), 415–422.

Ramos, G., Nghiem, D.X., Walterscheid, J.P., and Ullrich, S.E. (2002). Dermal application of jet fuel suppresses secondary immune reactions. *Toxicol. Appl. Pharmacol.,* 180(2), 136–144.

Riihimaki, V., and Pfaffli, P. (1978). Percutaneous absorption of solvent vapors in man. *Scand. J. Work Environ. Health,* 4, 73–85.

Ritchie, G., Still, K., Rossi, J., III, Bekkedal, M., Bobb, A., and Arfsten, D. (2003). Biological and health effects of exposure to kerosene-based jet fuels and performance additives. *J. Toxicol. Environ. Health B Crit. Rev.,* 6(4), 357–451.

Riviere, J.E., Brooks, J.D., Monteiro-Riviere, N.A., Budsaba, K., and Smith, C.E. (1999). Dermal absorption and distribution of topically dosed jet fuels Jet-A, JP-8, and JP-8(100). *Toxicol. Appl. Pharmacol.,* 160(1), 60–75.

Rogers, J.V., Garrett, C.M., and McDougal, J.N. (2003). Gene expression in rat skin induced by irritating chemicals. *J. Biochem. Mol. Toxicol.,* 17(3), 123–137.

Rogers, J.V., Gunasekar, P.G., Garrett, C.M., Kabbur, M.B., and McDougal, J.N. (2001a). Detection of oxidative species and low-molecular-weight DNA in skin following dermal exposure with JP-8 jet fuel. *J. Appl. Toxicol.,* 21(6), 521–525.

Rogers, J.V., Gunasekar, P.G., Garrett, C.M., and McDougal, J.N. (2001b). Dermal exposure to *m*-xylene leads to increasing oxidative species and low molecular weight DNA levels in rat skin. *J. Biochem. Mol. Toxicol.,* 15(4), 228–230.

Rogers, J.V., and McDougal, J.N. (2002). Improved method for in vitro assessment of dermal toxicity for volatile organic chemicals. *Toxicol Lett.,* 135(1-2), 125–135.

Rogers, J.V., Siegel, G.L., Pollard, D.L., Rooney, A.D., and McDougal, J.N. (2004). The cytotoxicity of volatile JP-8 jet fuel components in keratinocytes. *Toxicology,* 197(2), 113–121.

Rosenkilde, M.M., and Schwartz, T.W. (2004). The chemokine system — A major regulator of angiogenesis in health and disease. *APMIS,* 112(7-8), 481–495.

Sticherling, M., Bornscheuer, E., Schroder, J.M., and Christophers, E. (1991). Localization of neutrophil-activating peptide-1/interleukin-8-immunoreactivity in normal and psoriatic skin. *J. Invest. Dermatol.,* 96(1), 26–30.

Ullrich, S.E. (1999). Dermal application of JP-8 jet fuel induces immune suppression. *Toxicol. Sci.,* 52(1), 61–67.

Ullrich, S.E., and Lyons, H.J. (2000). Mechanisms involved in the immunotoxicity induced by dermal application of JP-8 jet fuel. *Toxicol. Sci.,* 58(2), 290–298.

Vere, R.A. (2003). Aviation Fuels. In *Modern Petroleum Technology, Part 2* (G.D. Hobson, Ed.), 5th ed. John Wiley & Sons, Chichester, pp. 723–771.

Wahlberg, J.E. (1993). Measurement of skin-fold thickness in the guinea pig. Assessment of edema-inducing capacity of cutting fluids, acids, alkalis, formalin and dimethyl sulfoxide. *Contact Dermatitis,* 28(3), 141–145.

Wang, S., Young, R.S., Sun, N.N., and Witten, M.L. (2002). In vitro cytokine release from rat type II pneumocytes and alveolar macrophages following exposure to JP-8 jet fuel in co-culture. *Toxicology,* 173(3), 211–219.

Weltfriend, S., Bason, M., Lammintausta, K., and Maibach, H. I. (1996). Irritant dermatitis (irritation). In *Dermatotoxicology* (F.N. Marzulli and H.I. Maibach, Eds.), fifth ed. Taylor & Francis, Washington, D.C., pp. 87–118.

Wigger-Alberti, W., Krebs, A., and Elsner, P. (2000). Experimental irritant contact dermatitis due to cumulative epicutaneous exposure to sodium lauryl sulphate and toluene: Single and concurrent application. *Br. J. Dermatol.,* 143(3), 551–556.

Wintermantel, T.M., Bock, D., Fleig, V., Greiner, E.F., and Schutz, G. (2005). The epithelial glucocorticoid receptor is required for the normal timing of cell proliferation during mammary lobuloalveolar development but is dispensable for milk production. *Mol. Endocrinol.,* 19(2), 340–349.

Wong, S.S., Vargas, J., Thomas, A., Fastje, C., McLaughlin, M., Camponovo, R., Lantz, R.C., Heys, J., and Witten, M.L. (2008). In vivo comparison of epithelial responses for S-8 versus JP-8 jet fuels below permissible exposure limit. *Toxicology,* 254(1-2), 106–111.

Yang, J.H., Lee, C.H., Monteiro-Riviere, N.A., Riviere, J.E., Tsang, C.L., and Chou, C.C. (2006). Toxicity of jet fuel aliphatic and aromatic hydrocarbon mixtures on human epidermal keratinocytes: evaluation based on in vitro cytotoxicity and interleukin-8 release. *Arch. Toxicol.,* 80(8), 508–523.

Zeiger, E., and Smith, L. (1998). The First International Conference on the Environmental Health and Safety of Jet Fuel. *Environ. Health Perspect.,* 106(11), 763–764.

10 The Effects of Aerosolized JP-8 Jet Fuel Exposure on the Immune System

A Review

David T. Harris

CONTENTS

OVERVIEW

Chronic jet fuel exposure could be detrimental to the health and well-being of exposed personnel, adversely affect their work performance, and predispose these individuals to increased incidences of infectious disease, cancer, and autoimmune disorders. Chronic exposure to jet fuel has been shown to cause liver dysfunction, emotional dysfunction, abnormal electroencephalograms, shortened attention spans, and decreased sensorimotor speed in exposed human subjects. Recently, several cancer clusters have been identified in proximity to military bases that use jet fuel, and have led to speculation that human exposures to jet fuel may directly participate in the development of malignancy. Currently, there are no mandatory standards for personnel exposure to jet fuels of any kind, let alone JP-8 jet fuel. Kerosene-based petroleum distillates (like JP-8) have been associated with hepatic, renal, neurological, and pulmonary toxicity in animal models and in human occupational exposures. The U.S. Department of Labor, Bureau of Labor Statistics, estimates that over 1.3 million workers were exposed to jet fuels in 1992. Thus, jet fuel exposure may have serious consequences for not only military personnel, but also for a significant number of civilian workers. Short-term (7 days) JP-8 jet fuel exposure has been shown to cause lung injury, as evidenced by increased pulmonary resistance, a decrease in bronchoalveolar lavage concentrations of Substance P (SP), increased lung/body wet weight ratios, and increased alveolar permeability. Long-term exposures, although demonstrating evidence of lung recovery, result in injury to secondary organs such as liver, kidneys, and spleen. The purpose of this review is to summarize the effects on the immune system that have been observed when animal and human subjects have been exposed to aerosolized jet fuels. A discussion of potential real-world consequences of such exposure and effects is discussed, as well as methods by which to ameliorate the effects of jet fuel exposure. Due to the observed immune effects, exposure of individuals to JP-8 (as well as any other) jet fuel may result in an increased risk of infectious disease and cancer. However, it may be possible to reverse or prevent many of these effects through the administration of SP. It is absolutely critical to ascertain and understand the potential consequences of immune function alterations as they pertain to the short-term and long-term health and well-being of exposed personnel, as well as their treatment before and after exposure.

A summary of the known effects of aerosolized JP-8 jet fuel exposure on the immune system is shown in Table 10.1.

INTRODUCTION

Exposure to environmental toxicants may have significant effects on the immune system. Significant changes in immune competence, even if short-lived, may have serious consequences for the exposed host that may affect susceptibility to infectious agents. Further, major alterations in immune function that are long lasting may result in an increased likelihood of the development or progression of cancer and infectious disease. Kerosene-based petroleum distillates such as jet fuel have been associated with hepatic, renal, neurological, and pulmonary toxicity in both animal models and human occupational exposures (Struwe et al., 1983; Gaworski et al., 1984; Dossing

TABLE 10.1
Summary of the Effects of Aerosolized JP-8 Exposure on the Immune System

Exposure to aerosolized JP-8 jet fuel adversely affects the immune system (mice) as evidenced by:

- JP8 exposure for 1 hour/day for 7 days is not overtly toxic to animals or tissues.
- Decreased immune organ weights
- Decreased T cell proliferation (function)
- Decreased NK cell, CTL and Th cell numbers and function
- Increased tumor incidences
- In utero effects (loss of male offspring and immune suppression)
- Increased morbidity after viral infections
- Doses as low as 100 mg/m^3 caused immune effects (thymus)
- JP-8 jet fuel exposure induced the secretion of immunosuppressive agents PGE_2 and IL-10 (mice and humans)
- Administration of Substance P protected/reversed the effects of JP-8 on the immune system (pM range).
- Similar immunotoxic effects were observed after exposure to either JP-8+100 or Jet A-1 jet fuels.
- S8 synthetic jet fuel exposure resulted in decreased immune organ weights, loss of viable immune cells, and decreased immune function

et al, 1985; Witten et al, 1990; Mattie et al, 1991; Chen et al., 1992; Witten et al., 1992; Pfaff et al., 1993, 1995). Chronic exposure to jet fuel has been shown to cause human liver dysfunction, emotional dysfunction, abnormal electroencephalograms, shortened attention spans, decreased sensorimotor speed (Knave et al., 1978; Struwe et al., 1983; Dossing et al., 1985), and elevations of immunosuppressive serum cytokines (Harris et al., 2007a). Furthermore, JP-8 exposure (at least in vitro) has been demonstrated to cause genotoxicity in both human liver and lymphocyte cultures (Grant et al., 2001; Jackman et al., 2002). These results are potentially worrisome with regard to the consequences for mutagenesis. JP-8 jet fuel exposure is known to cause lung injury, as evidenced by increased pulmonary resistance, decreased bronchoalveolar lavage concentrations of SP, increased lung/body wet weight ratios, and increased alveolar permeability (Hays et al., 1995; Pfaff et al., 1995). These results were observed with short-term exposures (7 days). Long-term exposures, although demonstrating some evidence of lung recovery, resulted in injury to secondary organs such as liver, kidney, and spleen (Pfaff et al., 1993; Hays et al., 1995; Pfaff et al., 1995). JP-8 jet fuel exposure at concentrations of 800 mg/m^3 resulted in lung injury as seen in red blood cell infiltration into the alveolar air spaces, distortion of the bronchial airways, and alterations of the type II alveolar epithelial cells (Pfaff et al., 1993; Hays et al., 1995; Pfaff et al., 1995). Both military and commercial aircraft workers are at risk of inhalation toxicity from jet fuels, as aircraft refueling results in the production of significant amounts of aerosolized jet fuel, despite vapor ventilation (Hays et al., 1995). The U.S. Department of Labor, Bureau of Labor Statistics, estimates that over 1.3 million workers were exposed to jet fuels in 1992. Thus, jet fuel exposure may not only have serious consequences for USAF personnel, but also may have potential harmful effects upon a significant number of civilian workers.

Currently, there are no absolute standards for military or private-industry personnel exposure to jet fuels, although the permissible exposure level (PEL) as recommended by the National Research Council is 350 mg/m^3 and the recommended short-term exposure limit is 1,800 mg/m^3 (Herrin et al., 2006).

Prior to widespread use, standard toxicological studies found that JP-8 jet fuel exposure resulted in minimal adverse effects in animals (Mattie et al, 1995; Cooper and Mattie, 1996). However, upon implementation there were numerous reports received of adverse health problems (e.g., skin and lung). Upon reexamination, immunological problems were found when animals were exposed either by gavage (Keil et al., 2004) or dermally exposed to JP-8 jet fuel (Ulrich and Lyons, 2000; Ramos et al., 2002, 2004). A suppressed immune system, particularly if the immunosuppression is long-lived, has been correlated with increased susceptibility to and severity of infectious diseases in both animals and people (Cohen et. al., 1982; Dunn, 1990). Due to the decision by the U.S. Air Force to implement the widespread use of JP-8 jet fuel in its operations, a thorough understanding of the mechanisms of its effects upon exposed personnel is critical and necessary, as well as potential therapies to prevent or counteract these effects.

Until recently (Harris et al., 1997a, 1997b, 1997c; Ulrich, 1999; Harris et al., 2000a, 2000b, 2000c; Ramos et al, 2002; Harris et al., 2003; Harris and Witten, 2003), there had been a paucity of data concerning the potential immunotoxicological effects of JP-8 jet fuel exposure. The results obtained in the studies from the Harris laboratory (Harris et al., 1997a, 1997b, 1997c; Harris et al., 2000a, 2000b, 2000c; Harris et al., 2003; Harris and Witten, 2003) and those of Ullrich et al. (Ullrich, 1999; Ramos et al., 2002) of the immunotoxicological effects of JP-8 exposure have shown that low-concentration, short-term exposure to JP-8 jet fuel results in a profound and significant alteration in immune function. If JP-8 exposure in humans results in similar alterations in immune function (either short term or long term) as seen in animals, it could have serious consequences for the exposed personnel. If an individual is significantly immunocompromised due to JP-8 exposure, it could result in an individual who is more susceptible to infectious disease, as well as an individual who may be predisposed to the development or progression (i.e., metastatic spread) of malignancy. Such effects could have serious consequences for USAF and civilian personnel, as well as their employers. It remains unknown if the immune system damage incurred after JP-8 jet fuel exposure is also subject to the same type of adaptive process (i.e., recovery) that appears to occur in the lung. Although this adaptive process in the lung tends to diminish the detrimental effects originally observed, there appear to be other effects (e.g., fibrosis) that may be equally, if not more, detrimental (Pfaff et al., 1993; Hays et al, 1995; Pfaff et al., 1995).

A portion of the effects of JP-8 exposure on the lung may be due, in part, to the depletion of SP from the lungs (De Sanctis et al., 1990; Witten et al., 1992), as hypothesized by Witten et al. Treatment of rats with capsaicin, an SP blocker, significantly acerbated the effects of JP-8 exposure (De Sanctis et al., 1990; Witten et al., 1992). Thus, SP may be instrumental in maintaining the integrity of the alveolar epithelium. SP is an 11-amino-acid peptide that is localized to the nerves in the airways of several species, including humans (Lundberg et al., 1984, 1987). SP preferentially activates NK-1 tachykinin receptors (Martling, 1987). When SP is administered in vivo by infusion or inhalation, it does not induce bronchoconstriction, in contrast to other tachykinins (Fuller et al., 1987). The effects of tachykinins (and possibly SP) are modulated by the lung epithelium. When the epithelium is damaged or missing, the effects of these modulators increase (Frossard et al., 1989). Although SP does not normally appear to cause bronchoconstriction, it does appear to be involved in mucus hypersecretion, vasodilation, and microvascular leakage (Belvisi et al., 1988).

The exposure procedure for the proposed experiments described in this chapter makes use of a simulated flightline exposure protocol to study the effects of aerosolized JP-8 jet fuel inhalation (Hays et al., 1995; Pfaff et al., 1995; Herrin et al., 2006). Animal exposure is performed via nose-only presentation while the animals are held in individual subject loading tubes. The tubes are nose-cone fitted to receiving adapters that originate from a common anodized aluminum exposure chamber. Nose-only exposure is also utilized to minimize ingestion of jet fuel during self-grooming. Animals are rotated on a daily basis through the 12 adapter positions on the exposure chamber. This procedure is done to minimize proximity to the jet fuel source as a variable in exposure concentration or composition. Exposure concentration is determined by a seven-stage cascade impactor, and is measured after each exposure (Hays et al., 1995; Pfaff et al., 1995). Mice are exposed for varying periods of time (i.e., hours to days). Previous experiments have demonstrated that exposure concentration during the 1-hour exposure period is constant, and that the exposure over the daily exposure period varies less than 10%, as measured by the SEM of the exposure concentration. In this fashion it is possible to obtain data that are relevant to the exposures incurred by USAF personnel. In addition, during the course of these studies, a real-time, in-line exposure system was developed that was capable of measuring both aerosol and vapor concentrations of jet fuel, as compared to the previous system that only assessed aerosol concentrations of jet fuel. Comparisons between the two exposure systems, as well as the effects of jet fuel source, will be discussed.

REVIEW OF THE EFFECTS ON THE IMMUNE SYSTEM

General Effects

My laboratory has observed that short-term (7 days) exposure of C57BL6 mice to low concentrations (100 to 1,000 mg/m^3) of aerosolized JP-8 jet fuel resulted in profound and significant alterations in the immune system. It was observed that even at exposure concentrations as low as 100 mg/m^3, detrimental effects on the immune system occurred (particularly in the thymus). Spleen and thymus organ weights, as well as total cell numbers recovered from each of the major immune system organs (spleen, thymus, lymph nodes, bone marrow, and peripheral blood), were significantly reduced (by as much as 50%). Further analyses revealed that T cell populations were lost and replaced by significant increases in inflammatory and B cell populations in these immune organs. Most worrisome was the observation that JP-8 exposure resulted in a significant depression of immune function. The suppression of immune function could not be overcome by the addition of exogenous growth factors (e.g., IL-2) known to stimulate immune function.

The detrimental effects of JP-8 exposure on the immune system were measurable after only a single day of (i.e., a single 1-hour/day) exposure. That is, it did not require the complete 7 days of exposure to observe significant changes in the immune system. Additional studies revealed that multiple toxicological effects on the immune system could be observed with as little as 1 hour of JP-8 exposure. That is, the damage to the immune system occurred rapidly and was cumulative in nature. Mice exposed to the full 7 days of JP-8 jet fuel exposure displayed a delayed ability to recover from the observed immune alterations. Overall, significant immunotoxicity occurred after a single 1-hour, 1,000-mg/m^3 JP-8 exposure and worsened over the next 6 days of 1-hour JP-8 exposure. Immune organ (thymus and spleen) weight loss occurred over the first 8 hours of exposures. Immune cell losses also occurred immediately after JP-8 exposure, although immune cells found in the bone marrow and peripheral blood did recover somewhat. In fact, the immediate effects of jet fuel exposure on peripheral blood immune cell numbers may be the best surrogate indicator for potential immunotoxicological damage from JP-8 exposure. Significantly, recovery of immune function was the most profoundly affected, taking more than 1 month to recover to normal levels. Immune function was also suppressed immediately (with a single, 1-hour exposure) after jet fuel exposure, was consistently suppressed during several days of exposure, and then continued to decrease. Immune function was not observed to recover during the exposure period. Based on these results, it would be expected that spleen and thymus immune cell numbers would eventually recover, but it would take several weeks to occur. Of interest were the observations that although peripheral blood immune cells were rapidly affected by JP-8 exposure, these cells recovered within the week, and that bone marrow cells were the most resistant to the effects of JP-8 exposure and the quickest to recover afterward. The finding that the levels of bone marrow cells overshot baseline levels may indicate a compensation mechanism of the immune system to the changes induced by JP-8. A similar compensatory tendency was seen with the spleen and peripheral blood, although not to the same extent. When animals were exposed to higher concentrations of JP-8 (e.g., 2,500 mg/m^3), the immune dysfunction persisted for the entire follow-up period of 4 weeks and did not reach 30% of normal values at this time. It is currently unknown how long it would take immune function to recover to normal levels, or even if full recovery is possible.

Thus, short-term, low-concentration exposures to JP-8 jet fuel resulted in long-lasting immune alterations. It appears that the immune system may be the most sensitive indicator of toxicological damage due to JP-8 exposure, as effects were seen at concentrations of jet fuel that did not evidence change in other biological systems (e.g., the lung).

Effects on Cellular Immunity

An in-depth analysis of the effects of JP-8 exposure on specific components of the cellular immune system was performed. Short-term (7 days, 1 hour/day), low-concentration (1,000 mg/m^3) jet fuel exposures were conducted in mice, and T cell and natural killer (NK) cell functions were analyzed 24 hours after the last exposure. The exposure regimen was found to almost completely ablate NK cell function, as well as significantly suppress the generation of lymphokine-activated killer (LAK) cell activity. Furthermore, JP-8 exposure suppressed the generation of cytotoxic T lymphocyte (CTL) cells from precursor T cells, and inhibited helper T cell activity. These findings demonstrated that JP-8 jet fuel exposure had significant detrimental effects on immune functions of exposed animals, and that chronic exposure to JP-8 jet fuel may have serious implications to the long-term health of exposed individuals with particular emphasis on viral infections and development of malignancy.

IL-10 and Prostaglandin E_2 Secretion

The above findings raised the following questions: How did JP-8-induce immunotoxicity that occurred so quickly and persisted for such extended periods of time? While investigating the mechanisms of JP-8's effects on the immune system, it was discovered that JP-8 exposure rapidly induced high levels of IL-10 and PGE_2 in the serum of exposed mice (Harris et al., 2007a). These immunosuppressive cytokines appear to be a contributory factor for the observed loss of immune function, but cannot entirely explain such findings, as prevention of PGE_2 increases (with COX-2 inhibitors) in vivo did not completely prevent immune dysfunction. Our findings were consistent with the observations reported by Ulrich and co-workers (Ullrich, 1999; Ullrich and Lyons, 2000; Ramos et al., 2002; Ramos et al., 2004), who found that dermal application of JP-8 resulted in suppression of immune function, with accompanying increases in serum PGE_2 and IL-10.

Both PGE_2 and IL-10 are known to have immunosuppressive actions (Ullrich and Lyons, 2000; Ramos et al., 2002) capable of inhibiting primary and secondary immune responses. Further, the production of these soluble immune agents was both temporally and functionally correlated with the loss (or recovery) of immune function in this model system. Aerosolized jet fuel exposure induced a rapid and persistent elevation in both PGE_2 and IL-10 serum levels, although the kinetics of the responses were somewhat different. IL-10 levels reached maximal levels within hours, decreased, and then remained persistently high for a period of at least 1 week. PGE_2 levels, however, did not reach maximal levels until after 4 days of exposure, declined thereafter, but remained significantly elevated. Significant increases in both responses could be observed within 1 to 2 hours following jet fuel exposure. When the decline in immune function was analyzed against the increasing levels of the immunosuppressive cytokines, it appeared that early increases in IL-10 levels might explain the rapid loss of immune function seen after jet fuel exposure, while the later rise in PGE_2 levels might explain the later losses in immune competence. Persistently elevated levels of both cytokines could be hypothesized to explain the persistent effects of exposure on the immune system.

Interestingly, treatment of exposed mice with a PGE_2 inhibitor (COX-2 inhibitor) was able to partially reverse the effects of exposure on immune function, but had no significant effects seen in terms of losses of immune organ weights and immune cell losses. Although PGE_2 levels did not return to baseline after treatment, the levels were such that immune function should have been comparable to control animals. PGE_2 secretion is known to be a function of macrophage stimulation (Ohnishi et al., 1991; Fadok et al., 1998), which in this setting seems logical in that the primary site of jet fuel exposure is in the lung, which contains a large number of macrophages (Chen et al., 1992). IL-10, however, is thought to be secreted by Th2 cells (Mosmann and Moore, 1991), B cells (O'Garra et al., 1992), and inflammatory macrophages, which are induced by JP-8 exposure. And neither of these cytokines is known to mediate the other effects attributable to JP-8 exposure—namely, the loss of immune organ weight and loss of immune cells. Therefore, increased levels of the immunosuppressive cytokines PGE_2 and IL-10 may explain some (but not all) of the effects of aerosolized jet fuel on the immune system; that is, there may be multiple mechanisms of action from JP-8 exposure—one for the effects on immune function, a different one for effects on immune organs and immune cells, and other mechanisms of action related to effects on the pulmonary and other body (neurological, skin, lung) systems (Witten et al., 1990, 1992; Chen et al, 1992; Pfaff et al., 1993; Hays et al., 1995; Pfaff et al., 1995).

New Exposure System Effects

Similar results have now been obtained when using a novel jet fuel exposure system based on real-time, in-line analysis of both aerosol and vapor-phase JP-8 concentrations during the exposure protocol (Wang et al., 2008). The new jet fuel exposure chamber consisted of a vapor/aerosol mixture that was generated using a Lovelace jet nebulizer (Model 01-100, In-Tox, Albuquerque, NM). Jet fuel vapor and aerosol concentrations were measured using an in-line, real-time total hydrocarbon (THC) analysis system (VIG Industries, Anaheim, CA). Measurements made using the THC system were more reliable than previous systems using the cascade impactor approach. Animals were exposed nose-only as before, and analyses revealed that aerosolized jet fuel made up 5% to 15% of the total hydrocarbon exposure, with vapor-phase JP-8 being the predominant constituent (Herrin et al., 2006).

Experiments were performed exposing mice to 1,000; 2,000; 4,000; and 8,000 mg/m^3 of JP-8 jet fuel for 1 hour/day for 7 days, using the new exposure apparatus with in-line monitoring of jet fuel concentrations (both aerosol and vapor). Comparisons were made to historical data obtained with the previous exposure system that only monitored aerosol concentrations of jet fuel during the exposure. The new exposure apparatus was found to produce aerosol concentrations that were approximately one eighth that of the previous exposure apparatus; that is, a total exposure to 1,000 mg/m^3 of jet fuel with the "new" apparatus was equivalent to an exposure concentration of approximately 125 mg/m^3 of aerosolized JP-8 with the "old" exposure apparatus.

However, similar immunotoxicological effects were observed regardless of the exposure system utilized. That is, regardless of exposure dose (1,000; 2,000; 4,000; or 8,000 mg/m^3 for 1 hour/day for 7 days using the new exposure apparatus), the immune cell viabilities for all immune organs (spleen, thymus, blood, and bone marrow) were between 90 and 100%, thus indicating that the exposures were not overtly toxic. However, spleen and thymus organ weights decreased (between 40 and 60%); immune cell numbers in bone marrow, thymus, spleen and peripheral blood decreased (between 20 and 70%), there were increased inflammatory cells observed; and most significantly, a decreased immune function was observed (between 40 and 50%). These results were reminiscent of those observed with JP-8 exposure using the previous exposure system. Thus, the detrimental effects of jet fuel exposure on the immune system may be correlated directly to total aerosol JP-8 concentration, regardless of total (aerosol and vapor) JP-8 concentration during the exposure period.

The Effects of Jet Fuel Source

The observed effects on the immune system was not limited to JP-8 jet fuel exposure, as similar effects were induced by exposure to JP-8+100, Jet A-1, and a synthetic jet fuel source. JP-8+100 jet fuel differs from JP-8 jet fuel in the addition of a package of fuel additives (the +100, which contains an antioxidant, a metal deactivator [chelating agent], a detergent, and a dispersant) designed to improve thermal stability and prevent engine coking, fouling, and sooting. Commercial Jet A-1 differs from JP-8 jet fuel in the composition of the major hydrocarbons that comprise the fuels. JP-8 contains primarily hydrocarbons of 14-, 15-, and 16-carbon chain lengths whereas Jet A-1 contains primarily 11-, 12-, and 13-carbon chain length hydrocarbons (White, 1999). Further, JP-8 differs from Jet A-1 in the addition of a deicer, an antistatic agent, an anticorrosive agent, and a lubricity agent. Significantly, exposure to any of the aerosolized fuel sources resulted in equivalent effects on the immune system of the exposed animals. That is, there was significant loss of immune organ weights, significant loss of viable immune cells recovered from the immune organs, a significant loss of the predominant cell population in the thymus responsible for the production of mature peripheral blood T cells, and a significant suppression of immune function. JP-8+100 and Jet A-1 jet fuels were equivalent to JP-8 jet fuel in terms of immunotoxicity. These results were significant in that it appears that civilian commercial aircraft personnel are at equivalent risk as military aircraft personnel. It does not appear that the addition of the fuel package to JP-8 jet fuel affects its toxicity. Neither does it appear that the long-chain hydrocarbon composition significantly determines or affects its immune toxicity. These results may imply that aerosolized hydrocarbon exposure in general has a deleterious effect on the immune system and should be minimized if not avoided. If the effects of JP-8+100 and Jet A-1 jet fuel exposure are as long-lived as that of JP-8 jet fuel exposure, then there may be considerable risk for a large population of workers from occupational jet fuel exposures. These exposures could possibly result in more total sick days, an increased frequency of workers becoming sick, and exposed personnel taking longer to recover.

The jet fuel termed S-8 is a synthetic fuel composed solely of synthetic aliphatic alkanes. It was constructed so as to eliminate the 18% polycyclic aromatic hydrocarbons found in JP-8 jet fuel as an attempt to determine if this polycyclic aromatic component was responsible for the observed toxicological effects observed in various physiological systems. JP-8 is normally composed of 81% aliphatic alkanes and respective isomers. We found that animals exposed to the S8 jet fuel at doses ranging from 1,000 to 8,000 mg/m^3 for 1 hour/day for 7 days showed no overt signs of toxicity at either the organismal or tissue level. S-8 exposure resulted in decreased immune organ weights, a loss of viable immune cells from all immune organs, and decreased immune function as assessed by the percentage of cytokine-secreting T cells. S-8 exposure had the greatest phenotypic effects on bone marrow B cells, thymic T cells,

TABLE 10.2
Preliminary Molecular Analysis of JP-8 Effects on the Immune System

Common Name	Gene Description	Fold Up-Regulated
Anxa11	Annexin A11	3–5 times
Casp8ap2	Caspase 8 associated protein 2	2–3 times
Cd14	Monocyte differentiation antigen	5 times
Rpl29	60S ribosomal protein L29	3–5 times
Rps24	40S ribosomal protein S24	3–5 times
Rpl18	60S ribosomal protein L18	2–3 times
Rps16	40S ribosomal protein S16	2–3 times
Rpl18A	60S ribosomal protein L18A	2–3 times

and peripheral blood T cells. Thus, the S-8 and JP-8 jet fuels seem comparable in terms of effects on the immune system (Wang et al., manuscript in preparation).

Genomic Analyses

Preliminary experiments have been performed in an attempt to ascertain the molecular effects of jet fuel exposure on the immune system of mice. In this fashion it is hoped that an elucidation of the mechanisms of action of JP-8 exposure can be obtained. It was observed that in vivo exposure of mice to JP-8 (1,000 mg/m^3) caused changes in gene expression profiles in the spleen (as determined by microarray analyses) less than 0.5 hours following exposure. With increasing time after a single exposure, the spleen cells responded with changes in expression in a variety of genes, including genes for cell growth and maintenance, catalytic activity, binding, and cell communication. For example, spleen microarray analyses revealed that some apoptotic and ribsomal protein genes were up-regulated at 1 hour post exposure. Pertinent examples of gene expression changes are shown in Table 10.2.

REMEDIAL EFFECTS OF SUBSTANCE P THERAPY

General Effects

Previous studies have shown that JP-8-induced pulmonary dysfunction was associated with a decrease in levels of the neuropeptide SP in lung lavage fluids (De Sanctis et al., 1990; Witten et al., 1992). It was found that administration of aerosolized SP was able to protect exposed animals from such JP-8-induced pulmonary changes. Treatment of rats with capsaicin, an SP blocker, significantly acerbated the effects of JP-8 exposure (De Sanctis et al., 1990; Witten et al., 1992). Interestingly, treatment of mice with the neuropeptide SP immediately after JP-8 exposure (1 μM, 15 minutes) was able to reverse or prevent many of the observed immune effects as well. That is, SP administration was able to protect or reverse the effects of JP-8 jet fuel exposure on immune organ weights, immune organ numbers, and immune function (Harris et al., 1997c; Harris et al., 2000c). It was observed that SP administration could protect JP-8 exposed animals from losses of viable immune cell numbers, but not losses in immune organ weights. SP appeared to act on all immune cell populations equally as analyzed by flow cytometry, as no one immune cell population appeared to be preferentially protected by SP. Also, SP administration was capable of protecting JP-8-exposed animals from loss of immune function at all concentrations of JP-8 utilized (250 to 2,500 mg/m^3). Significantly, SP only needed to be administered for 15 minutes after JP-8 exposure, and was active at both 1 μM and 1 nM concentrations. The role of SP in this process was confirmed by the finding that administration of SP inhibitors worsened the effects of JP-8 exposure in the immune system. SP may be instrumental in maintaining the integrity of the alveolar epithelium; it is an 11 amino acid peptide that is localized to the nerves in the airways of several species, including humans (Lundberg et al., 1984, 1987). SP preferentially activates the NK-1 tachykinin receptors (Martling, 1987). Although SP does not normally appear to cause bronchoconstriction, it does appear to be involved in mucus hypersecretion, vasodilation, and microvascular leakage (Belvisi et al., 1988). Hypothetically, JP-8 exposure may cause a loss of epithelial integrity that, in combination with SP depletion, results in the observed lung injury. How these observations are related to the observed immune system damage is unclear at this point, as it is not known whether JP-8 acts directly or indirectly on the immune system. It should be noted that although there is considerable published information regarding the effects of SP on the immune system, its role as it relates to administration by inhalation (Marriott and Bost, 2001) and in immunotoxicity is unexplored. It is known, however, that neuropeptides of various kinds do indeed have the ability to modulate the immune system (Marriott and Bost, 2001). SP fibers are known to innervate immune organs such as the lymph nodes (Felten et al., 1992); and that T, B, macrophage, and mast cells express receptors for SP (Stanisz et al., 1987). Administration of SP results in lymphocyte proliferation (Payan et al., 1983; Payan et al., 1984; Scicchitano et al., 1988) and activates macrophages (Stanisz et al., 1987). In vivo, SP increases mucosal synthesis of IgA, is elevated in chronically inflamed tissues, and even enhances inflammatory responses (Stanisz et al., 1987; Scicchitano et al., 1988; Bienenstock et al., 1989). Our results have indicated that SP has a beneficial effect in reversing or protecting the immune system from the detrimental effects of JP-8 exposure, a significant observation that may be able to be transitioned to the clinical setting.

Pre or Post Exposure Effects

Additional SP experimentation revealed that it was possible to wait for 1 or 6 hours post-JP-8 exposure to administer SP and reverse the effects seen in the immune system (Harris et al., 2000c). Furthermore, administration of SP 15 minutes prior to JP-8 exposure could prevent the JP-8-induced immunotoxicity. Administration of SP either 1 or 6 hours prior to JP-8 exposure was not effective. Thus, it appeared that SP required time for its protective actions to take effect; that is, if one pretreated animals with SP for 15 minutes before JP-8 exposure, it was ineffective. However, both 1- and a 6-hour pre JP-8 exposure treatment were effective in preventing immune organ weight loss and viable immune cell loss. These results might imply that SP exerted its effects indirectly (rather than directly) through the production of some metabolite or secretion of some factor, and that its effects on the immune system could occur long range. There was a time limit to the protective effects of SP in that although a 1- and a 6-hour pretreatment were sufficient to prevent organ weight loss and immune cell loss, it was not sufficient to prevent immune dysfunction. Only SP given 15 minutes before JP-8 exposure was able to prevent immune dysfunction, although it had limited effects on the other two parameters measured. This result implied that SP may act in two ways: (1) an immediate effect that is short-lived and protects competent immune function, and (2) a long- lived (or an effect requiring longer time to produce or become active) effect that protects against organ weight and cell loss. Treatment of exposed animals either 15 minutes or 1 hour post JP-8 exposure, but not 6 hours post exposure, was sufficient to reverse any spleen or thymus organ weight loss. This effect could be explained by recruitment of immune cells during the early treatment time-periods or clonal expansion, or it could be due to loss or emigration of cells at later time periods. It is interesting to note that SP administration did not alter any particular immune cell population—whether administered pre or post JP-8 exposure. Thus, the effects of SP on the restoration of immune organ weights, immune cell numbers, and immune function cannot be attributed to the recovery or expansion of one particular population of cells after SP administration. Similar results were found in terms of the effects of SP post JP-8 treatment on cell loss (except for the spleen). A 6-hour SP post JP-8 treatment did indeed prevent spleen cell, but not thymus cell, loss. These findings may indicate that it takes longer for effects to be seen in the spleen, that the mechanisms are different, or that the active factors are longer lived in the spleen. In terms of immune function, all of the SP post JP-8 exposure treatments were quite effective, although the 1-hour time point was the most effective. These results may imply that SP could counteract or prevent the production of a JP-8-induced factor that damages immune function, and that this factor either takes time to produce or requires time to exert its effects. SP either might nullify the factor in some fashion or might act to make immune cells less susceptible to the factor.

Dose Responses

Finally, it is worth noting the observation that a 1-nM concentration of SP was almost as effective in protecting the immune system from JP-8 exposure as a 1-μm concentration, in that it implied that this mediator had potent effects that might be easily transitioned from in vitro to in vivo applications. It is feasible to speculate that doses such as the ones described could easily be administered via a hand-held nebulizer or ''puffer- bottle.'' In that fashion, it might be possible to provide at-risk personnel with the means to protect themselves against the effects of JP-8 exposure.

POTENTIAL REAL-WORLD CONSEQUENCES

A series of studies were conducted in an effort to correlate the observed immunotoxicological effects of aerosolized JP-8 jet fuel exposure in animals with what might be expected to occur in a "real-life" setting for exposed personnel. Therefore, the effects of jet fuel exposure on fetal development, on susceptibility to viral infections, and on the development of malignancy were analyzed.

In Utero Exposures

As described above, work from this laboratory and other groups had indicated that exposure to JP-8 jet fuel, even at relatively low concentrations for short periods of time, was detrimental to the immune system (Harris et al., 1997a, 1997b; Harris et al., 2000a; Ullrich and Lyons, 2000; Ramos et al., 2003, 2004). At least a portion of these effects was attributed to the increased secretion of prostaglandin E_2 in exposed animals (Ullrich and Lyons, 2000; Harris et al., 2007). As increasing numbers of women are employed in the military, and as prostaglandins are known to adversely affect the maintenance of pregnancy (Jain and Mishell, 1994), we determined the effects of JP-8 exposure on pregnant animals. Two different scenarios were utilized. In the first scenario mice were exposed starting during the first trimester of pregnancy to simulate female employees who might be exposed to jet fuel during their pregnancy. The first trimester of pregnancy is generally when the greatest growth of the fetus occurs and where adverse effects of jet fuel exposure might be most easily observed. In the second instance, the mice were exposed to JP-8 starting during the last trimester of pregnancy to simulate inadvertent exposure of a pregnant employee or a worker that continued in her job regardless of the concerns of exposure. The last trimester of pregnancy is generally thought to be less susceptible to environmental influences, and thus fewer adverse effects might be observed. Regardless of when exposure occurred, jet fuel exposure began post-implantation, during the fetal period of gestation, and stopped at birth. Thus, any effects observed in the offspring could not be attributed to events occurring prior to implantation and development of the embryo, nor to post-natal effects of JP-8 exposure. It was observed that JP-8 exposure of the mothers during pregnancy resulted in a long lasting immunosuppression in the mothers, even long after delivery of the offspring (Harris et al., 2008). It is known that pregnancy itself results in some immunosuppression in the pregnant animal (Harris et al., 1994), and is thought to be required for maintenance of the haploidentical fetus. Pregnancy-induced immunosuppression is generally antigen-specific in nature (directed towards the tissue transplantation antigens of the fetus, possibly as a result of transplacental crossover of fetal cells into the maternal circulation; Bianchi et al., 1996), rather than global as observed in the current study. However, the level of immunosuppression observed was greater than that seen during and immediately after pregnancy, and persisted for a longer period of time. It might be of concern that such immune compromise could predispose the pregnant individual to infectious complications both during and after pregnancy, both in terms of the mother and the offspring. Similarly, Kiel et al. (2004) have previously reported that JP-8 exposure by the gavage route in mice also resulted in immune changes. In utero exposure by daily gavage of the dam from day 6 to 15 of gestation also resulted in immune changes in the newborn mice (Kiel et al., 2003). Taken together, these studies seem to indicate that the developing immune system of the fetus is particularly sensitive to the effects of in utero jet fuel exposure. The most significant and troublesome findings were those involving the newborn animals. In utero JP-8 jet fuel exposure, regardless of when during the pregnancy the exposure occurred, significantly suppressed immune organ weights, immune organ cell numbers, and immune function—even 6 to 8 weeks post birth. Even more significantly, in utero exposure to JP-8 resulted in fewer live births from the exposed mothers, and fewer viable male pups being born. Thus, JP-8 exposure exerted a strong epigenetic effect on the development of the future immune system of the unborn animals (Harris et al., 2008). However, there was also a genetic component involved as the adverse effects of JP-8 exposure were correlated to the level of adverse effects exhibited by the JP-8-exposed mothers. It has generally been accepted that male fetuses are more fragile during pregnancy (Astolfi and Zonta, 1999), but specific sex targeting by an environmental agent is unusual, particularly as the effects on the male fetuses are exhibited both early and late during pregnancy. Currently, it is not known if the immunosuppression seen in the surviving pups is permanent or whether immune dysfunction (e.g., atopy) might result as the animals age. Thus, it seems reasonable to recommend that females of reproductive age (whether pregnant or not) avoid exposure to JP-8 (and potentially other types of jet fuel; see Harris et al., 2000c), both in terms of dermal and aerosol exposures. Once exposed to JP-8, the jet fuel, being a hydrocarbon, is prone to depot in fatty tissues that tend to accumulate during pregnancy. Thus, past exposures to jet fuel may pose a hazard to the developing fetus, even many years after the last maternal exposure.

Influenza

As described in detail above, JP-8 aerosol exposure results in significant rapid and cumulative immunotoxicity (Harris et al., 1997a,b,c) that is long lasting and slow to recover. Further studies revealed that aerosolized JP-8 jet fuel exposure resulted in rapid and sustained secretion of the immunosuppressive factors IL-10 and prostaglandin E_2 (PGE_2) (Harris et al., 2007c). Both PGE_2 and IL-10 are known to have immunosuppressive actions (Ullrich and Lyons, 2000; Ramos et al., 2002). Because the presence of an intact and robust immune system is crucial to the control of infectious disease and, in particular, viral infections, it could be hypothesized that aerosolized JP-8 exposure would affect the severity of viral infections. We utilized a murine-adapted influenza virus model for these studies, and JP-8 jet fuel was given at doses mimicking flightline personnel exposures. The jet fuel exposures were arranged to occur 1 week prior to viral infection. Animals were then observed for signs of morbidity and mortality, and at 2 weeks post–infection, the surviving animals were sacrificed for immunological assessment. JP-8 exposure at the levels used in these studies has been shown repeatedly to be non-lethal (Witten et al., 1992; Hays et al., 1995; Harris et al., 1997a,b,c; Wong et al., 2004). It was found that JP-8 exposure worsened the outcomes of influenza infection, as evidenced by increased overall mortality of exposed and infected mice, as well as increased morbidity in surviving mice (Harris et al., 2008). These findings could be attributed to suppressed immune (cellular) responses to infection, as evidenced by decreased immune organ weights, decreased immune cell viability, decreased numbers of both helper and cytotoxic T cells, and suppressed immune function. It was significant to note that a large immune response (i.e., to influenza infection), in combination with a significant inflammatory response (i.e., jet fuel exposure), could result in significant toxicity to the immune system that might operate independently of the immunosuppressive effects of jet fuel exposure. The effects of JP-8 exposure on the immune response to viral infection would seem to be a logical extension of the immune-compromised state that results from JP-8 exposure. The data would seem to suggest that JP-8 exposure suppressed the immune responses necessary for clearance of the viral infection; impaired overall recovery of the infected animal; and, if persistent, could result in increased mortality (Harris et al., 2008). Therefore, efforts should be considered to minimize JP-8 exposure, particularly in those individuals at greater risk for exposure to infectious agents (e.g., personnel stationed in tropical environs).

Cancer

Knowing that aerosolized JP-8 exposure was immunotoxic and resulted in a rapid and sustained secretion of the immunosuppressive factors IL-10 and PGE_2 (Harris et al., 2007a), and because the presence of an intact and robust immune system is crucial to the control of malignancy, we investigated whether aerosolized JP-8 exposure would affect either the development or the spread of cancer in an animal model. The B16 melanoma tumor model was utilized for these studies. B16 tumor cells can be injected either subcutaneously to study solid tumor development, or the cells can be injected intravenously to examine metastasis. JP-8 exposure, given at doses mimicking flightline personnel exposures, were arranged to occur either at the time of tumor induction or 1 week prior to tumor induction. These situations were chosen to mimic the effects of jet fuel exposure on tumor development and tumor incidence, as well as any effects on tumor metastasis. It was observed that JP-8 exposure increased the number of lung tumor foci an average of 560% above control mice. However, prior exposure of the animals to JP-8 for 1 week before tumor inoculation resulted in an average 870% increase in tumor foci. These data indicated that JP-8 exposure increased both the incidence and metastasis of cancer in this model system (Harris et al., 2008). Exposure to jet fuel induced a rapid state of long-lasting immune compromise that accelerated tumor development and metastatic spread. When the tumor cells were injected subcutaneously to model the development of a solid tumor, an interesting difference was observed. Exposure of the mice to JP-8 at the time of tumor induction did not affect or alter the incidence, growth rate, or ultimate size of these tumors. Similar results were seen when the mice were exposed to JP-8 for 1 week prior to tumor injection. However, the JP-8-exposed mice succumbed to their disease at a faster rate than control animals. The decreased survival seemed to be attributable to increased metastatic spread of the tumor as evidenced by increased lung metastases. Thus, in this model system, JP-8 exposure does not significantly alter tumor incidence or development, but once again increases tumor metastases (Harris et al., 2008). The effects on tumor development would seem to be a logical extension of the immune-compromised state that results from JP-8 exposure. Overall, it seemed that JP-8 exposure has a more significant effect on tumor metastasis than on tumor development, which if persistent could result in increased mortality. Therefore, efforts should be considered to minimize JP-8 exposure, particularly in those individuals at greater risk for malignancy (e.g., cigarette smokers).

OTHER

In concluding this chapter, it should be noted that the effects of occupational JP-8 exposure on military personnel have been preliminarily investigated. Peripheral blood was obtained in a blinded study (performed in collaboration with Dr. T. Risby, Johns Hopkins University) from exposed military personnel (National Guard personnel), immune cells were isolated, and immune analyses performed. Extremely low-level JP-8 exposures (less than 50 mg/m^3) resulted in decreased total leukocyte numbers in the peripheral blood, increased peripheral blood neutrophils immediately after exposure, increased peripheral blood eosinophil numbers immediately after exposure, and increased plasma PGE_2 levels. These findings are similar to those in mice, which reported increased in inflammatory cells after jet fuel exposure as well as the secretion of immunosuppressive cytokines. Preliminary analyses indicated that there might be more susceptible subpopulations within the human population as a whole, again indicative of a genetic component influencing levels of susceptibility to JP-8's effects on the immune system (see the "In Utero" results described above).

ACKNOWLEDGMENTS

The work reviewed in this chapter would not have been possible without the expert technical assistance of Miss D. Sakiestewa, Miss J. Wang, and Mr. D. Titone, and the fruitful collaboration with Dr. M. Witten.

We most gratefully acknowledge the foresight, intellectual direction, and financial assistance provided by Dr. W. Kozumbo and the AFOSR.

REFERENCES

Astolfi, P., and Zonta, L.A. (1999) Reduced male births in major Italian cities. *Hum. Reproduct.,* 14(12): 3116–3119.

Belvisi, M.G., Chung, K.F., Jackson, D.M., and Barnes, P.J. (1988) Opioid modulation of non-cholinergic neural bronchoconstriction in guinea-pig in vivo. *Br. J. Pharmacol.,* 95: 413–418.

Bianchi, D.W., Zickwolf, G.K., Weil, G., Sylvester, S., and DeMaria, M.A. (1996) Male fetal progenitor cells persist in maternal blood for as long as 27 years postpartum. *Proc. Natl. Acad. Sci. USA*, 93(2): 705–708.

Bienenstock, J., Croitoru, K., Ernst, P.B., Stead, R.H., and Stanisz, A. (1989) Neuroendocrine regulation of mucosal immunity. *Immunolog. Invest.,* 18(1-4): 69–76.

Chen, H., Witten, M.L., Pfaff, J.K., Lantz, R.C., and Carter, D. (1992). JP-8 jet fuel exposure increases alveolar permeability in rats. *FASEB J.,* 6: A1064.

Cohen, J., Pinching, A.J., Rees, A.J., and Peters, D.K. (1982) Infection and immunosuppression. A study of the infective complications of 75 patients with immunologically-mediated disease. *Q. J. Med.,* 201: 1–15.

Cooper, J.R., and Mattie, D.R. (1996) Developmental toxicity of JP-8 jet fuel in the rat. *J. Appl. Toxicol.,* 16: 197–200.

De Sanctis, G.T., App, E.M., Trask, J.K., De Sanctis, B.I., Remmers, J.E., Green, F.H., Man, S.F., and King, M. (1990) Resorptive clearance and transepithelial potential difference in capsaicin-treated F344 rats. *J. Appl. Physiol.,* 68: 1826–1832.

Dossing, M., Loft, S., and Schroeder, E. (1985) Jet fuel and liver function. *Scand. J. Work Environ. Health,* 11: 433–437.

Dunn, D.L. (1990) Problems related to immunosuppression. Infection and malignancy occurring after solid organ transplantation. *Crit. Care Clin.,* 6(4): 955–977.

Fadok, V.A., Bratton, D.L., Konowal, A., Freed, P.W., Westcott, J.Y., and Henson, P.M. (1998) Macrophages that have ingested apoptotic cells in vitro inhibit proinflammatory cytokine production through autocrine/paracrine mechanisms involving TGF-β, PGE_2, and PAF. *Clin. Invest.,* 101(4): 890–898.

Felten, D.L., Felten, S.Y., Bellinger, D.L., Lorton, D. (1992) Noradrenergic and peptidergic innervation of secondary lymphoid organs: role in experimental rheumatoid arthritis. *Eur. J. Clin. Invest.,* 22(Suppl. 1): 37–41.

Frossard, N., Rhoden, K.J., and Barnes, P.J. (1989) Influence of epithelium on guinea pig airway responses to tachykinins: Role of endopeptidase and cyclooxygenase. *J. Pharmacol. Exp. Ther.,* 248: 292–298.

Fuller, R.W., Maxwell, D.L., Dixon, C.M., McGregor, G.P., Barnes, V.F., Bloom, S.R., and Barnes, P.J. (1987) Effect of Substance P on cardiovascular and respiratory function in subjects. *J. Appl. Physiol.,* 62: 1473–1479.

Gaworski, G., MacEwen, J., Vernot, E., Bruner, R., and Cowen, M. (1984) Comparison of the subchronic inhalation toxicity of petroleum and oil shale JP-5 jet fuel. In *Advances in Modern Environmental Toxicology* (H.N. MacFarland and C.E. Holdsworth, Eds.). Vol. 6. Princeton Scientific Publishers, Inc., Princeton, NJ, pp. 33–47.

Grant, G.M., Jackman, S.M., Kolanko, C.J., and Stenger, D.A. (2001) JP-8 jet fuel-induced DNA damage in H4IIE rat hepatoma cells. *Mutat. Res.,* 490: 67–75.

Harris, D.T., Schumacher, M.J., LoCascio, J., Booth, A., Bard, J., and Boyse, E.A. (1994) Immunoreactivity of umbilical cord blood and post-partum peripheral blood with regard to HLA-haploidentical transplantation. *Bone Marrow Transplant.,* 14: 63–68.

Harris, D.T., Sakiestewa, D., Robledo, R.F., and Witten, M. (1997a). Immunotoxicological effects of JP-8 jet fuel exposure. *Toxicol. Indus. Health,* 13(1): 43–55.

Harris, D.T., Sakiestewa, D., Robledo, R.F., and Witten, M. (1997b) Short-term exposure to JP-8 jet fuel results in long term immunotoxicity. *Toxicol. Indus. Health,* 13(5): 559–570.

Harris, D.T., Sakiestewa, D., Robledo, R., and Witten, M. (1997c) Protection from JP-8 jet fuel induced immunotoxicity by administration of aerosolized Substance P. *Toxicol. Indus. Health,* 13(5): 571–588.

Harris, D.T., Sakiestewa, D., Robledo, R.F., Young, R.C., and Witten, M. (2000a) Effects of short term JP-8 jet fuel exposure on cell-mediated immunity. *Toxicol. Indus. Health,* 16: 78–84.

Harris, D.T., Sakiestewa, D., Robledo, R.F., Young, R.S., and Witten, M. (2000b) Effects of short term JP-8 jet fuel exposure on cell-mediated immunity. *Toxicol. Indus. Health,* 16: 78–84.

Harris, D.T., Sakiestewa, D., Titone, D., Robledo, R.F., Young, R.S., and Witten, M. (2000c) Substance P as prophylaxis for JP-8 jet fuel-induced immunotoxicity. *Toxicol. Indus. Health,* 16: 253–259.

Harris, D.T., Sakiestewa, D., Titone, D., Robledo, R.F., Young, R.S., and Witten, M. (2000d) Jet fuel induced immunotoxicity. *Toxicol. Indus. Health,* 16: 261–265.

Harris, D.T., Sakiestewa, D., and Witten, M.L. (2003) JP-8 jet fuel exposure results in immediate immunotoxicity, which is cumulative over time. *Toxicol. Indus. Health,* 18: 77–83.

Harris, D.T., and Witten, M.L. (2003) Aerosolized Substance P protects against cigarette-induced lung damage and tumor development. *Cell. Molec. Biol.,* 49(2): 151–157.

Harris, D.T., Sakiestewa, D., Titone, D., and Witten, M. (2007a) JP-8 jet fuel exposure induces high levels of IL-10 and PGE2 secretion and is correlated with loss of immune function. *Toxicol. Indus. Health,* 23: 223–230.

Harris, D.T., Sakiestewa, D., He, X., Titone, D., Witten, M. (2007b) Effects of in utero JP-8 jet fuel exposure on the immune systems of pregnant and newborn mice. *Toxicol. Indus. Health,* 23(9): 545–552.

Harris, D.T., Sakiestewa, D., Titone, D., He, X., Hyde, J., and Witten, M. (2007c) JP-8 Jet fuel exposure potentiates tumor development in two experimental model systems. *Toxicol. Indus. Health,* 23(10): 617–623.

Harris, D.T., Sakiestewa, D., Titone, T., He, X., Hyde, J., and Witten, M. (2008) JP-8 jet fuel exposure suppresses the immune response to viral infections. *Toxicol. Indus. Health,* 24(4): 209–216.

Hays, A.M., Parliman, G., Pfaff, J.K., Lantz, R.C., Tinajero. J, Tollinger B, Hall, J.N., and Witten, M.L. (1995). Changes in lung permeability correlates with lung histology in a chronic exposure model. *Toxicol. Indus. Health,* 11(3): 325–326.

Herrin, B.R., Haley, J.E., Lantz, R.C., and Witten, M.L. (2006) A reevaluation of the threshold exposure level of inhaled JP-8 in mice. *J. Toxicol. Sci.,* 31(3): 219–228.

Jackman, S.M., Grant, G.M., Kolanko, C.J., Stenger, D.A., and Nath, J. (2002) DNA damage assessment by comet assay of human lymphocytes exposed to jet propulsion fuels. *Environ. Molec. Mutagen.,* 40: 18–23.

Jain, J.K., and Mishell, D.R. (1994) A comparison of intravaginal Misoprostol with prostaglandin E2 for termination of second-trimester pregnancy. *N. Eng. J. Med.,* 331: 290–293.

Keil, D., Warren, D.A., Jenny, M.J., EuDaly, J.G., Smythe, J., and Peden-Adams, M.M. (2003) Immunological function in mice exposed to JP-8 jet fuel in utero. *Toxicol. Sci.,* 76: 347–356.

Keil, D., Dudley, A., EuDaly, J., Dempsey, J., Butterworth, L., Gilkeson, G., and Peden-Adams, M. (2004) Immunological and hematological effects observed in B6C3F1 mice exposed to JP-8 jet fuel for 14 days. *J. Toxicol. Environ. Health A,* 67(14): 1109–1129.

Knave, B., Olson, B.A., Elofsson, S., Gamberale, F., Isaksson, A., Mindus, P., Persson, H.E., Struwe, G., Wennberg, A., and Westerholm, P. (1978) Long-term exposure to jet fuel. II. A cross-sectional epidemiologic investigation on occupationally exposed industrial workers with special reference to the nervous system. *Scand. J. Work Environ. Health,* 4: 19–45.

Lundberg, J.M., Hökfelt, T., Martling, C.R., Saria, A., and Cuello, C. (1984) Substance P-immunoreactive sensory nerves in the lower respiratory tract of various mammals including man. *Cell Tissue Res.,* 235: 251–261.

Lundberg, J.M., et al. In Kaliner, M.A. and P.J. Barnes (Eds.), *The Airways: Neural Control in Health and Disease,* Marcel Dekker (New York), p. 417, 1987.

Marriott, L., and Bost, K.L. (2001) Substance P receptor mediated macrophage responses. *Adv Exp. Med. Biol.,* 49: 247–255.

Martling, C.R., Theodorsson-Norheim, E., and Lundberg, J.M. (1987) Occurrence and effects of multiple tachykinins; Substance P, neurokinin A and neuropeptide K in human lower airways. *Life Sci.,* 40: 1633–1643.

Mattie, D.R., Alden, C.L., Newell, T.K., Gaworski, C.L., and Flemming, C.D. (1991). A 90-day continuous vapor inhalation toxicity study of JP-8 jet fuel followed by 20 or 21 months of recovery in Fischer 344 rats and C56BL/6 mice. *Toxicol. Pathol.,* 19(2): 77–87.

Mattie, D.R., Marit, G.B., Flemming, C.D., and Cooper, J.R. (1995) The effects of JP-8 jet fuel on male Sprague-Dawley rats a 90-day exposure by oral gavage. *Toxicol. Indus. Health,* 11: 423–435.

Mosmann T.R., and Moore KW. (1991) The role of IL-10 in crossregulation of TH1 and TH2 responses. *Immunol. Today,* 12(3): A49–53.

O'Garra, A., Chang, R., Go, N., Hastings, R., Haughton, G., and Howard, M. (1992) Ly-1 B (B-1) cells are the main source of B cell-derived interleukin 10. *Eur. J. Immunol.,* 22(3): 711–717.

Ohnishi, H., Lin, T.H., Nakajima, I., and Chu, T.M. (1991) Prostaglandin E2 from macrophages of murine splenocyte cultures inhibits the generation of lymphokine-activated killer cell activity. *Tumour Biol.,* 12(2): 99–110.

Payan, D.G., Brewster, D.R., and Goetzl, E.J. (1983) Specific stimulation of human T lymphocytes by Substance P. *J. Immunol.,* 131: 1613–1615.

Payan, D.G., Brewster, D.R., Missirian-Bastian, A., and Goetzl, E.J. (1984) Substance P recognition by a subset of human T lymphocytes. *J. Clin. Invest.,* 14: 1532–1539.

Pfaff, J., Parliman, G., Parton, K., Lantz, R., Chen, H., Hays, A., and Witten, M. (1993) Pathologic changes after JP-8 jet fuel inhalation in Fischer 344 rats. *FASEB J.,* 7: A408.

Pfaff, J., Parton, K., Lantz, R.C., Chen, H., Hays, A., and Witten, M.L. (1995) Inhalation exposure to JP-8 jet fuel alters pulmonary function and Substance P levels in Fischer 344 rats. *J. Appl. Toxicol.,* 15(4): 249–256.

Ramos, G., Nghiem, D.X., Wlaterscheid, J.P., and Ullrich, S.E. (2002) Dermal application of jet fuel suppresses secondary immune reactions. *Toxicol. Appl. Pharmacol.,* 180(2): 136–144.

Ramos, G., Kazimi, N., Nghiem, D.X., Walterscheid, J.P., and Ullrich, S.E. (2004) Platelet activating factor receptor binding plays a critical role in jet fuel-induced immune suppression. *Toxicol. Appl. Pharmacol.,* 195(3): 331–338.

Scicchitano, R., Biennenstock, J., and Stanisz, A.M. (1988) In vivo immunomodulation by the neuropeptide substance P. *Immunology,* 63: 733–735.

Stanisz, A.M., Scicchitano, R., Dazin, P., Bienenstock, J., and Payan, D.G. (1987) Distribution of Substance P receptors on murine spleen and Peyer's patch T and B cells. *J. Immunol.,* 139: 749–754.

Struwe, G., Knave, B., and Mindus, P. (1983). Neuropsychiatric symptoms in workers occupationally exposed to jet fuel-a combined epidemiologic and causistic study. *Acta Psychiat. Scand.,* 303(S): 55–67.

Ullrich, S.E. (1999) Dermal application of JP-8 jet fuel induces immune suppression. *Toxicol. Sci.,* 52: 61–67.

Ullrich, SE., and Lyons, HJ. (2000) Mechanisms involved in the immunotoxicity induced by dermal application of JP-8 jet fuel. *Soc. Tox.,* 58: 290–298.

Wang, J., Badowski, M., and Harris, D.T. A comparison of the immunotoxicological effects of JP8 versus S8 jet fuels. Manuscript in preparation.

White, R.D. (1999) Refining and blending of aviation turbine fuels. *Drug Chem. Toxicol.,* 22(1): 143–153.

Witten, M.L., Leeman, S.E., Lantz, R.C., Joseph, P.M., Burke, C.H., Jung, W.K., Quinn, D., and Hales, CA. (1990) Chronic jet fuel exposure increases lung Substance P (SP) concentration in rabbits. In *Substance P and Related Peptides: Cellular and Molecular Physiology* (S.E. Leeman, Ed). International Symposium, University of Massachusetts and the New York Academy of Science, New York, p. 29.

Witten, M.L., Pfaff, J., Lantz, R.C., Parton, K.H., Chen, H., Hays, A., Kage, R., and Leeman, S.E. (1992) Capsaicin pretreatment before JP-8 jet fuel exposure causes a large increase in airway sensitivity to histamine in rats. *Regulatory Peptides,* S1: 176.

Wong, S.S., Hyde, J., Sun, N.N., Lantz, R.C., and Witten, M.L. (2004) Inflammatory responses in mice sequentially exposed to JP-8 jet fuel and influenza virus. *Toxicology,* 197(2): 138–146.

11 The Involvement of Poly(ADP-ribosyl)ation in Defense against JP-8 Jet Fuel and Other Chemical Toxicants

Luis A. Espinoza, Aria Attia, Eric M. Brandon, and Mark E. Smulson

CONTENTS

INTRODUCTION

The poly(ADP-ribose) polymerase 1 (PARP-1) enzyme catalyzes the covalent long-chain poly(ADP-ribosyl)ation of various nuclear proteins utilizing β-nicotinamide adenine dinucleotide (NAD^+) as a substrate, with PARP-1 itself being the major target of the modifications. However, many other nuclear DNA-binding proteins are also modified by active PARP-1, a protein that may be actuated by DNA damage. After PARP-1 is activated, it binds to single- or double-stranded DNA ends via its two zinc

TABLE 11.1
Cellular Events and Exogenous Stimuli Associate with PARP-1 Activation

Agents that Activate Poly(ADP-ribosyl)ation	Process where DNA Strand Breaks Occur	Poly(ADP-ribose) Acceptors Proteins
Aging	Apoptosis	DNA polymerase α/β
Alkylating agents	Chromosomal aberrations	Endonucleases
Apoptosis inducers	Differentiation	Histones
Ionizing radiation	DNA repair	p53
Oxidizing agents	DNA replication	PARP-1
Topoisomerase inhibitors	Gene expression	PCNA
	Genetic instability	Topoisomerase I and II

fingers that recognize DNA breaks independent of the DNA sequence. Following these events, PARP-1 catalyzes a sequential transfer reaction of ADP-ribose units from NAD^+ to various nuclear proteins, forming a protein-bound polymer of ADP-ribose units. The covalent poly(ADP-ribosyl)ation of nuclear DNA-binding proteins in eukaryotic cells is a post-translational modification reaction related in part to the modulation of chromatin structure and function in DNA-damaged and apoptotic cells (Table 11.1). DNA strand breakage can occur during DNA replication or during DNA repair in response to exposure of cells to genotoxic stressors. One of the earliest cellular events that follows these circumstances is the poly(ADP-ribosyl)ation of an array of DNA-binding proteins that are localized predominantly adjacent to the DNA strand breaks [1]. This chapter presents evidence that constituents of JP-8 jet fuel may be genotoxic stressors that induce, in a dose-dependent manner, different levels of PARP-1 activation. This, in turn, causes differential responses in diverse types of cells.

JP-8, a jet fuel used by U.S. and NATO forces, has many chemical constituents that may be toxic if absorbed into the body by any of several routes. Exposure to JP-8 has been associated with numerous symptoms, including fatigue, headache, and skin irritation [2, 3]. The immune system is one of the major target organs for JP-8 toxicity. Studies with laboratory animals have shown that exposure to JP-8 in vapor, aerosol, or neat form resulted in a marked decrease in immune cell numbers and organ weight, as well as suppression of immune function [4–6], including effects on pulmonary function [7, 8].

This chapter focuses predominantly on the role of PARP-1 activation in response to the possibly toxic effects of JP-8 and its capacity of PARP-1 to modulate the events leading to cell apoptosis. The gene expression profiles after JP-8 exposure from primary cells, established cell lines, and from animal tissues are also discussed. These findings will be correlated with the different types of responses of cells and tissue to the toxic effects of JP-8. Additionally, involvement of PARP-1 in the process of the inflammatory response (Table 11.2), with the modulation of redox transcription factors and the production of pro-inflammatory factors, are discussed.

TABLE 11.2
Approaches Used to Identify Changes in Gene Profiles of Cell Lines and Tissues in Response to JP-8

Macroarray	Microarray	
Clontech	IntegriDerm	Affymetrix
Atlas Human Apoptosis and Stress cDNA expression arrays	DermArray GeneFilters DNA microarray	GeneChip Rat Genome U34 Array A
439 probes	5,171 probes	8,000 probes
Radioactive-labeled probes	Radioactive-labeled probes	Biotin-labeled probes

CELL TYPE SPECIFICITY OF JP-8 TOXICITY

It was initially observed that an 80-μg/ml dilution of JP-8 fuel induced the formation of various biochemical markers of apoptosis, such as activation of caspase-3 activity, which was assessed by AMC released from the caspase-3 substrate DEVD-AMC (N-acetyl-Asp-Glu-Val-Asp-7-amino-4-methylcoumarin), as early as 4 hours after the addition of JP-8 [9]. PARP-1 cleavage was noted at 12 hours, and lamin B1 cleavage initiates were detected at approximately 6 hours. By 24 hours, 80% of human histiocytic lymphoma (U937) and the acute T cell leukemia (Jurkat) cells treated with the JP-8 fuel experienced apoptotic death. Furthermore, most of the biological and morphological characteristics of the apoptotic cascade, such as cell rounding, membrane blebbing, fragmentation, detachment from the substrate, and cellular fragmentation, were also observed in Jurkat T cells exposed to JP-8. Most of these events, that is, caspase-3 activation, cleavage of poly(ADP ribose) polymerase 1 (PARP-1), chromatin condensation, membrane blebbing, the release of cytochrome c from mitochondria into the cytosol, and fragmentation of genomic DNA, were also detected in rat lung epithelial (RLE-6TN) cells during a 24-hour period of JP-8 exposure [9]. However, this concentration of JP-8 did not induce toxicity in skin fibroblasts and keratinocytes [10]. In contrast, higher dilutions of JP-8 (200 μg/ml) induced markers of necrotic cell death in primary and immortalized human keratinocytes. The latter cells did not display caspase activation, PARP-1 cleavage, or nuclear fragmentation. The induction of necrosis by JP-8 in primary skin fibroblasts derived from either wild-type or PARP knockout ($PARP^{-/-}$) mice indicated that JP-8 toxicity in these skin cells is mediated by a PARP-independent pathway [10]. Together, these observations suggest that the type of cell death induced by JP-8 is cell-type specific.

JP-8 toxicity induction in alveolar type II epithelial cell line RLE-6TN cells was directly correlated with depletion of the intracellular concentration of glutathione (GSH). The mechanism by which JP-8 induces cell death in epithelial cells was associated with an oxidative environment, resulting in a substantial generation of reactive oxygen species (ROS). This was concomitant with the reduction in GSH content and its precursor *N*-acetyl-cysteine (NAC) in a time-dependent manner [11]. The relevant role of cellular antioxidants was established when pre-incubation of RLE-6TN cells with NAC or GSH consistently blocked JP-8-induced depletion in GSH content. The protective effect of thiol antioxidants against JP-8 toxicity in epithelial cells was also associated with a reduction in caspase-3 activation and mitochondrial dysfunction, thus preventing the degradation of structural and anti-apoptotic proteins. These findings led to the suggestion that disruption of mitochondrial integrity during apoptosis is a consequence of the modification of intracellular Ca^{2+} homeostasis by JP-8 that may result in the activation of Ca^{2+}- and Mg^{2+}-dependent endonucleases. This process can be responsible for the occurrence of internucleosomal DNA fragmentation in RLE-6TN cells exposed to JP-8 [9].

JP-8-INDUCED CHANGES IN GENE EXPRESSION PROFILES OF HUMAN CELLS

Transcriptional Profiling of the JP-8 Response in Lymphocytes

Our laboratory used (Table 11.2) cDNA expression macroarrays (Clontech) to gain insight into gene expression changes that might contribute to JP-8 toxicity [12]. This assay identified altered expression in stress response, or apoptosis-related genes to JP-8 in cultured lymphocytes (Jurkat cells). Of the 439 total apoptosis, or stress response-related genes analyzed, 26 genes were identified whose expression level was either up- or down-regulated by a factor ≥2, comparing JP-8- and vehicle-treated cells. Among this group, 16 genes were up-regulated and 10 genes were down-regulated. Next, the impact of some of the more relevant genes in the activation of stress and apoptosis cascades is considered (Table 11.3).

The abundance of caspase-3 mRNA in Jurkat cells was significantly increased by exposure of cells to JP-8, providing further support for the notion that caspase-3 (and presumably PARP-1 inactivation) plays an important role in JP-8-induced apoptosis in these cells, as well as in other cell lines, in response to genotoxic stressors [9]. Observation of the up-regulation of caspase-9 in Jurkat cells was also consistent with the role of this protease, together with Apaf-1 (apoptotic protease activating factor 1), in the activation of caspase-3 [13]. Cleavage of vimentin, an intermediate filament protein that contributes to the mechanical stability of cells, by caspase-9 is thought

TABLE 11.3
JP-8-Induced Dysfunction on Gene Expression Pathways in Human Cells and in Rat Lung Tissues

Jurkat Cells	Human Keratinocytes	Rat Lung Tissue
Apoptosis-related proteins	Apoptosis-resistant proteins	Antioxidant-related proteins
Cell cycle-related proteins	Cell signaling and ligands	Chaperons, heat shock response
Metabolic enzymes and DNA repair proteins	Cytoskeleton and ECM	Cytoskeleton and ECM
Stress response-related proteins	Growth promoting proteins	Cell signaling, ligands
Transcriptional activators or repressors	Metabolic enzymes	Metabolic enzymes
	Stress-resistance proteins	

to facilitate the disruption of cellular integrity during apoptosis [14–22]. The cleavage of this protein has been observed in Jurkat cells exposed to various apoptotic stimuli. Because stable transfection of these cells with a caspase-resistant form of vimentin resulted in a delay of apoptosis initiation [15], JP-8 induction of the down-regulation of vimentin gene expression in Jurkat cells is consistent with the depletion of its encoded protein in the progression of apoptosis.

Exposure of Jurkat cells to JP-8 resulted in an increase in the abundance of p38 MAPK (mitogen-activated protein kinases) mRNA. Activation of p38 MAPK in response to a stimulus contributes to cell cycle arrest at the G_1–S phase transition [23]. It is also implicated in the regulation of the transcriptional response mediated by p53, caspase-3 activation, and PARP-1 cleavage during programmed cell death triggered by nitrogen oxide (NO) accumulation induced by genotoxic agents [18, 24]. Increased expression of GAS1, a plasma membrane glycoprotein, has been associated with growth arrest and prevention of DNA synthesis in several tumor cell lines [25]. GAS1 has also been shown to induce the expression of a variety of apoptotic mediators [26], and it is thought to have an active role in excitotoxic neuronal death [22]. Accordingly, our observations of an increase in the abundance of mRNAs for the transcription factors E2F-1 and E2F-5, as well as that for DP-1 induced by JP-8 in Jurkat cells, suggest augmented sensitivity of these cells to undergo cell death. Increased expression of E2F-1 (E2F transcription factor 1) in response to DNA damage has been associated with apoptosis induction [27, 28]. Additionally, E2F-5 contributes to the inhibition of cell proliferation after it is transported to the nucleus in complexes with transcriptional repressors that inhibit the expression of cell cycle genes in response to serum deprivation [20]. Our combined work in this field suggests that changes on transcripts that modulate cell proliferation may contribute to apoptosis initiation in Jurkat cells treated with JP-8.

It was also observed that JP-8 increased the abundance of mRNAs for the heat shock proteins HSC70 and HSP90, as well as that of the antioxidant enzyme GSTO1 in Jurkat cells [12]. An increased expression of these proteins has also been observed in several models of oxidative stress in response to the generation of ROS and the depletion of cellular thiols [17, 19]. However, if the level of cellular stress exceeds the capacity of heat shock proteins for cellular defense, these proteins may be unable to protect against cell death induction [29]. Proteomic analysis revealed that the expression of GST in the kidney was increased in rats exposed to JP-8 [30]. We have also established a direct relationship between the generation of ROS and depletion of GSH in response to JP-8 and the effects of this fuel on mitochondria and downstream events that disrupt cell morphology [11]. Consequently, these findings indicate that the toxic effects of JP-8 in an immune cell evoke a series of events leading to the activation of genes that may trigger irreversible cell death by apoptosis, and ultimately result in decreased immune function in related cells and organ types.

Transcriptional Profiling of the JP-8 Response in Keratinocytes

Because percutaneous absorption (dermal exposure) is one of the most prevalent routes of occupational exposure to JP-8, the response of primary human keratinocytes to JP-8 treatment in vitro is an important parameter to be considered. Because a dose of 80 μg/ml of JP-8 did not induce cell death in keratinocytes after 24 hours of treatment, a 7-day treatment was tested to determine if the longer exposure at 80 μg/ml of JP-8 would result in cell death in keratinocytes [31]. This experiment revealed that prolonged incubations with this dose of JP-8 had no impact on the cell viability of keratinocytes. In contrast, a much higher dose (200 μg/ml) of JP-8 that reliably killed the cells induced a necrotic cell death in approximately 40% of keratinocyte cells after 24 hours of treatment. The (lower) 80 μg/ml of JP-8 dilution was chosen for microarray experiments to explore the molecular mechanism activated by keratinocytes in response to treatment with JP-8 during shorter and longer periods. Cells were harvested at 24 hours or at 7 days after JP-8 exposure. Both of these time periods were considered determinants to analyze the variation of two sets of regulated genes using the DermArray Gene Filter DNA (Table 11.2), a microarray that was created specifically for dermatology research.

Consistently, the expression of several genes with obvious relevance to keratinocyte cells and potential functional relevance was examined (Table 11.3). For example, the under-regulation of the kinase MLCK (myosin light chain kinase) has been associated with apoptosis induction. Although high levels of MLCK were detected in stressed animals, and MLCK has been associated with inflammation and apoptosis [32], the mechanism of how MLCK increases in programmed cell death remains unknown. However, protein recruitment of MLCK and caspase-8 activation in a tumor necrosis factor (TNF)α-induced apoptosis model appears to provide to the cytoskeleton a rearrangement that allows the assembly of the TNFR1 death domains [33]. Consistent with the under-expression of MLCK, the low levels of TNFR1 detected in our microarray data suggest that activation of this pathway is efficiently controlled by keratinocytes in response to JP-8 exposure.

Also noteworthy were the elevated levels of SERPINE2, a serine proteinase inhibitor gene that has been shown be a negative prognostic marker in several human cancers [34]. It is interesting that PAI-1 (SERPINE1), closely related to SERPINE2, is believed to play a key role in tumor invasion, depending on its concentration and in conjunction with expression levels of uPA (PLAU) and the uPA receptor (PLAUR) on cells [35]. These interactions have also been observed in a full-thickness wound mode that promotes keratinocyte migration by blocking certain integrin recognition sites in the ECM (erythema chronicum migrans) and facilitating uPAR internalization and recycling [36]. High levels of PLAU and PLAUR in keratinocytes under hypoxia are also considered critical in increased migration that mimics normal wound healing [37], while PAI-1 is believed to be one of the major stress-regulated genes in response to various physiological and pathological stresses [38]. This suggests that SERPINE2 may be also involved in the stress response against a type of insult-inducing oxidative stress [37]. Furthermore, the observed up-regulation of the genes BACAH and PDHA1C implies that induction of several detoxifying enzymes is important in the process of metabolism of JP-8 components in skin cells. However, the reduced levels of SDHC and ABFB indicate that some enzymes are probably not activated or perhaps may even be inhibited after JP-8 exposure [31].

The cultures also over-expressed keratin K18 and vimentin, an intermediate filament protein that contributes to the mechanical stability of cells and is degraded in response to various inducers of apoptosis [39]. We proposed that increased levels of

K18 and vimentin in our study might be implicated in protection against cell stress (i.e., JP-8), cell signaling, and apoptosis, as evidenced in keratinocytes and other cell types [40, 41].

These findings suggest that repeated exposure to JP-8 may induce altered cell growth. This may account for the disruption of the skin barrier function, skin irritation, and alteration of the skin structure [42, 43]. Our results also show that JP-8 induced in keratinocytes a modification in the expression of genes with diverse functions principally associated with proliferation and stress. The altered gene profile observed in keratinocytes is also relevant to a rational explanation for the resistance of these cells to JP-8 toxicity. Although the fuel is composed of hundreds of hydrocarbons, in this study we focused on determining whether correlations exist between toxic responses to JP-8 and global changes in the overall genetic expression profiles of normal human keratinocytes exposed to this complex compound [31].

Transcriptional Profiling of the Rat Lung Response to JP-8

To compare the in vitro results with intact animal expression changes, gene expression was analyzed in rats exposed by inhalation to JP-8 [44]. For this purpose, we utilized Affymetrix chip microarrays (Table 11.2). The gene expression profiles of lung tissue isolated from rats exposed to JP-8 at 171 or 352 mg/m^3 were compared with those of lung tissue derived from control animals exposed to air alone [44]. JP-8 significantly affected the expression of 56 genes at the lower doses and 66 genes at the higher doses, respectively. The expression of most (48/56, 86%) of the genes affected by JP-8 at 171 mg/m^3 was observed to be down-regulated, whereas the expression of 42% (28/66) of the genes affected by JP-8 at 352 mg/m^3 was up-regulated. The affected genes were classified according to their primary biological function and were divided into groups for the low and high doses of JP-8, respectively (Table 11.3). The diversity of the biological functions of the affected genes likely reflects the chemical complexity of the fuel.

The lungs of rats exposed to JP-8 at the occupationally relevant dose of 352 mg/m^3 manifested a 5.5-fold increase in expression of the gene for γ-synuclein, a centrosomal protein that plays an important role in the regulation of cell growth and is present at a relatively low level in normal lung tissue. γ-Synuclein was found to promote cancer cell survival and inhibit apoptosis induced by cellular stress or chemotherapeutic drugs through modulation of intracellular signaling by mitogen-activated protein kinase (MAPK) pathways [45]. The modification of the cytoskeleton by overexpression of γ-synuclein in lung cells might be explained by the increased lung permeability [46], as well as the morphological changes associated with impaired barrier function (perivascular edema) and the early signs of pulmonary fibrosis (apparent from an increase in the number and size of surfactant-secreting lamellar bodies) [47, 48] induced by inhalation of JP-8.

The expression of genes whose products contribute to the cellular response to oxidative stress or chemical toxicants, including glutathione *S*-transferases (Gsta1, GstaYc2) and cytochromes P450 (CYP2C13, Cyp2e1), was also prominently increased in lung tissue from rats exposed to JP-8 at the higher dose. These antioxidants almost certainly play key roles in the bioactivation of JP-8 components contributing to reduction in oxidative stress in the rat lung tissue when exposed to the fuel. Other proteomics analyses have also shown that the expression of GSTs is markedly increased in lung tissue from mice and in brain tissue from rats exposed to JP-8 [49, 50]. The expression of these detoxifying genes has been shown to be induced by Nrf2, a transcription factor that protects against oxidative or chemically induced cellular damage in the lung [51].

In contrast, no changes in the expression of any of the apoptosis-related genes represented on the microarray were induced by JP-8 at either dose, which is consistent with the results of the histological analysis showing minimal cell damage in the lungs of the JP-8-treated rats. The abundance of mRNAs for various structural proteins, including α-actin, β-actin, and myosin heavy chain 7 (Myh7), increased upon exposure to JP-8 at 352 mg/m^3, whereas that of the mRNA for high-molecular-weight microtubule-associated protein 2 (HMW-MAP2), a protein important in microtubule assembly, decreased. The altered expression of Myh7 has been associated with chronic lung hyperinflation in humans [52] as well as with severe emphysema in rats [53]. Proteomics analyses have also shown that repeated inhalation of JP-8 alters the expression of several proteins related to the maintenance of cell structure, in addition to those related to detoxification and cell proliferation [49, 54]. The reduced expression of the HMW-MAP2 gene has been detected in whole brain tissue of rats exposed to JP-8 [55] and has also been proposed as a specific marker for pulmonary carcinoid tumor and small cell carcinoma [56]. It is therefore plausible to propose that persistent altered expression of extracellular matrix or cytoskeleton genes on cells exposed to JP-8 may induce morphological and functional changes. These events may increase the risk of development of diseases such as pulmonary fibrosis [46–48].

Expression of the gene for the inositol 1,4,5-trisphosphate receptor (InsP3R1), a ligand-gated Ca^{2+} channel that mediates the release of Ca^{2+} from intracellular stores [57], was reduced by a factor of 10 in the lungs of rats exposed to JP-8 at 352 mg/m^3. We have previously shown that JP-8-induced fragmentation of DNA during apoptosis in RLE-6TN cells is Ca^{2+} dependent [9, 11]. Treatment with exogenous GSH (glutathione) or the thiol-containing antioxidant *N*-acetyl cysteine protected against the toxic effects of JP-8 and other chemical toxicants in several cell lines [11, 58]. Oxidized glutathione, a product of oxidative stress, has been shown to reduce the Ca^{2+} content of inositol 1,4,5-trisphosphate-sensitive stores in pulmonary endothelial cells [57]. Furthermore, high levels of intracellular Ca^{2+} are implicated in airway hyper-responsiveness and allergic pulmonary inflammation [59]. Finally, the expression of the genes for aquaporin 1 (Aqp1) and aquaporin 4 (Aqp4), proteins that mediate water transport in various tissues including the lungs, was increased and decreased, respectively, in the lungs of rats exposed to JP-8 at 352 mg/m^3.

Exposure to JP-8 has been associated with deleterious effects on pulmonary function, including an increase in pulmonary resistance, permeability, and the accumulation of inflammatory cells [8, 60]. Repetitive exposure to this fuel also induces long-term effects on the immune system, as evidenced by reduced numbers of viable immune cells, decreased weight of immune organs, and loss of immune function [61, 62]. Indeed, we have shown that JP-8 induces a series of events in lymphocytes that lead to the activation of genes triggering irreversible cell death by apoptosis [12]. In this regard, the expression of immunoglobulin genes in lung tissue was down-regulated by JP-8 in our studies [44]. Such effects on the immune system may increase the susceptibility of personnel exposed to JP-8 to infectious agents as well as confer a predisposition to the development of autoimmune disease as a result of immune suppression [61].

The integration of our data with previous proteomic analysis performed in lung tissue from mice [54], using approximately the same dose of JP-8, revealed a substantial agreement on genes associated with the extracellular matrix, such as actin and myosin genes, which were up-regulated as a consequence of JP-8 treatment. Increased expression of these contractile filament genes may imply a protective response of cells to the oxidative stress induced by jet fuel. Proteomic data obtained from lungs of animals exposed to higher doses of JP-8 (1,000 and 2,500 mg/m^3) [49], compared to that used in the present study, still demonstrated a remarkable correspondence in the up-regulation of certain detoxification genes such as GSTs, which are known to play an important role in the defense mechanisms against the possible toxic effects of JP-8 components. These findings clearly indicate that short-term exposure of rats to JP-8, under conditions that mimic the occupational exposure of U.S. Air Force personnel, induced marked changes in genes whose functions are related to defense against toxicant-induced oxidative stress, and may also be responsible for protection of lung cells against the effects of JP-8 in vivo.

POLY(ADP-RIBOSE) POLYMERASE-1 IS ACTIVATED BY JP-8 IN PULMONARY CELLS

Effects of JP-8 on Cell Viability of Respiratory Epithelial Cells

To determine the effects of nonlethal doses (8 μg/ml) of JP-8 in pulmonary cells, RLE-6TN cells were used. These cells originated from rat alveolar type II epithelial cells (AEII) and preserve many of the original phenotypic characteristics [63]. This nonlethal JP-8 dose had no impact on cell survival even at longer exposure times (24, 36, and 48 hours) compared to that observed in control cells exposed with the equivalent dilution of ethanol for each time course.

Because JP-8 toxicity may be mediated by the generation of ROS and depletion of intracellular GSH [11], we decided to monitor changes in the GSH content and glutathione *S*-transferases Pi (GST-P) protein expression in RLE-6TN cells exposed to the nonlethal dose of 8 μg/ml JP-8. GST-P is part of a superfamily of enzymes responsible for the detoxification of electrophiles by conjugation with the nucleophilic thiol reduced GSH [64]. The immunoblot analysis showed that GST-P expression is increased in a time-dependent manner in cells exposed to JP-8. A significant increase in the intracellular amount of GSH was also observed during the 24 to 48 hours of treatment. The increase in GSH and GST-P levels indicates that both antioxidants may act to directly limit the potentially nocive effects of ROS accumulation. The same dilution of JP-8 did not modify GSH content in NR8383 cells, a rat pulmonary macrophage cell line. Compared to the (alcohol alone) 48-hour control, a slight increase in GSH levels was observed. In contrast to that observed in RLE-6TN cells, no alterations in GST-P protein amount were triggered by the fuel in NR8383 cells. The protein expression of catalase, an antioxidant enzyme, showed a significant increase starting at 6 hours after treatment with JP-8, with a marked augment during the 12- to 48-hour period. In contrast, Mn SOD (superoxide dismutase), another antioxidant defense enzyme, evidenced reduction in a time-dependent manner after 1 hour of treatment. These observations indicate that even when the normal or augmented level of antioxidants elicits a protective effect in pulmonary cells exposed to JP-8, changes in the expression of some of these molecules (i.e., Mn SOD protein) may stimulate the production of ROS. Accumulation of these signaling molecules may induce the activation of redox-sensitive transcription factors, including NF-κB and AP-1 that are involved in mediating inflammatory response.

Effects of JP-8 on Phosphorylation and Degradation of IκBα in Epithelial Cells

To gain insight into the role of pulmonary cells in the regulation of lung inflammation, we determined whether NF-κB is activated in RLE-6TN and NR8383 cultures exposed to JP-8 (8 μg/ml) [64, 65]. NF-κB activation in several models is preceded by the phosphorylation of the inhibitory IκB-α protein [66]. Treatment of both cell lines with JP-8 resulted in phosphorylation and degradation of IκB-α in a time-dependent manner, thereby allowing the mobilization of NF-κB subunits to the nucleus. Untreated RLE-6TN cells harvested at 48 hours had negligible or very low levels of phospho-IκB-α. Because the nuclear localization of p65 and p50 is preceded by the phosphorylation and degradation of IκBα, the slight reduction in IκB-α levels in a time-dependent manner that was observed in the total protein extract [67] suggested that JP-8 promotes the phosphorylation and degradation of IκB-α, which is a critical step toward the activation of NF-κB. In NR8383 cells, JP-8 induced the progressive phosphorylation of IκB-α protein, which is consistent with the proteolytic degradation of IκB-α when compared with untreated control macrophages that showed negligible or very low levels of phospho-IκB-α. The slight decline in the phospho-IκB-α level from 24 to 48 hours may reflect the increased amount of the detoxifying enzymes, catalase, GSH, and GST-P. This defense mechanism displayed by alveolar macrophages against the numerous potential oxidants present in JP-8 fuel may be induced to attenuate the expression of cytokine-induced activation of NF-κB in NR8383 cells (Figure 11.1). These results suggest that JP-8 promotes in pulmonary macrophages a consistent phosphorylation and degradation of IκB-α, thus inducing NF-κB-dependent gene expression (i.e., inflammatory proteins).

JP-8 Prolongs NF-κB Activation of Pulmonary Cells

A gel shift assay showed that a prolonged, nontoxic concentration of JP-8 directly induced a marked augmentation on DNA-binding activity of NF-κB in rat pulmonary epithelial cells. In nuclear extracts from RLE-6TN cells exposed to JP-8, the supershift analysis showed that the mobility of the NF-κB DNA complex was only super-shifted by anti-p65 antibody but not by the anti-p50 antibody. These data

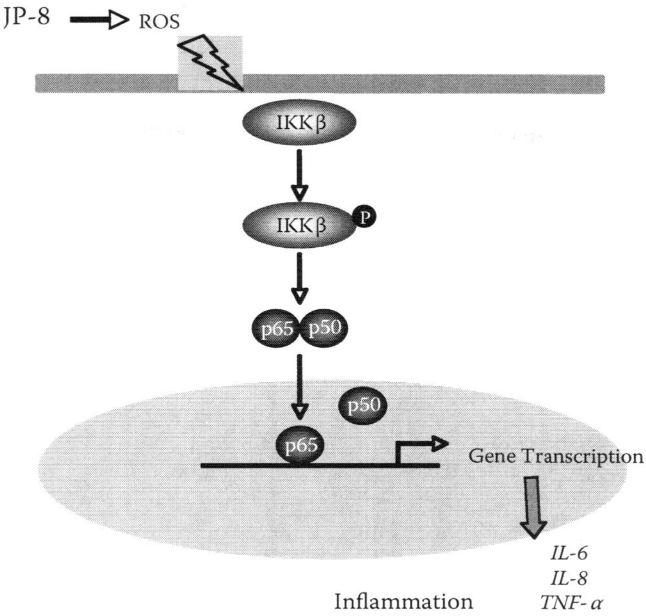

FIGURE 11.1 JP-8 induces NF-κB activation. A schematic representation of the phosphorylation and degradation of IκB-α, which is a critical step to the activation of NF-κB in pulmonary cells exposed to JP-8. The generation of ROS may have a key role in the activation of this redox transcription factor that finally may influence the expression of proinflammatory factors.

suggest that only the p65 subunit is involved in NF-κB activation in RLE-6TN cells treated with JP-8 [64]. The translocation of p65 and p50 NF-κB subunits into the nucleus was confirmed by Western blot analysis. A marked increase in the DNA-binding activity of NF-κB to its consensus sequences (Figure 11.1) was also evident in nuclear extracts isolated from NR8383 cells treated with fuel [65]. The binding activity of NF-κB activity is initiated as early as 1 hour after treatment, compared to the untreated control cells, and it consistently increased during the subsequent 12- to 48-hour period. We also observed that JP-8 promotes the rapid and increased binding activation of AP-1, which is also involved in the expression and production of proinflammatory molecules such as IL-1 and TNFα [68].

Effect of JP-8 on the Expression of Proinflammatory Mediators

The IL-8 gene, whose expression is regulated by NF-κB, plays an important role in several types of inflammatory responses such as chronic obstructive pulmonary disease (COPD) associated lung inflammation, by inducing the chemotaxis of polymorphonuclear leukocytes [69]. Total RNA isolated from untreated and treated RLE-6TN cells with JP-8 (8 μg/ml) subjected to semiquantitative RT-PCR analysis showed that JP-8 increased the expression of IL-8 gene transcripts at 24 hours, with a stable expression of IL-8 mRNA through the 36- and 48-hour exposure periods [64]. Because TNFα-induced signal transduction in respiratory cells is regulated through NF-κB activation, changes in the expression levels of this proinflammatory mediator were also determined by semiquantitative reverse transcription polymerase chain reaction (RT-PCR). We observed a time-dependent increase in TNFα transcripts with a pronounced induction at 48 hours of treatment. To determine whether JP-8 also up-regulates the expression of IL-8 and TNFα transcripts in vivo, we analyzed the expression of these inflammation mediators in the lung tissues of rats previously exposed for 7 days at doses that mimic the level of occupational exposure in humans. We found that JP-8 induced in rat lung tissues the up-regulation of both IL-8 and TNFα transcripts, which appear to be stable during the entire course of rat lung exposure [64].

IL-6 is a pleiotropic cytokine induced together with the alarm cytokine TNFα in acute-phase reactions. JP-8 treatment of RLE-6TN cells markedly stimulated IL-6 mRNA expression at 24-, 36- and 48-hour time points, compared to untreated control cells where IL-6 levels were almost undetectable [64]. The levels of IL-6 became more pronounced in epithelial cells 48 hours after exposure to the fuel. At this time exposure, RLE-6TN cells also exhibited a prominent induction of TNFα. These data suggest that the induction of IL-6 in RLE-6TN cells may be an alternative mechanism to regulate the production of proinflammatory mediators induced by JP-8 fuel. The RT-PCR analysis for IL-6 mRNA in lung tissue extracts of rats exposed to JP-8 by nose-only aerosol administration revealed that IL-6 levels were also substantially increased after a prolonged treatment with JP-8. In general, the RT-PCR data obtained from lung tissues of animals exposed to JP-8 treatment for 7 days consistently supported those observed in the cell culture system [64].

JP-8 Induces PARP-1 Activation

ROS-induced DNA damage, as may occur with significant JP-8 exposure, leads to the activation of PARP-1 [64]. Active PARP-1 then binds to DNA strand breaks and, using NAD^+, covalently modifies [poly(ADP-ribosyl)ation] nuclear proteins such as histones, adjacent to these breaks, with PAR polymers. This action is the signal mechanism for DNA repair enzymes. The amount of PAR formed in living cells with DNA damage is commensurate with the extent of DNA damage. Increased activation of PARP-1 due to excessive damage to the DNA drastically depletes levels of intracellular NAD^+ and adenosinetriphosphate (ATP), which are proposed to induce cell death. The time-course treatment studies revealed that the JP-8 dose of 8 μg/ml induces the progressive poly(ADP-ribosyl)ation auto modification of PARP-1 in RLE-6TN and NR8383 cells as early as 1 hour after treatment, and was continuously activated during the time-course treatment [64, 65]. PARP-1 activation was also evidenced in NR8383 cells treated with the 4-μg/ml dose at 1 hour, with a pronounced auto modification of this enzyme at the 24-hour time point. Because enhanced PARP-1 activation has been associated with initiation of apoptosis [70], it was of interest that PARP-1 activation did not trigger cell death in NR8383 cells treated with either dose of JP-8. In support of these data, pulmonary cells exposed to the fuel did not show evidence of PARP-1 and pro-caspase-3 cleavage. These findings are in agreement with the absence of death in RLE-6TN and NR8383 cells [64, 65].

PARP-1 has been indicated as a co-activator of NF-κB in inflammatory disorders [71]. We have found that activation of PARP-1 was concomitant with the prolonged activation of NF-κB in RLE-6TN cells exposed to JP-8-mediated oxidative stress. The time-course treatment revealed that JP-8 induces poly(ADP-ribosyl)ation automodification of PARP-1. This prolonged poly(ADP-ribosyl)ation of PARP-1 is in agreement with the increased and prolonged NF-κB activation post JP-8 exposure, compared with untreated control cells. While several proteins, including histones, topoisomerase 1, and p53, appear to be modified, the major acceptor by far was PARP-1 itself (Figure 11.2). Although variations in PARP-1 activation were observed at different time points, PARP-1 protein level was constant and nondegraded by caspase-3-like activity throughout the experiment. No cell death was observed at any time point, as evidenced by the constant presence of intact pro-caspase-3 and no caspase-3 activation [64]. A number of possibilities for the variation of PARP activity may be explained by the fact that prolonged treatment with JP-8 would promote at some point significant consumption of NAD^+ (and ATP), which would be required for cell viability. Additionally, considerable evidence indicates that both PARP-1 activation, as well as poly(ADP-ribose) (PAR) accumulation, is proliferation and cell cycle dependent. We earlier showed that, in synchronized HeLa and WI-38 cells [72], PARP-1 mRNA is high in the mid S phase but is significantly higher in the G2 phase. This was confirmed in nuclear runoff transcription studies. Others have shown that in synchronized intact cells there are two major points of PAR accumulation in cells:

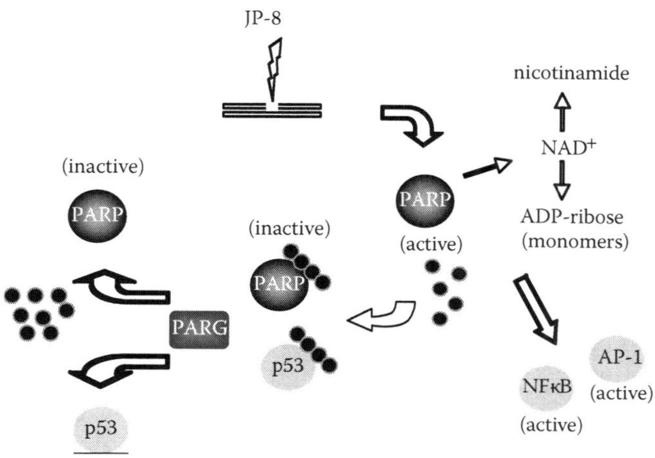

FIGURE 11.2 Scheme of proposed poly(ADP-ribosyl)ation cycle. JP-8-induced DNA damage activates PARP-1. This enzyme catalyzes the formation of long-branched poly(ADP-ribose) polymers using NAD^+ as a substrate on proteins acceptor to ADP-ribosylation various proteins and itself. Alternatively, active PARP-1 mediates signaling pathways by co-activating several transcription factors. The poly(ADP-ribose) glycohydrolase (PARG) hydrolyzes the polymer of poly(ADP-ribose) from the poly(ADP-ribosyl)ated proteins to produce free ADP-ribose residues.

one in mid S and a second at the G2 phase, due to synthesis as well as possibly less PAR turnover of the poly(ADP-ribose) glycohydrolase (PARG) [73]. Poly(ADP-ribosyl)ation is a phenomenon that contributes to various physiologic and pathophysiologic events associated with DNA strand breakage, repair of DNA damage, gene expression, and apoptosis [70, 74]. Hence, either apoptosis or necrosis associated with persistent PARP-1 activation and NAD^+/ATP depletion is indicative of DNA strand breakage. This was consistently supported by our findings using apoptotic doses (80 μg/ml) of JP-8 exposure and the generation of ROS-inducing DNA damage in lung epithelial cells [11]. However, this damage to the DNA appears to be insufficient to induce cell death using threshold levels of JP-8 exposure [64, 65].

CONCLUSIONS

Inhalation is thought to be the primary route of human exposure to JP-8 and, for that reason, the epithelial surface is an important target for possible toxicity. Inhalation exposure to JP-8 for prolonged periods or to elevated doses may result in deleterious effects on pulmonary functions, as well as histopathological changes in the lungs. The generation of high levels of ROS after exposure to JP-8 sufficient to induce DNA damage appears to be the key event that induces PARP-1 activation. The level of PARP-1 activation appears to be directly associated with the JP-8 concentration and time of exposure, which can impact differently on signaling pathway activation, especially those that converge on *redox*-sensitive *transcription factors*. Increased doses can also induce two different modes of cell death: apoptosis or necrosis. Concurrently, low doses at repetitive exposure promote the expression of pro-inflammatory mediators that may induce long-term effects on the tissues affected, thus increasing the susceptibility of personnel exposed to the fuel to the development of pulmonary diseases.

REFERENCES

[1] Espinoza, L.A. and Crotti, L.B. *Roles of PARP-1 and p53 in the Maintenance of Genome Integrity*. Hauppauge, NY: Nova Science Publishers, 2007.

[2] Smith, L.B., Bhattacharya, A., Lemasters, G., Succop, P., Puhala, E., 2nd, Medvedovic, M., and Joyce, J. Effect of chronic low-level exposure to jet fuel on postural balance of US Air Force personnel. *J. Occup. Environ. Med.*, 39, 623, 1997.

[3] Zeiger, E., and Smith, L. The First International Conference on the Environmental Health and Safety of Jet Fuel. *Environ. Health Perspect.,* 106, 763, 1998.

[4] Harris, D.T., Sakiestewa, D., Robledo, R.F., and Witten, M. Immunotoxicological effects of JP-8 jet fuel exposure. *Toxicol. Indus. Health*, 13, 43, 1997.

[5] Harris, D.T., Sakiestewa, D., Titone, D., Robledo, R.F., Young, R.S., and Witten, M. Jet fuel-induced immunotoxicity. T*oxicol. Indus. Health*, 16, 261, 2001.

[6] Ramos, G., Nghiem, D.X., Walterscheid, J.P., and Ullrich, S.E. Dermal application of jet fuel suppresses secondary immune reactions. *Toxicol. Appl. Pharmacol.*, 180, 136, 2002.

[7] Pfaff, J., Parton, K., Lantz, R.C., Chen, H., Hays, A.M., and Witten, M.L. Inhalation exposure to JP-8 jet fuel alters pulmonary function and Substance P levels in Fischer-344 rats. *J. Appl. Toxicol.*, 15, 249, 1995.

[8] Robledo, R.F., Young, R.S., Lantz, R.C., and Witten, M.L. Short-term pulmonary response to inhaled JP-8 jet fuel aerosol in mice. *Toxicol. Pathol.*, 28, 656, 2000.

[9] Stoica, B.A., Boulares, A.H., Rosenthal, D.S., Iyer, S., Hamilton, I.D., and Smulson, M.E. Mechanisms of JP-8 jet fuel toxicity. I. Induction of apoptosis in rat lung epithelial cells. *Toxicol. Appl. Pharmacol.*, 171, 94, 2001.

[10] Rosenthal, D.S., Simbulan-Rosenthal, C.M., Liu, W.F., Stoica, B.A., and Smulson, M.E. Mechanisms of JP-8 jet fuel cell toxicity. II. Induction of necrosis in skin fibroblasts and keratinocytes and modulation of levels of Bcl-2 family members. *Toxicol. Appl. Pharmacol.,* 171, 107, 2001.

[11] Boulares, A.H., Contreras, F.J., Espinoza, L.A., and Smulson, M.E. Roles of oxidative stress and glutathione depletion in JP-8 jet fuel-induced apoptosis in rat lung epithelial cells. *Toxicol. Appl. Pharmacol.,* 180, 92, 2002.

[12] Espinoza, L.A., and Smulson, M.E. Macroarray analysis of the effects of JP-8 jet fuel on gene expression in Jurkat cells. *Toxicology*, 189, 181, 2003.

[13] Zou, H., Li, Y., Liu, X., and Wang, X. An APAF-1.cytochrome c multimeric complex is a functional apoptosome that activates procaspase-9. *J. Biol. Chem.,* 274, 11549, 1999.

[14] Nakanishi, K., Maruyama, M., Shibata, T., and Morishima, N. Identification of a caspase-9 substrate and detection of its cleavage in programmed cell death during mouse development. *J. Biol. Chem.,* 276, 41237, 2001.

[15] Belichenko, I., Morishima, N., and Separovic, D. Caspase-resistant vimentin suppresses apoptosis after photodynamic treatment with a silicon phthalocyanine in Jurkat cells. *Arch. Biochem. Biophys.*, 390, 57, 2001.

[16] Byun, Y., Chen, F., Chang, R., Trivedi, M., Green, K.J., and Cryns, V.L. Caspase cleavage of vimentin disrupts intermediate filaments and promotes apoptosis. *Cell Death Differ.*, 8, 443, 2001.

[17] Chen, H.W., Chien, C.T., Yu, S.L., Lee, Y.T., and Chen, W.J. Cyclosporine A regulates oxidative stress-induced apoptosis in cardiomyocytes: Mechanisms via ROS generation, iNOS and Hsp70. *Br. J. Pharmacol.,* 137, 771, 2002.

[18] Cheng, A., Chan, S.L., Milhavet, O., Wang, S., and Mattson, M.P. p38 MAP kinase mediates nitric oxide-induced apoptosis of neural progenitor cells. *J. Biol. Chem.*, 276, 43320, 2001.

[19] Dilworth, C., Bigot-Lasserre, D., and Bars, R. Spontaneous nitric oxide in hepatocyte monolayers and inhibition of compound-induced apoptosis. *Toxicol. In Vitro*, 15, 623, 2001.

[20] Fujita, N., Furukawa, Y., Itabashi, N., et al. Differences in E2F subunit expression in quiescent and proliferating vascular smooth muscle cells. *Am. J. Physiol. Heart. Circ. Physiol.*, 283, 204, 2002.

[21] Ledo, F., Kremer, L., Mellstrom, B., et al. Ca2+-dependent block of CREB-CBP transcription by repressor DREAM. *Embo. J.*, 21, 4583, 2002.

[22] Mellstrom, B., Cena, V., Lamas, M., et al. Gas1 is induced during and participates in excitotoxic neuronal death. *Mol. Cell. Neurosci.*, 19, 417, 2002.

[23] Zhang, W., and Liu, H.T. MAPK signal pathways in the regulation of cell proliferation in mammalian cells. *Cell. Res.*, 12, 9, 2002.

[24] Mandir, A.S., Simbulan-Rosenthal, C.M., Poitras, M.F., et al. A novel in vivo post-translational modification of p53 by PARP-1 in MPTP-induced Parkinsonism. *J. Neurochem.*, 83, 186, 2002.

[25] Del Sal, G., Ruaro, E.M., Utrera, R., et al. Gas1-induced growth suppression requires a transactivation-independent p53 function. *Mol. Cell. Biol.*, 15, 7152, 1995.

[26] Lee, K.K., Leung, A.K., Tang, M.K., et al. Functions of the growth arrest specific 1 gene in the development of the mouse embryo. *Dev. Biol.*, 234, 188, 2001.

[27] Simbulan-Rosenthal, C.M., Rosenthal, D.S., Iyer, S., et al. Involvement of PARP and poly(ADP-ribosyl)ation in the early stages of apoptosis and DNA replication. *Mol. Cell. Biochem.*, 193, 137, 1999.

[28] Polager, S., Kalma, Y., Berkovich, E., et al. E2Fs up-regulate expression of genes involved in DNA replication, DNA repair and mitosis. *Oncogene*, 21, 437, 2002.

[29] Sharp, F.R., Massa, S.M., and Swanson, R.A. Heat-shock protein protection. *Trends Neurosci.*, 22, 97, 1999.

[30] Witzmann, F.A., Carpenter, R.L., Ritchie, G.D., et al. Toxicity of chemical mixtures: proteomic analysis of persisting liver and kidney protein alterations induced by repeated exposure of rats to JP-8 jet fuel vapor. *Electrophoresis*, 21, 2138, 2000.

[31] Espinoza, L.A., Li, P., Lee, R.Y., et al. Evaluation of gene expression profile of keratinocytes in response to JP-8 jet fuel. *Toxicol. Appl. Pharmaco*l., 200, 93, 2004.

[32] Kaneko, K., Satoh, K., Masamune, A. et al. Myosin light chain kinase inhibitors can block invasion and adhesion of human pancreatic cancer cell lines. *Pancreas*, 24, 34, 2002.

[33] Petrache, I., Birukov, K., Zaiman, A.L., et al. Caspase-dependent cleavage of myosin light chain kinase (MLCK) is involved in TNF-alpha-mediated bovine pulmonary endothelial cell apoptosis. *FASEB J.*, 17, 407, 2003.

[34] Buchholz, M., Biebl, A., Neebetae, A., et al. SERPINE2 (protease nexin I) promotes extracellular matrix production and local invasion of pancreatic tumors *in vivo*. *Cancer Res.*, 63, 4945, 2003.

[35] Devy, L., Blacher, S., Grignet-Debrus, C. et al. The pro- or antiangiogenic effect of plasminogen activator inhibitor 1 is dose dependent. *Faseb. J.*, 16, 147, 2002.

[36] Jones, J.M., Cohen, R.L., and Chambers, D.A. Collagen modulates gene activation of plasminogen activator system molecules. *Exp. Cell. Res.*, 280, 244, 2002.

[37] Daniel, R.J., and Groves, R.W. Increased migration of murine keratinocytes under hypoxia is mediated by induction of urokinase plasminogen activator. *J. Invest. Dermatol.*, 119, 1304, 2002.

[38] Yamamoto, K., Takeshita, K., Shimokawa, T., et al. Plasminogen activator inhibitor-1 is a major stress-regulated gene: implications for stress-induced thrombosis in aged individuals. *Proc. Natl. Acad. Sci. U.S.A.*, 99, 890, 2002.

[39] Morishima, N. Changes in nuclear morphology during apoptosis correlate with vimentin cleavage by different caspases located either upstream or downstream of Bcl-2 action. *Genes. Cells.,* 4, 401, 1999.

[40] Sanchez, A., Alvarez, A.M., Lopez Pedrosa, J.M., Roncero, C., et al. Apoptotic response to TGF-β in fetal hepatocytes depends upon their state of differentiation. *Exp. Cell. Res.*, 252, 281, 1999.

[41] Omary, M.B., Ku, N.O., Liao, J., et al. Keratin modifications and solubility properties in epithelial cells and *in vitro*. *Subcell. Biochem.*, 31, 105, 1998.

[42] Kabbur, M.B., Rogers, J.V., Gunasekar, P.G., et al. Effect of JP-8 jet fuel on molecular and histological parameters related to acute skin irritation. *Toxicol. Appl. Pharmacol.*, 175, 83, 2001.

[43] Kanikkannan, N., Locke, B.R., and Singh, M. Effect of jet fuels on the skin morphology and irritation in hairless rats. *Toxicology*, 175, 35, 2002.

[44] Espinoza, L.A., Valikhani, M., Cossio, M.J., et al. Altered expression of γ-synuclein and detoxification-related genes in lungs of rats exposed to JP-8. *Am. J. Respir. Cell. Mol. Biol.*, 32, 192, 2005.

[45] Pan, Z.Z., Bruening, W., Giasson, B.I., et al. Gamma-synuclein promotes cancer cell survival and inhibits stress- and chemotherapy drug-induced apoptosis by modulating MAPK pathways. *J. Biol. Chem.*, 277, 35050, 2002.

[46] Hays, A.M., Parliman, G., Pfaff, J.K., et al. Changes in lung permeability correlate with lung histology in a chronic exposure model. *Toxicol. Indus. Health*, 11, 325, 1995.

[47] Robledo, R.F., and Witten, M.L. Acute pulmonary response to inhaled JP-8 jet fuel aerosol in mice. *Inhal. Toxicol.*, 10, 531, 1998.

[48] Hays, A.M., Lantz, R.C., and Witten, M.L. Correlation between in vivo and in vitro pulmonary responses to jet propulsion fuel-8 using precision-cut lung slices and a dynamic organ culture system. *Toxicol. Pathol.*, 31, 200, 2003.
[49] Witzmann, F.A., Bauer, M.D., Fieno, A.M., et al. Proteomic analysis of simulated occupational jet fuel exposure in the lung. *Electrophoresis,* 20, 3659, 1999.
[50] Lin, B., Ritchie, G.D., Rossi, J., 3rd, et al. Identification of target genes responsive to JP-8 exposure in the rat central nervous system. *Toxicol. Indus. Health*, 17, 262, 2001.
[51] Lee, J.M., Calkins, M.J., Chan, K., et al. Identification of the NF-E2-related factor-2-dependent genes conferring protection against oxidative stress in primary cortical astrocytes using oligonucleotide microarray analysis. *J. Biol. Chem.*, 278, 12029, 2003.
[52] Mercadier, J.J., Schwartz, K., Schiaffino, S., et al. Myosin heavy chain gene expression changes in the diaphragm of patients with chronic lung hyperinflation. *Am. J. Physiol.*, 274, L527, 1998.
[53] Kim, D.K., Zhu, J., Kozyak, B.W., et al. Myosin heavy chain and physiological adaptation of the rat diaphragm in elastase-induced emphysema. *Respir. Res.*, 4, 1, 2003.
[54] Drake, M.G., Witzmann, F.A., Hyde, J., et al. JP-8 jet fuel exposure alters protein expression in the lung. *Toxicology*, 191, 199, 2003.
[55] Lin, B., Ritchie, G.D., Rossi, J., 3rd, et al. Gene expression profiles in the rat central nervous system induced by JP-8 jet fuel vapor exposure. *Neurosci. Lett.,* 363, 233, 2004.
[56] Liu, Y., Sturgis, C.D., Grzybicki, D.M., et al. Microtubule-associated protein-2: A new sensitive and specific marker for pulmonary carcinoid tumor and small cell carcinoma. *Mod. Pathol.*, 14, 880, 2001.
[57] Henschke, P.N., and Elliott, S.J. Oxidized glutathione decreases luminal Ca2+ content of the endothelial cell in (1,4,5)P3-sensitive Ca2+ store. *Biochem. J.*, 312(Pt. 2), 485, 1995.
[58] Han, S., Espinoza, L.A., Liao, H., et al. Protection by antioxidants against toxicity and apoptosis induced by the sulphur mustard analog 2-chloroethylethyl sulphide (CEES) in Jurkat T cells and normal human lymphocytes. *Br. J. Pharmacol.,* 142, 229, 2004.
[59] Ubl, J.J., Grishina, Z.V., Sukhomlin, T.K., et al. Human bronchial epithelial cells express PAR-2 with different sensitivity to thermolysin. *Am. J. Physiol. Lung Cell. Mol. Physiol.*, 282, L1339, 2002.
[60] Pfaff, J.K., Tollinger, B.J., Lantz, R.C., et al. Neutral endopeptidase (NEP) and its role in pathological pulmonary change with inhalation exposure to JP-8 jet fuel. *Toxicol. Ind. Health*, 12, 93, 1996.
[61] Harris, D.T., Sakiestewa, D., Robledo, R.F., et al. Short-term exposure to JP-8 jet fuel results in long-term immunotoxicity. *Toxicol. Indus. Health,* 13, 559, 1997.
[62] Harris, D.T., Sakiestewa, D., Robledo, R.F., et al. Effects of short-term JP-8 jet fuel exposure on cell-mediated immunity. *Toxicol. Indus. Health*, 16, 78, 2000.
[63] Driscoll, K.E., Carter, J.M., Iype, P.T., et al. Establishment of immortalized alveolar type II epithelial cell lines from adult rats. *In Vitro Cell. Dev. Biol. Anim.*, 31, 516, 1995.
[64] Espinoza, L.A., Tenzin, F., Cecchi, A.O., et al. Expression of JP-8-induced inflammatory genes in AEII cells is mediated by NF-kappaB and PARP-1. *Am. J. Respir. Cell. Mol. Biol.,* 35, 479, 2006.
[65] Espinoza, L.A., Smulson, M.E., and Chen, Z. Prolonged poly(ADP-ribose) polymerase-1 activity regulates JP-8-induced sustained cytokine expression in alveolar macrophages. *Free Radic. Biol. Med.*, 42, 1430, 2007.
[66] Huang, T.T., Kudo, N., Yoshida, M., et al. A nuclear export signal in the N-terminal regulatory domain of IkappaBalpha controls cytoplasmic localization of inactive NF-kappaB/IkappaBalpha complexes. *Proc. Natl. Acad. Sci. U S A.,* 97, 1014, 2000.
[67] Hayden, M.S., and Ghosh, S. Signaling to NF-kappaB. *Genes Dev.*, 18, 2195, 2004.

[68] Rahman, I., Marwick, J., and Kirkham, P. Redox modulation of chromatin remodeling: impact on histone acetylation and deacetylation, NF-kappaB and pro-inflammatory gene expression. *Biochem. Pharmacol.,* 68, 1255, 2004.
[69] Lezcano-Meza, D., and Teran, L.M. Occupational asthma and interleukin-8. *Clin. Exp. Allergy,* 29, 1301, 1999.
[70] Simbulan-Rosenthal, C.M., Rosenthal, D.S., et al. PARP-1 binds E2F-1 independently of its DNA binding and catalytic domains, and acts as a novel coactivator of E2F-1-mediated transcription during re-entry of quiescent cells into S phase. *Oncogene,* 22, 8460, 2003.
[71] Hassa, P.O., and Hottiger, M.O. The functional role of poly(ADP-ribose)polymerase 1 as novel coactivator of NF-kappaB in inflammatory disorders. *Cell. Mol. Life Sci.*, 59, 1534, 2002.
[72] Bhatia, K., Kang, V.H., Stein, G.S., et al. Cell cycle regulation of an exogenous human poly(ADP-ribose) polymerase cDNA introduced into murine cells. *J. Cell. Physiol.*, 144, 345, 1990.
[73] Kidwell, W.R., and Mage, M.G. Changes in poly(adenosine diphosphate-ribose) and poly(adenosine diphosphate-ribose) polymerase in synchronous HeLa cells. *Biochemistry*, 15, 1213, 1976.
[74] Menissier-de Murcia, J., Molinete, M., Gradwohl, G., et al. Zinc-binding domain of poly(ADP-ribose)polymerase participates in the recognition of single strand breaks on DNA. *J. Mol. Biol.*, 210, 229, 1989.

12 Evaluation of Methods Used to Generate and Characterize Jet Fuel Vapor and Aerosol for Inhalation Toxicology Studies

Raphaël T. Tremblay, Sheppard A. Martin, and Jeffrey W. Fisher

CONTENTS

INTRODUCTION

Studies of human toxicological effects from jet fuel exposure rely heavily on animal data. Inhalation effects are usually studied by exposing animals to laboratory-generated jet fuel atmospheres [34]. This chapter surveys and discusses the various methods and experimental techniques used to generate and quantify atmospheres composed of jet fuel vapor and aerosol for animal exposures. Included in these discussions are the strengths and drawbacks of the surveyed methodology and the potential impact of the selected methodology on the interpretation of results. While most of the discussed studies were funded by the U.S. Air Force Office of Scientific Research (AFOSR), relevant studies supported by other funding sources are also included. Based on our laboratory experience with aerosolized fuel atmospheres and a review of the literature, this chapter first discusses the general issues associated with generating an atmosphere of jet fuel, and secondly considers specific methods for the generation and characterization of jet fuel vapor and aerosol atmospheres, as appropriate.

ISSUES ASSOCIATED WITH THE GENERATION OF A JET FUEL ATMOSPHERE

Aviation fuels are complex mixtures containing hundreds if not thousands of different hydrocarbon compounds, and may also include a number of performance additives. Individual fuel blends (e.g., JP-8 or JP-5) may differ substantially in composition from batch to batch because these blends are defined only by general physical and chemical specifications [34]. Although jet fuel blend compositions may vary among batches, the most important considerations for controlling the chemical composition and concentration of generated aerosolized fuel exposures include the (1) type of generation system and (2) operational settings of the generation system.

A jet fuel atmosphere, of course, differs significantly from the chemical composition of the neat fuel used to generate the atmosphere. Physical properties of the chemical constituents of the neat fuel, particularly vapor pressure, are of significant importance for the generation of a jet fuel atmosphere. Regardless of the method used to generate a fuel atmosphere, higher vapor pressure compounds are enriched in the vapor phase compared to the neat fuel and, conversely, lower vapor pressure compounds are enriched in the droplet or aerosol phase (if present). Without going into the details of vapor or aerosol fuel generation, the process can be most simply described as the continuously preferential transfer of the more volatile compounds from the liquid to the gas/vapor phase. At the beginning of the generation of the fuel atmosphere, all compounds are in the liquid phase (aerosol or liquid layer); the more volatile compounds are preferentially evaporated to the gas phase, and a liquid or aerosol droplet enriched in heavier (less volatile) compounds is left behind. The extent of this process is primarily controlled by several "controlled" parameters of the experimental methodology, including generation heating temperature, residence time (travel time between the nebulizer and the exposure chamber through heated or unheated tubing), and the temperature and input rate of room air. Ultimately, nearly all of the lighter and more volatile fuel components will be in the gas phase, leaving only the heavier constituents in the liquid phase (or aerosol). This complex and dynamic process impacts the (1) gas/vapor phase composition, (2) aerosol composition (if present), (3) overall aerosol to gas/vapor phase ratio, and finally (4) compound-specific aerosol ratio. These descriptive parameters may be rapidly altered as the controlled instrumental parameters are varied. For example, in our laboratory, a decrease in the total aerosol fraction from 52% to 36% was induced by simply increasing the residence time (travel time between the nebulizer and the exposure chamber) from 7 to 61 seconds (Unreported data).

This phenomenon of preferential evaporation leads to widely different exposure characteristics for different inhalation chambers, suggesting possible difficulty in comparing toxicology data reported by laboratories using different generation systems. Typically, at no point during the fuel exposure is the vapor or aerosol composition absolutely predictable in terms of representing the known neat fuel composition. Table 12.1 presents an example of the measured composition of vapor and aerosol phases generated for a JP-8 inhalation exposure, and is compared to the composition of neat JP-8. In this example, the volatile compounds (methylcyclohexane, toluene, 2-methylheptane, and octane) account for nearly 10% of the vapor phase, although they only represent about 3% of the neat fuel. A similar but reverse discordance between the neat fuel composition and aerosol droplet composition is observed for low vapor pressure compounds in the aerosol. The last six compounds in Table 12.1 (dodecane and other larger-molecular-weight compounds) account for more than 14% of the aerosol composition, while they account for only about 8.5% of the neat fuel.

As can be seen in Table 12.2, the aerosol-to-gas/vapor phase ratios also vary widely among the different fuel components. Table 12.2 shows this ratio for a series of detected compounds ordered by molecular weight for a targeted 1000-mg/m^3 exposure to S-8 synthetic jet fuel made through the Fischer-Tropsch process. In this example, over 90% of the octane is in the gas phase while 70% to 80% of the heavier hydrocarbons are in the aerosol phase.

An issue not fully addressed in this chapter is the relevance of the generated laboratory jet fuel atmospheres to real-life occupational human exposures. Exposure assessment has been discussed in detail by the National Academy of Science [34]. In general, the jet fuel is expected to partition between the vapor and aerosol in occupational settings in a manner similar to that predicted by animal models.

GENERATION OF A JET FUEL ATMOSPHERE

The most commonly used generation techniques employed to create a vapor or vapor/aerosol atmosphere for animal studies is discussed in the following sections. Table 12.3 presents a summary of selected jet fuel inhalation studies along with the techniques used.

HEATED TUBE

The use of a heated tube to generate a jet fuel vapor-only atmosphere for animal studies is a very simple and common technique [3, 5, 9, 22, 25–29, 32, 35, 41, 44, 47, 49, 53, 54]. While the actual laboratory setup can vary, the liquid fuel generally

TABLE 12.1
Measured Composition of Vapor and Aerosol from Aerosolized JP-8 and Neat JP-8

Fuel Constituent	Vapor	Aerosol	Neat
Benzene	0.04	0.01	0.35
2-Methylhexane	0.34	0.16	0.13
n-Heptane	<0.01	<0.01	1.16
Methylcyclohexane	4.39	0.90	2.06
Toluene	1.11	0.34	0.30
2-Methylheptane	1.43	0.35	0.24
n-Octane	3.00	0.94	0.44
2,5-Dimethylheptane	0.42	0.07	0.04
Ethylbenzene	0.61	0.35	0.23
2-Methyloctane	1.85	0.76	0.42
m-Xylene	1.47	0.99	1.00
3-Methyloctane	1.48	0.49	0.45
o-Xylene	0.96	0.76	0.42
n-Nonane	3.75	1.50	1.00
2-Methylnonane	1.11	0.60	0.25
1,3,5-Trimethylbenzene	0.39	0.49	0.36
3-Methylnonane	0.93	0.54	0.67
o-Ethyltoluene	0.52	0.62	0.44
1,2,4-Trimethylbenzene	1.32	1.90	1.45
Butylcyclohexane	0.44	0.39	1.11
2-Methyldecane	0.74	0.88	0.60
n-Decane	3.84	2.71	2.19
n-Undecane	2.83	5.11	3.14
n-Dodecane	1.56	6.00	3.12
Naphthalene	0.03	0.13	0.12
n-Tridecane	0.54	4.79	2.37
2-Methylnaphthalene	0.01	0.02	0.28
n-Tetradecane	0.66	1.82	1.62
n-Pentadecane	0.05	1.51	0.91

Source: Unpublished data, using exposure system and characterization methods described in Tremblay, R.T., Martin, S.A., Fechter, L.D., et al. *The Toxicologist CD — J. Soc. Toxicol.*, 102, Abstract 948, 2008; and Martin, S.A., Flynt, K., Kendrick, C., et al. Abstract #948. *The Toxicologist CD — J. Soc. Toxicol.,* 102, Abstract 1100, 2008.

Note: Data presented as fraction (%) of each phase. Error ±10%.

TABLE 12.2
Concentrations (mg/m³) of a Series of Hydrocarbons in a 1,000-mg/m³ Atmosphere of Aerosolized S-8

Constituent	Gas Phase	Aerosol	Aerosol Fraction (%)
2-Methylheptane	8.87	0.19	2
4-Methylheptane	4.01	0.03	1
3-Methylheptane	13.15	<0.01	<1
n-Octane	40.45	3.35	8
2,5-Dimethylheptane	19.59	2.06	10
2,3-Dimethylheptane	5.54	<0.01	<1
2-Methyloctane	72.46	10.61	13
3-Methyloctane	46.04	6.65	13
3,3-Diethylpentane	1.88	0.06	3
n-Nonane	47.44	9.51	17
2,2-Dimethyloctane	6.11	0.45	7
3,3-Dimethyloctane	10.37	0.67	6
2-Methylnonane	19.82	6.52	25
3-Ethyloctane	4.90	0.99	17
3-Methylnonane	22.10	6.37	22
4-Methyldecane	7.76	3.27	30
2-Methyldecane	7.50	4.13	36
3-Methyldecane	8.38	4.59	35
n-Decane	30.84	10.65	26
n-Undecane	14.21	10.26	42
n-Dodecane	6.47	10.16	61
n-Tridecane	1.79	8.37	82
n-Tetradecane	1.64	3.33	67
n-Pentadecane	0.52	1.43	73

Source: From Tremblay, R.T., Martin, S.A., Fechter, L.D., et al. *The Toxicologist CD — J. Soc. Toxicol.*, 102, Abstract 1100, 2008.
Note: Error of ±10 %.

enters through the top of a vertical heated tube, while fresh air enters through the bottom. The tube, most often made of glass, can be filled with glass beads to substantially increase the evaporative surface area in some cases (e.g., [53]). The tube is heated to increase the speed and rate of evaporation of the fuel. The fraction of the fuel that does not evaporate is collected and measured from a port at the base of the tube. This technique has the advantage of providing consistent vapor composition (beyond the tube warm-up period). It can be referred to as a "single-pass generation system" because the liquid fuel not evaporated is discarded. Tube temperature can be varied to control at least part of the chemical composition of the atmosphere and also the total fuel concentration in the atmosphere. A potential problem with use of the heated tube system is the formation of aerosol from condensation in the delivery

TABLE 12.3
Summary of Animal Exposure Studies to Jet Fuel Vapor and Aerosol, Including the Fuels Tested, Generation System, and Exposure Characterization Techniques Used

Study	Fuel(s)	Generation System	Quantification of Vapor and Aerosol	Note
		Vapor Only		
Carpenter et al., 1976 [5]	Kerosene	Heated tube	Vapor: GC-FID (off-line, direct inject)	
Bruner et al., 1993; MacEwen and Vernot, 1981, 1982; Wall et al., 1990 [3, 47, 84, 85]	JP-4	Heated tube	Vapor: THC	
MacNaughton and Uddin, 1984 [29]	JP-4	Heated tube	Vapor: THC	Aerosol monitored with particle counter (PC)
Gaworski et al., 1984 [9]	JP-5	Heated tube	Vapor: THC	Aerosol monitored with PC
Wall et al., 1990; MacEwen and Vernot, 1984, 1985 [27, 28,47]	JP-5 JP-8	Heated tube	Vapor: THC	Aerosol monitored with PC
Starek and Vojtisek, 1986 [45]	Kerosene	Not specified (appears to be vapor)	Vapor : GC	
Newton et al., 1991 [35]	JP-4	Heated tube	Vapor: THC	
Mattie et al., 1991 [32]	JP-8	Heated tube	Vapor: THC	Aerosol monitored with PC
Nordholm et al., 1999 [36]	JP-4	Evaporator tower	Vapor: IR	GC characterization of vapor
Witzmann et al., 2000b [53]	JP-8	Heated tube	Vapor: IR	Aerosol monitored with real time aerosol monitor
Rossi et al., 2001 [44]	JP-5 JP-8	Heated tube	Vapor: IR	

Lin et al., 2001, 2004; Witzmann et al., 2003; Ritchie et al., 2001 [22, 23, 41, 51]	JP-8	Heated tube	Vapor: IR	
Campbell and Fisher, 2007 [4]	JP-8	Air bubbled through fuel	Vapor: online GC-FID	Aerosol filtered out with glass wool before exposure chamber
		Vapor and Aerosol		
Pfaff et al., 1995 [37]	JP-8	Devilbiss PulmoSonic Nebulizer	Vapor: none Aerosol: cascade impactor	Concentration calculated from weights of impactor plates. Aerosol composition by desorption of plate followed by GC analysis.
Robledo and Witten, 1998; Robledo et al., 2000; Wang et al., 2001; Witzmann et al., 1999, 2000a, b; Baldwin et al., 2001, 2007; Drake et al., 2003; Harris et al., 1997a, b, c, 2000a, b, c, 2002, 2007; McGuire et al., 2000; Wong et al., 2004; Pfaff et al., 1995, 1996 [1, 2, 7, 11–18, 33, 37, 38, 42, 43, 48, 51–53, 55]	JP-8	Devilbiss Ultra-Neb Nebulizer	Vapor: none Aerosol: cascade impactor	Same as above; reported aerosol to vapor ratio of 1.5
Whitman and Hinz, 2001 [49]	JP-4 JP-8 JP-8+100	Heated tube for JP-4 vapor only (reported aerosol at high concentration); collison nebulizer for aerosol	Vapor: GC-FID of charcoal tube extract Aerosol: GC-FID of filter extract	Vapor monitored by IR when no aerosol (no filter); aerosol monitored by PC Some detailed FID work with extract from tubes and glass filters; size by impactor
Whitman and Hinz, 2004 [50]	JP-5 JPTS JP-7 DFM JP-10	Collison nebulizer and heated tube (JP-5); single-pass spray atomizer for others	Same as above	Same as above

Continued

TABLE 12.3 (*Continued*)

Summary of Animal Exposure Studies to Jet Fuel Vapor and Aerosol, Including the Fuels Tested, Generation System, and Exposure Characterization Techniques Used

Study	Fuel(s)	Generation System	Quantification of Vapor and Aerosol	Note
Herrin et al., 2006 [19]	JP-8	Lovelace jet nebulizer	Total: THC Aerosol: cascade impactor	
Fechter et al., 2007 [8]	JP-8	Single-pass spray atomizer	Total: THC	Some aerosol characterization on filter and polyurethane foam.
Martin et al., 2008; Tremblay et al., 2008 [31, 46]	JP-8 S-8	Single-pass Savillex jet nebulizer	Vapor: charcoal tube + thermal desorption in GC-MS Aerosol: same as vapor (see text for details)	Online GC-FID for monitoring Size by impactor
Naval Health Research Center Detachment/ Environmental Health Effects Laboratory exposure system [Personal communication from James Reboulet at NHRC/EHEL, 2008]		Single-pass Misonix spray nebulizer	Vapor: IR Aerosol: filter weight	Filtered aerosol prior to IR; calibration performed with vapor standards of similar chemical composition

Note: DFM, Diesel Fuel Marine; S-8, Synthetic-8; JPTS, Jet Propellant Thermally Stable; THC, Total Hydrocarbon Analyzer; IR, Infrared Spectroscopy; GC-FID, Gas Chromatography-Flame Ionization Detection; PC, Particle Counter; GC-MS, Gas Chromatography-Mass Spectrometry.

system once the vapors cool, although this has not been shown to be an issue except for very high concentrations [49].

Bubbler System

Bubbling air through jet fuel is another very simple way to generate fuel vapors [4]. Air is usually bubbled through a glass frit. This technique is very well suited for single compound exposures, but has drawbacks when used with complex mixtures. First, this system "recycles" the fuel because the same liquid is used for a long period of time, as opposed to the single-pass system discussed previously. Rapid depletion of the lighter (more volatile) chemical constituents of the mixture would be expected, leading to a non-consistent vapor composition over time. Another problem is the possible creation of aerosols from bubbles that burst at the fuel surface. While the aerosol quantity generated is typically minimal, nevertheless the aerosols must be filtered if not desired in the exposure.

Atomizers

In recent years, atomizer systems have grown in use and popularity [8, 19, 31, 49, 50]. Atomization is the conversion of a liquid into an aerosol using compressed gas. These atomization systems are often referred as a jet nebulizer, spray nebulizer, or spray atomizer system. These systems use Bernoulli's principle to entrain and shear-off a liquid using pressurized gas. The liquid then breaks apart into aerosol droplets. Some systems (e.g., Misonix, Inc., Farmingdale, NY) generate a sonic field with the incoming compressed air, which breaks the liquid stream into smaller droplets than a conventional atomizer. While these systems use a sonic field, they are more like compressed-air nebulizers in their operation than ultrasonic nebulizers. Such a system is presently in use at the Naval Health Research Center Detachment/ Environmental Health Effects Laboratory NHRC/EHEL) [personal communication from James Reboulet, NHRC/EHEL, 2008]. While the basic principles remain the same among all atomizers, a single difference in design may greatly affect the end results. Systems like the Collison nebulizer [49, 50] and the Lovelace jet nebulizer [19] impact the generated aerosols onto a hard surface to reduce the amount and size of the droplets. Most of the generated droplets impacting that surface are recycled back into the starting liquid. Lighter fuel constituents are rapidly depleted in this cycle and consequently may lead to an inconsistent exposure composition. Figure 12.1 shows that concentrations of light fuel components are rapidly depleted when using a Collison nebulizer. These systems are better suited for single compounds and nonvolatile mixtures (components with vapor pressures similar to that of decane or lower) and have the advantage of generating very little waste. Some atomizers (e.g., single-pass) do not recycle the impacted liquid (if impacted at all) and therefore produce an exposure with a more constant composition. Fechter and co-workers [8] designed such a single-pass jet nebulizer based on a previously published design [24]. A disadvantage of these systems is that they require a higher fuel use than recycling systems. However, small-scale single-pass jet nebulizers with low consumption rates (approximately 10 to 15 ml/hour) are now available commercially

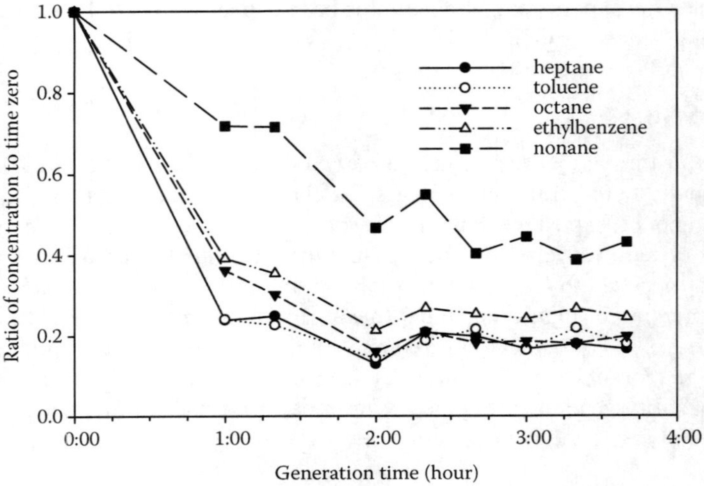

FIGURE 12.1 Vapor phase relative concentration for a series of hydrocarbons from a JP-8 exposure using a Collison nebulizer, as a function of time. (*Source:* From Tremblay, R.T., Martin, S.A., Fechter, L.D., et al. *The Toxicologist CD — J. Soc. Toxicol.*, 102, Abstract 1100, 2008.)

(Savillex, Minnetonka, Minnesota). These small-volume, single-pass nebulizers are very well suited for small-scale, nose-only animal exposures [31]. The Misonix nebulizer used at NHRC/EHEL (Naval Health Research Center / Environmental Health Effects Laboratory) is a single-pass system [Personal communication from James Reboulet, 2008].

Ultrasonic Nebulizer

Ultrasonic nebulizers have been used extensively by at least one jet fuel toxicity research group [1, 2, 7, 11–18, 33, 37, 38, 42, 43, 48, 51–53, 55]. These nebulizers are widely available in the medical market and are commonly used to deliver drugs [39]. Generally, the fuel sample is placed inside a containment vessel, which is placed in a water bath and subjected to ultrasound. Aerosols are generated above the liquid fuel surface and are entrained away using gas flow. A major problem with such a system is the potential for the aerosol to deposit back into the liquid fuel, therefore possibly recycling the fuel repeatedly from the liquid to the aerosol phase. Similar to the Collison and Lovelace nebulizers, this recycling can lead to the rapid depletion of the lighter hydrocarbons and enrichment of the liquid over time. The impact of this issue depends on the frequency at which the fuel is replaced from the containment vessel during exposure.

Conclusions on Generation Systems

While many factors affect the choice of a generation system, production of an exposure atmosphere with a constant chemical composition is obviously essential. Of the discussed generation systems, heated tubes and single-pass jet nebulizers provide the best systems for vapor and vapor/aerosol exposures, respectively. Both generation systems provide a constant composition, which is crucial when generating complex laboratory inhalation atmospheres. These systems are therefore preferred for generation of exposures from jet fuels or other complex mixtures. Another important concern is how representative are the experimental atmospheres produced by different exposure systems for typical human exposures to jet fuel vapor and aerosols. It is assumed that human fuel exposures vary from occasional inhalation of very low levels of fuel vapor (e.g., when visiting an airport), to daily exposure to moderate levels (e.g., when employed as a fuel handler or aircraft worker), to occasional exposure to extremely high levels of vapor and aerosol (e.g., when working as an aircraft fuel cell cleaner or jet mechanic). In each of these cases, analytical air sampling systems can be used to characterize minimum, average, and maximum human exposures of fuel vapor and aerosol. To properly evaluate these human data, it is important that laboratory animal data are collected using systems that reliably produce constant chemical composition atmospheres.

CHAMBER ATMOSPHERE CHARACTERIZATIONS

Proper characterization of the generated fuel exposures is essential for the interpretation and comparison of toxicology studies. Characterization methods range from simply measuring the total jet fuel concentration, to more complex quantifications of the aerosol and vapor compositions. Techniques to quantify and characterize fuel vapors are discussed first, followed by techniques applicable to aerosols. Finally, aerosol particle size characterization is discussed.

Gas-Phase Characterization

Total Hydrocarbon (THC) Analyzer

Quantification of the total chamber concentration is often performed using a total hydrocarbon (THC) analyzer [3, 8,29, 32, 35, 19, 47, 82–85]. These systems have the advantage of being online, providing results every few seconds or minutes, and requiring very little attention during exposures. THC systems are based on flame ionization detection (FID), a detection technique commonly used in gas chromatography (GC). The incoming hydrocarbons are burned in a hydrogen/air (sometimes hydrogen/oxygen) flame, resulting in positively charged carbon ions and electrons. The carbon ions are collected on a negative electrode, inducing a measurable current. The response is proportional to the amount of ionized carbon. The detector is insensitive to noncombustible and non-carbon-containing compounds such as water, nitrous oxide, etc. However, the THC technique has two potential biases. First, the main characteristic of using FID is that compositionally similar compounds may result in very similar detector responses. That is, the composition of a mixture does not need to be known in order to derive the carbon content of its hydrocarbon constituents. This suggests that quantifying a mixture consisting of hydrocarbons from different chemical families can be problematic. Compounds that have oxidized carbons (alcohol, carbonyl, halogen, etc.) will not produce as many ionized carbons as their nonoxidized carbon counterparts, leading to a lower signal per carbon. This bias may be minimal for jet fuel analyses, as such compound comparisons are generally insignificant. However, aromatic hydrocarbon compounds found in jet fuel can induce a bias. Unsaturated compounds, including aromatic hydrocarbons, do not give the same signal intensity per carbon compared to straight-chain alkane compounds. Most of the signal intensity differences are less than 5%, although some compounds have a larger bias than 5%. For example, ethylbenzene, a significant JP-8 constituent, has been shown to give the response of 6.88 carbon atoms while it is actually composed of 8 carbons (a 15% difference) [20]. This analytical issue leads to underestimation of the amount of hydrocarbons measured.

A second issue with use of THC analysis is the bias created when converting the carbon content to a mass content. The carbon-to-mass ratio for the measured compound or mixture must be known. Saturated alkanes have very similar constant carbon-to-mass ratio (<2%), and calibration with any saturated hydrocarbon (usually a gas or very volatile hydrocarbon) is sufficient to accurately convert the results to a mass basis. Bias arises with unsaturated hydrocarbons (alkenes and aromatics) because they have a higher carbon-to-mass ratio than saturated hydrocarbons (5% to 10% difference). While this ratio can be calculated for any compound, it needs to be measured experimentally when dealing with complex mixtures because it cannot be calculated accurately. This issue can lead to overestimating the concentration of hydrocarbon present.

These two bias issues together are related to THC calibration errors, the most important errors associated with the use of the system. While the THC system can be routinely calibrated for carbon response using a single hydrocarbon, the carbon-to-mass ratio of the measured vapors needs to be known to convert the results back to a mass basis. The only way to accomplish this calibration is to generate a known vapor concentration of the same chemical composition as the one that will be measured during the exposure. This is technically challenging, and only one group has published such a technique [40].

Most of the cited studies have used hexane or heptane for calibration, while some studies did not specify how the calibration was performed. The bias mainly affects aromatics, which are an important part of neat JP-8, may provide significant human and animal toxicity in jet fuel exposures, and are potentially a major part of the vapors in an inhalation chamber as shown in Table 12.1. It is rather challenging to attempt to evaluate the impact of using a single saturated hydrocarbon for the calibration of a system used to measure numerous other hydrocarbons. Fortunately, the two previously discussed THC system biases (variable signal response per carbon and variable carbon-to-molecular weight ratios) act in opposite manners. Nonetheless, we believe that even a significant concentration of aromatic compounds in a jet fuel mixture is not likely to bias the results by much more than 5% to 10%.

All the studies referenced in Table 12.3 using THC analysis were measuring a vapor-only atmosphere, except for two studies where aerosols were detected [8, 19]. Although THC instruments are not designed to analyze droplets, they can sometimes analyze aerosol droplets, depending on their engineering design and operational settings. The primary challenge for adequate detector response is the proper delivery of the aerosol droplets as a vapor to the flame. Tubing diameter and flow must be selected to generate laminar flows at the tubing entry point inside the chamber. Without laminar flow, aerosol droplets will not be sampled quantitatively. Furthermore, THC instruments are typically set up 1 to 3 meters from the actual chamber. Particles can potentially impact inside the transfer line. High-molecular-weight components (low vapor pressure) are likely to stay inside the tubing and not evaporate if they impact. Finally, THC instruments have complex tubing systems with pumps, valves, and other components that are likely to interact with the aerosol. Fechter et al. [8] used an oven-heated THC analyzer where all of these components were heated to 160°C. Operating at such a temperature would allow for the nearly complete volatilization of the aerosol upon entry into the THC analyzer. Herrin et al. [19] did not list the model used, but only the company, a company that has both oven-heated and non–oven-heated THC systems. The use of a THC analyzer without an oven or at lower temperature will generally underestimate the total hydrocarbon quantity. The extent of this bias depends on the amount of aerosol present, the system used, and the operational settings.

Gas Chromatography-Flame Ionization Detection

Similar to THC analysis, gas chromatography coupled with flame ionization detection (GC-FID) has been used to obtain total hydrocarbon concentration in fuel atmospheres [4, 5, 31]. The calibration of this system suffers the same issues as THC analysis because it is based on the same detection technique. Sample introduction can be performed offline where an air sample is collected from the chamber using a gas-tight syringe and injected into the GC-FID apparatus. This offline technique is only suitable for gas-phase exposure, as the collection of aerosol using a syringe is not possible. This system can also be used online with an automated collection/injection system. It is usually made of a heated transfer line with continuous flow and a fixed-volume loop that can be used to inject a sample into the GC-FID apparatus. This technique allows sample collection/analysis every few minutes. The use of a heated transfer line allows the use of this technique with aerosols. The transfer system (loop) also must be kept at the same or higher temperature to minimize condensation.

Infrared Spectroscopy

Infrared (IR) spectroscopy is also a very common technique used to obtain a total concentration [22, 23, 36, 40, 41, 44, 53, 54]. The technique is based on the absorption of energy in the mid-infrared region (4,000 to 400 cm^{-1}; 30 to 1.4 μm) by covalent bonds. The absorption is observed at specific wavelengths that correspond to the vibration and rotation of bonds. Functional groups absorb in different regions according to their composition and structure, which makes IR spectroscopy a very good identification technique. The C–H bond is of most interest for fuel, and absorption of this bond is between 2,850 and 3,000 cm^{-1} for an sp^3 carbon (alkanes), above 3,000 cm^{-1} for an sp^2 carbon (alkenes and aromatics), and near 3,300 cm^{-1} for an sp carbon (alkynes). In the context of this work, IR spectroscopy is a vapor-phase-only technique and is not suitable for quantifying aerosols. In fact, droplets must be removed if present. All the studies presented above were vapor-phase-only exposure or the aerosol was filtered prior to entering the IR spectrometer.

Calibration of complex mixtures, such as fuel, is a problem with IR spectroscopy. Similar to the THC technique, calibration must be performed with a standard of exact composition. Again, such a technique has only been achieved by one group [40]. All the other studies mentioned above used a single compound (hexane vapor) to calibrate the IR spectrometer in the range of 3.4 to 3.5 μm (2,941 to 2,857 cm^{-1}), which is absorption from the C–H stretch of saturated hydrocarbons. There are two issues with this calibration technique. First, that range of wavelengths is only responsive to C–H bonds from saturated hydrocarbons. The only other bond that absorbs in this range is the O–H bond from a carboxylic acid, which is not expected to be significant for fuel vapors. Unsatured hydrocarbons do not absorb in this range and therefore are not quantified. This is especially significant for aromatic compounds, which are a significant part of petroleum-based fuels and even more important in the gas phase. The other issue is the choice of one compound to calibrate the instrument. The absorbance per mass can vary by 10% within saturated hydrocarbons. These two issues together can lead to significant underestimation. Jet A vapors have been shown to absorb about 40% less per weight than heptane at 3.39 μm [21]. This agrees with Reboulet et al. [40], who found that Jet A vapors were about 60% less absorbing than those of hexane per weight. The main reason for these differences comes from measuring only bond absorptions from saturated C–H bonds, while non-saturated hydrocarbons are not measured.

Considering the above, it is likely that the use of IR spectroscopy to quantify jet fuel vapors has led to the underestimation of fuel exposure concentrations. The extent of this bias is difficult to estimate but a figure of 50%, or so, appears to be a good estimate [21, 40].

Adsorbent Tubes

Charcoal tubes have been used quantitatively to sample gas-phase compounds [46, 49, 50]. Samples must be individually collected, which limits the number of data points that can be obtained over the course of a study. However, this method allows the analyses to be performed at a later time. Extraction of the samples can be done by chemical or thermal desorption of the hydrocarbons from the charcoal. Thermal desorption is far superior to chemical (solvent) desorption, because it eliminates possible preferential desorption issues and potential loss of volatile components from the extract. It also has the advantage of being less time consuming than other techniques.

Gas chromatography along with flame ionization detection (GC-FID) or mass spectrometry (GC-MS) is usually the chosen technique for quantification. Calibration issues are similar to THC, as discussed earlier.

The main advantage and power of the adsorbent tube technique is its ability to provide compositional information. Some studies have used this technique to obtain limited, but significant compositional characterization of the chamber gas-phase atmosphere [4, 6, 10, 36, 44, 46, 49, 50]. Calibration for a series of compounds of interest can be performed with authentic standards, and accurate compound concentrations in the gas phase can be obtained. This is particularly important for pharmacokinetic studies where compound-specific exposure concentrations are needed. It is also a useful tool to relate the generated exposure to occupation exposure, as one or more chemical markers are often chosen to track the occupational fuel exposure.

Aerosol Characterization

The techniques presented focus on the quantification of gas-phase compounds. While some (e.g., THC analysis and online GC-FID) can provide the total combined concentration of the vapor and aerosol content, none of these methods can determine the aerosol concentration separately. Techniques that focus on measuring aerosol composition and concentration are discussed below.

Filters

Filtering air for particles is one of the oldest and simplest techniques to collect aerosols. Filters can then be weighted or chemically desorbed for quantification. This sampling technique along with chemical desorption has been used to quantify the aerosol fraction in some studies [49, 50]. It has also been used to chemically characterize the aerosol [6, 8, 10]. While this technique is fairly standard in the environmental pollution and industrial hygiene fields, its application to jet fuel exposure is not without limitations. The primary concern is the volatilization of the fuel droplets (aerosol) from the filter during and after the collection phase. As discussed earlier, jet fuel contains a series of volatile compounds, and proper sampling of the aerosol is extremely challenging. Filters (glass fiber and quartz fiber) do not have the ability to retain the compounds once sampled, like charcoal, which is commonly used for gas-phase compounds. Using this method, Whitman and Hinz [49, 50] reported that their total measured concentrations (gas phase and aerosol) were below the expected nominal concentrations. For JP-8, the difference was about 30% to 45% between the measured and nominal total concentrations. It is likely that they underestimated the aerosol concentrations, considering that they used filters to collect and quantify the aerosol. Our laboratory found that aerosol concentrations for JP-8 and S-8 exposures, calculated from weighing filters, were 30% to 35% of the actual aerosol concentration as measured by the tube differential method, as is discussed below [46]. An aerosol fraction of about 8% would have been calculated using the filter data, while the actual fraction was about 23%. It is likely that studies using filters to collect and measure aerosol have significantly underestimated the amount of aerosol present in the chamber atmosphere. Underestimating the aerosol concentration in a chamber is less important for exposures where limited amounts of aerosol are present.

Adsorbent Tube

Researchers in our laboratory have developed a method using a dual charcoal setup to also determine aerosol compositions [46]. One charcoal tube is collected behind a filter to remove aerosol (i.e., sampling gas phase only), while a second tube is collected directly from the chamber without a filter (i.e., sampling gas phase and aerosol). Both tubes are analyzed using thermal desorption onto a GC-MS (gas chromatography-mass spectrometry). The mathematical difference between the two tubes represents the aerosol fraction. This method has significant advantages over the use of filters. First, aerosols are sampled on an adsorbing media (charcoal), as opposed to a filter without such adsorbing capacity, thus reducing loss by evaporation. Second, only a small sample amount is needed, as the thermal desorption coupled with mass spectrometry (MS) is a very sensitive technique. This reduces the sampling time and therefore reduces sampling artifacts. Finally, the thermal desorption technique uses no solvent, which removes any possible preferential solvent desorption issues and also allows its use off-site. This technique works best when a significant concentration of aerosol is present, because the aerosol amount/concentration is calculated from the differential of two large numbers (vapor only and vapor/aerosol total).

Optical Aerosol Measurement

Particle counters or real-time aerosol monitors have been used to monitor aerosol levels, but without providing quantification [9, 29, 32, 47, 49, 50, 53, 84, 85]. Most of these studies used this technique to ensure only that no aerosol content was present. Calibration of these systems is very complex and many assumptions must be made to convert the signal (particle count or light scattering intensity) to a total mass/volume concentration. They also offer no chemical composition information. Nonetheless, these instruments have the potential to accurately measure fuel aerosol without disturbing the vapor-aerosol equilibrium, which is a significant problem with other techniques. Some of these instruments can also provide droplet size information, which is almost impossible to obtain accurately using any other method. It is perhaps surprising to observe how infrequently these systems are used in a quantitative manner for fuel research. This is most likely due to the complexity of operation and calibration; the high aerosol concentrations, which require dilution apparatus; and the high cost of the instrumentation. Of all the techniques presented in this chapter, optical aerosol measurement techniques can potentially offer the most accurate aerosol quantitation.

Aerosol Size Determination

While optical measurement techniques have potential, the most common instrument for aerosol size characterization in this field of research is the cascade impactor. Impactors differentiate particle size based on their aerodynamic diameters using geometry and gas flow. Particles larger than a given aerodynamic diameter impact a surface while particles smaller than that given diameter will stay in the gas stream. The cut point (aerodynamic diameter) is defined as the diameter where 50% of the particle impacts. A cascade impactor consists of a number of individual impactors in series with reducing cut sizes. The resulting impactor provides a series of samples, each from a different size range. These samples can then be weighted or extracted for further analysis.

Cascade impactors rely on either a large pressure drop across the impactor or small jets (micro-orifice) to create the fast gas flow needed to separate the smallest particles. Pressures down to about 70% of ambient pressure can be found in lower stages (manufacturer data from In-Tox Products, Moriarty, NM). This can lead to significant evaporation of the aerosol droplets in the gas stream and also, once collected, on the impaction plates. MOUDI impactors rely on micro-orifices to generate the necessary conditions for particle separation while keeping the internal pressure above 90% of ambient pressure [30]. In either case, significant aerosol loss occurs during sampling from particles not impacting on the plates and evaporation. Impactors should only be used to obtain size information and some limited compositional information, and not to calculate aerosol concentrations. Our laboratory group has found that aerosol concentrations derived from the weight collected on the impactor plates underestimate the aerosol concentration by at least 50% for JP-8 exposure. Unfortunately, a series of studies has used cascade impactors to derive chamber aerosol exposure concentrations, and did not concurrently measure vapor concentrations [1, 2, 7, 11–18, 33, 37, 38, 42, 43,48, 51, 52, 55]. In one study, a THC analyzer was used to compare the impactor-derived data with the data collected from the THC analyzer [19]. While they did not report the data from both instruments, they reported a correction factor of 8.1, which can be used to convert the aerosol concentration published in their prior work to a total atmosphere concentration (vapor and aerosol). The correction factor appears to have been calculated from biological responses (e.g., lung injury) and not from the concentration difference obtained from a comparison of the data from the impactor and THC techniques. While the aerosol concentrations reported in these studies might be converted to a total hydrocarbon concentration, it is clear that the aerosol concentrations reported have dramatically underestimated the actual concentration present. This issue must be considered when interpreting toxicology findings from a number of published studies.

CONCLUSIONS

The generation and characterization of jet fuel atmospheres for animal research is complex. The fuel itself represents a challenge because it is made up of hundreds of compounds with different chemical, biological, and physical properties. The generation and characterization of an experimental atmosphere is complicated by the fact that the chemical constituents of the "end-result" exposure may vary substantially from those found in the neat fuel.

A series of methods to generate vapor and aerosol from jet fuel were discussed. The heated tube method appears to be the best suited for vapor-only exposure, while single pass atomization is the key for aerosol/vapor exposure. Both methods can deliver exposures with constant chemical composition.

Characterization of the exposures is also a challenge, especially when aerosols are present. THC and GC-FID appear to be the best-suited instruments to quantify fuel vapor. While IR can be as useful, calibration with a single compound can lead to more bias than with THC. In either case, the use of an identical calibration standard is the best way to achieve accurate results. Aerosol characterization is still rarely sufficient. Collection on filters and impactors has led to numerous errors reported in the literature and underestimates the aerosol exposure level. While online instruments have been used to monitor aerosol levels, they have generally not been used for quantification. These instruments have the greatest potential to provide accurate data because they can measure the aerosol.

Finally, the generation and characterization issues discussed in this chapter must be taken into account when interpreting results from published studies. Many toxicology studies have limited reporting on their analytical methods, which complicates the interpretation of the data. There is a need for better descriptions of how complex fuel exposures are generated and characterized in future studies reported in the scientific literature.

REFERENCES

[1] Baldwin, C.M., Figueredo, A.J., Wright, L.S., et al. Repeated aerosol-vapor JP-8 jet fuel exposure affects neurobehavior and neurotransmitter levels in a rat model. *J. Toxicol. Environ. Health, Part A*, 70, 1203, 2007.

[2] Baldwin, C.M., Houston, F.P., Podgornik, M.N., et al. Effects of aerosol-vapor JP-8 jet fuel on the functional observational battery, and learning and memory in the rat. *Arch. Environ. Health,* 56, 216, 2001.

[3] Bruner, R.H., Kinkead, E.R., Oneill, T.P., et al. The toxicologic and oncogenic potential of JP-4 jet fuel vapors in rats and mice-12-month intermittent inhalation exposures. *Fundamen. Appl. Toxicol.*, 20, 97, 1993.

[4] Campbell, J.L., and Fisher, J.W. A PBPK modeling assessment of the competitive metabolic interactions of JP-8 vapor with two constituents, *m*-xylene and ethylbenzene. *Inhalat. Toxicol.,* 19, 265, 2007.

[5] Carpenter, C.P., Geary, D.L., Myers, R.C., et al. Petroleum hydrocarbon toxicity studies. II. Animal and human response to vapors of deodorized kerosene. *Toxicol. Appl. Pharmacol.*, 36, 443, 1976.

[6] Dietzel, K.D., Campbell, J.L., Bartlett, M.G., et al. Validation of a gas chromatography/mass spectrometry method for the quantification of aerosolized Jet Propellant 8. *J. Chromatogr. A,* 1093, 11, 2006.

[7] Drake, M.G., Witzmann, F.A., Hyde, J., et al. JP-8 jet fuel exposure alters protein expression in the lung. *Toxicology,* 191, 199, 2003.

[8] Fechter, L.D., Gearhart, C., Fulton, S., et al. JP-8 jet fuel can promote auditory impairment resulting from subsequent noise exposure in rats. *Toxicol. Sci.,* 98, 510, 2007.

[9] Gaworski, C., MacEwen, J., Vernot, E., et al. Comparison of the subchronic inhalation toxicity of petroleum and oil shade JP-5 jet fuels. In *Advances in Modern Environmental Toxicology, Vol. 7. Renal Effects of Petroleum Hydrocarbons*, Mehlman, M., Hemstreet III, G., Thorpe, J., et al., Eds., Princeton Scientific Publisher, Princeton, NJ, 1984, pp. 33–48.

[10] Gregg, S.D., Campbell, J.L., Fisher, J.W., et al. Methods for the characterization of Jet Propellent-8: vapor and aerosol. *Biomed. Chromatogr.*, 21, 463, 2007.

[11] Harris, D.T., Sakiestewa, D., Robledo, R.F., et al. Immunotoxicological effects of JP-8 jet fuel exposure. *Toxicol. Ind. Health,* 13, 43, 1997a.

[12] Harris, D.T., Sakiestewa, D., Robledo, R.F., et al. Protection from JP-8 jet fuel induced immunotoxicity by administration of aerosolized substance P. *Toxicol. Ind. Health*, 13, 571, 1997b.

[13] Harris, D.T., Sakiestewa, D., Robledo, R.F., et al. Short-term exposure to JP-8 jet fuel results in long-term immunotoxicity. *Toxicol. Indus. Health,* 13, 559, 1997c.

[14] Harris, D.T., Sakiestewa, D., Robledo, R.F., et al. Effects of short-term JP-8 jet fuel exposure on cell-mediated immunity. *Toxicol. Indus. Health*, 16, 78, 2000a.

[15] Harris, D.T., Sakiestewa, D., Titone, D., et al. Jet fuel-induced immunotoxicity. *Toxicol. Indus. Health,* 16, 261, 2000b.

[16] Harris, D.T., Sakiestewa, D., Titone, D., et al. Substance P as prophylaxis for JP-8 jet fuel-induced immunotoxicity. *Toxicol. Indus. Health*, 16, 253, 2000c.

[17] Harris, D.T., Sakiestewa, D., Titone, D., et al. JP-8 jet fuel exposure rapidly induces high levels of IL-10 and PGE2 secretion and is correlated with loss of immune function. *Toxicol. Indus. Health,* 23, 223, 2007.

[18] Harris, D.T., Sakiestewa, D., Titone, D., et al. JP-8 jet fuel exposure results in immediate immunotoxicity, which is cumulative over time. *Toxicol. Indus. Health,* 18, 77, 2002.

[19] Herrin, B.R., Haley, J.E., Lantz, R.C., et al. A reevaluation of the threshold exposure level of inhaled JP-8 in mice. *J. Toxicol. Sci.*, 31, 219, 2006.

[20] Kallai, M., Veres, Z., and Balla, J. Response of flame ionization detectors to different homologous series. *Chromatographia*, 54, 511, 2001.

[21] Klingbeil, A.E., Jeffries, J.B., and Hanson, R.K. Temperature- and pressure-dependent absorption cross sections of gaseous hydrocarbons at 3.39 mu m. *Meas. Sci. Technol.,* 17, 1950, 2006.

[22] Lin, B.C., Ritchie, G.D., Rossi, J., et al. Identification of target genes responsive to JP-8 exposure in the rat central nervous system. *Toxicol. Indus. Health,* 17, 262, 2001.

[23] Lin, B.C., Ritchie, G.D., Rossi, J., et al. Gene expression profiles in the rat central nervous system induced by JP-8 jet fuel vapor exposure. *Neurosci. Lett.,* 363, 233, 2004.

[24] Liu, B.Y.H., and Lee, K.W. Aerosol generator of high stability. *Am. Ind. Hyg. Assoc. J.*, 36, 861, 1975.

[25] MacEwen, J., and Vernot, E. Toxic Hazards Research Unit Annual Technical Report. AMRL-TR-81-126., Aerospace Medical Research Laboratory, Wright-Patterson Air Force Base, Dayton, OH, 1981.

[26] MacEwen, J., and Vernot, E. Toxic Hazards Research Unit Annual Technical Report. AMRL-TR-82-62, Aerospace Medical Research Laboratory, Wright-Patterson Air Force Base, Dayton, OH, 1982.

[27] MacEwen, J., and Vernot, E. Toxic Hazards Research Unit Annual Technical Report. AMRL-TR-84-001, Aerospace Medical Research Laboratory, Wright-Patterson Air Force Base, Dayton, OH, 1984.

[28] MacEwen, J., and Vernot, E. Toxic Hazards Research Unit Annual Technical Report. AMRL-TR-85-058, Aerospace Medical Research Laboratory, Wright-Patterson Air Force Base, Dayton, OH, 1985.

[29] MacNaughton, M., and Uddin, D. Toxicology of mixed distillate and high energy synthetic fuels. In *Advances in Modern Environmental Toxicology, Vol. 7. Renal Effects of Petroleum Hydrocarbons,* Mehlman, M., Hemstreet III, G., Thorpe, J., and Weaver, N., Eds., Princeton Scientific Publisher, Princeton, NJ, 1984, pp. 121–132.

[30] Marple, V., Rubow, K., and Behm, S.A. microorifice uniform deposit impactor (MOUDI): description, calibration, and use. *Aerosol Sci. Technol.*, 16, 434, 1991.

[31] Martin, S.A., Flynt, K., Kendrick, C., et al. Inhalation kinetics of jet fuel components in the rat, Abstract #948. *The Toxicologist CD – J. Soc. Toxicol.,* 102, 2008.

[32] Mattie, D.R., Alden, C.L., Newell, T.K., et al. A 90-day continuous vapor inhalation toxicity study of JP-8 jet fuel followed by 20 or 21 months of recovery in Fischer-344 rats and C57BL/6 mice. *Toxicol. Pathol.,* 19, 77, 1991.

[33] McGuire, S., Bostad, E., Smith, L., et al. Increased immunoreactivity of glutathione-S-transferase in the retina of Swiss Webster mice following inhalation of JP8+100 aerosol. *Arch. Toxicol.,* 74, 276, 2000.

[34] National Research Council of the National Academies, Ed. *Toxicologic Assessment of Jet-Propulsion Fuel 8.* The National Academies Press, Washington, D.C., 2003

[35] Newton, P.E., Becker, S.V., and Hixon, C.J. Pulmonary function and particle deposition and clearance in rats after a 90-day exposure to shale-oil-derived jet fuel JP-4. *Inhalat. Toxicol.,* 3, 195, 1991.

[36] Nordholm, A.F., Rossi, J., Ritchie, G.D., et al. Repeated exposure of rats to JP-4 vapor induces changes in neurobehavioral capacity and 5-HT/5-HIAA levels. *J. Toxicol. Environ. Health-Part A,* 56, 471, 1999.

[37] Pfaff, J., Parton, K., Lantz, R.C., et al. Inhalation exposure to JP-8 jet fuel alters pulmonary-function and substance-P levels in Fischer-344 rats. *J. Appl. Toxicol.*, 15, 249, 1995.

[38] Pfaff, J.K., Tollinger, B.J., Lantz, R.C., et al. Neutral endopeptidase (NEP) and its role in pathological pulmonary change with inhalation exposure to JP-8 jet fuel. *Toxicol. Indus. Health*, 12, 93, 1996.

[39] Rau, J. Design principles of liquid nebulization devices currently in use. *Respir. Care,* 47, 1257, 2002.

[40] Reboulet, J., Cunningham, R., Gunasekar, P.G., et al. Development of an Infrared Spectrophotometric Method for the Analysis of Jet Fuel Using a Loop Calibration Technique. Naval Health Research Center-Environmental Health Effects Laboratory, Unpublished Report, Wright-Patterson AFB, OH, 2007.

[41] Ritchie, G.D., Rossi, J., Nordholm, A.F., et al. Effects of repeated exposure to JP-8 jet fuel vapor on learning of simple and difficult operant tasks by rats. *J. Toxicol. Environ. Health, Part A,* 64, 385, 2001.

[42] Robledo, R.F., and Witten, M.L. Acute pulmonary response to inhaled JP-8 jet fuel aerosol in mice. *Inhalat. Toxicol.,* 10, 531, 1998.

[43] Robledo, R.F., Young, R.S., Lantz, R.C., et al. Short-term pulmonary response to inhaled JP-8 jet fuel aerosol in mice. *Toxicol. Pathol.*, 28, 656, 2000.

[44] Rossi, J., Nordholm, A.F., Carpenter, R.L., et al. Effects of repeated exposure of rats to JP-5 or JP-8 jet fuel vapor on neurobehavioral capacity and neurotransmitter levels. *J. Toxicol. Environ. Health, Part A*, 63, 397, 2001.

[45] Starek, A., and Vojtisek, M. Effects of kerosene hydrocarbons on tissue metabolism in rats. *Polish J. Pharmacol. Pharm.*, 38, 461, 1986.

[46] Tremblay, R.T., Martin, S.A., Fechter, L.D., et al. A novel method for the chemical characterization of generated jet fuel vapor and aerosol for animal studies, Abstract 1100. *The Toxicologist CD — J. Soc. Toxicol.*, 102, 2008.

[47] Wall, H., Vingegar, A., and Kinkead, E. Toxic Hazards Research Unit Annual Technical Report. AMRL-TR-90-063, Aerospace Medical Research Laboratory, Wright-Patterson Air Force Base, Dayton, OH, 1990.

[48] Wang, S.J., Young, R.S., and Witten, M.L. Age-related differences in pulmonary inflammatory responses to JP-8 jet fuel aerosol inhalation. *Toxicol. Indus. Health*, 17, 23, 2001.

[49] Whitman, F.T., and Hinz, J.P. Sensory Irritation Study in Mice: JP-4, JP-8, JP-8+100, USA Force and Exxon Mobil Biomedical Sciences, Inc., 1, 2001.

[50] Whitman, F.T., and Hinz, J.P. Sensory Irritation Study in Mice: JP-5, JP-TS, JP-7, DFM, JP-10, USA Force and Exxon Mobil Biomedical Sciences, Inc., 1, 2004, p. 1.

[51] Witzmann, F.A., Bauer, M.D., Fieno, A.M., et al. Proteomic analysis of simulated occupational jet fuel exposure in the lung. *Electrophoresis,* 20, 3659, 1999.

[52] Witzmann, F.A., Bauer, M.D., Fieno, A.M., et al. Proteomic analysis of the renal effects of simulated occupational jet fuel exposure. *Electrophoresis*, 21, 976, 2000a.

[53] Witzmann, F.A., Carpenter, R.L., Ritchie, G.D., et al. Toxicity of chemical mixtures: Proteomic analysis of persisting liver and kidney protein alterations induced by repeated exposure of rats to JP-8 jet fuel vapor. *Electrophoresis,* 21, 2138, 2000b.

[54] Witzmann, F.A., Bobb, A., Briggs, G.B., et al. Analysis of rat testicular protein expression following 91-day exposure to JP-8 jet fuel vapor. *Proteomics,* 3, 1016, 2003.

[55] Wong, S.S., Hyde, J., Sun, N.N., et al. Inflammatory responses in mice sequentially exposed to JP-8 jet fuel and influenza virus. *Toxicology*, 197, 2004.

13 Genetic Damage in the Blood and Bone Marrow of Mice Treated with JP-8 Jet Fuel

Vijayalaxmi

CONTENTS

INTRODUCTION

The jet fuel JP-8 is used primarily by the U.S. Department of Defense to power aircraft and land vehicles, and contains complex mixtures of aliphatic, aromatic, and other substituted naphthalene hydrocarbon compounds [Riviere et al., 1999]. The health effects of exposures to JP-8 on cardiovascular, developmental, immune, renal, neuronal, respiratory tract, and reproductive systems as well as carcinogenicity were reviewed recently [NRC, 2003]. In some investigations, "potential" adverse effects were reported in the respiratory, immune, and nervous systems in animals as well as in humans. With respect to genotoxicity, earlier in vitro investigations failed to demonstrate mutagenic effects induced by jet fuels. However, observations made in recent studies suggest significant increases in DNA single-strand breaks in mammalian cells exposed in vitro and in vivo to petroleum derivatives, engine exhausts, and jet fuels, including JP-8 [Brusick and Matheson, 1978; McKee et al., 1994; Lemasters et al., 1997, 1999; Pitarque et al., 1999; Grant et al., 2000, 2001; Rogers et al., 2001; Jackman et al., 2002]. Considering the apparently contradictory reports related to the genotoxic effects of JP-8 and other related jet fuels, the National Research Council (NRC) recommended that future studies in animals should include evaluation of in vivo genotoxicity [NRC, 2003].

A series of investigations was conducted in mice treated on the skin with JP-8 to determine the extent of genetic damage. A rodent micronucleus test was used to measure the incidence of MN in polychromatic erythrocytes (PCE) in blood and bone marrow tissues. The data in experimental mice were compared with those in untreated controls as well as in positive control animals. The investigations were described in detail in earlier publications [Vijayalaxmi et al., 2004, 2006], and the data are summarized in this chapter.

MATERIALS AND METHODS

All investigations and experimental procedures described below were conducted in collaboration with Dr. Stephen E. Ullrich at the Department of Immunology, M.D. Anderson Cancer Center (Houston, Texas). The protocols for handling the animals were reviewed and approved by its Institutional Animal Care and Use Committee. JP-8 (Lot# 3509) was supplied by the Operational Toxicology Branch, Air Force Research Laboratory, Wright-Patterson Air Force Base (Dayton, Ohio).

Adult C3H/HeNCr (MTV-) mice were obtained from the National Cancer Institute Frederick Cancer Research Facility Animal Production Area (Frederick, Maryland). They were acclimatized for 7 days and then randomized to different groups:

Group 1: Untreated controls.

Group 2: Two hours before the start of the experiment(s), an area of approximately 8 cm^2 on the backs of all mice was shaved and treated on the skin with JP-8. Undiluted JP-8 was applied directly to the shaved dorsal skin with a micropipette. The mice were then caged individually in a chemical fume hood for the next 3 hours, at the end of which the residual fuel had been either absorbed through the skin or had evaporated. The animals were then returned to standard housing in the animal facility. The dose of JP-8, treatment schedule, and euthanization of mice were different in different experiments (see below). The highest single dose used was 300 μl (240 mg/mouse) because, in previous studies, this dose was found to be immunotoxic [Ullrich, 1999; Ramos et al., 2002].

Groups 3: Single intraperitoneal injection of cyclophosphamide (40 mg/kg body weight), a known genotoxic agent (positive control).

Peripheral blood and bone marrow smears were prepared, fixed in absolute methanol, and air-dried. Coded slides were then sent to the Department of Radiation Oncology, University of Texas Health Science Center (UTHSC) in San Antonio, where all slides were stained with acridine orange (0.01 mg/ml in 0.2M sodium phosphate buffer, pH 7.4). Dr. Vijayalaxmi examined one complete set of slides at UTHSC, while a duplicate set was mailed to the Environmental Carcinogenesis Division, U.S. Environmental Protection Agency (U.S. EPA) (Research Triangle Park, North Carolina) where Dr. Andrew D. Kligerman examined the slides. Thus, two independent investigators (the UTHSC and U.S. EPA) assessed the extent of genetic damage, assessed from the examination of 2,000 consecutive PCE in each mouse and in each tissue (Heddle et al., 1984). The results were decoded after complete microscopic evaluation by both investigators. Because there were no significant differences in MN indices recorded by both investigators, the results were pooled and subjected to various statistical analyses using the SAS user guide [SAS, 1977]. Statistical significance was taken at a level of $p < 0.05$ [Vijayalaxmi et al., 2004, 2006].

JP-8 Treatment(s) and Tissue Collection Schedule(s) for Group 2 Mice

Experiment 1. Single treatment and tissue collection at different times.

(a) 300 μl. Blood smears were prepared at 0, 24, 48, and 72 hours following treatment with JP-8. Bone marrow smears were prepared when the animals were euthanized at 72 hours following the treatment.

Experiment 2. Three treatments [Tice et al., 1990]. 50 μl (total 150 μl), 100 μl (total 300 μl), or 300 μl (total 900 μl) each on 3 consecutive days. Blood and bone marrow smears were prepared at 24 hours after the last treatment.

Experiment 3. Single or weekly treatments.

(a) 300 μl single treatment. Blood smears were prepared at 72 hours after treatment (as in Experiment 1).

(b) 300 μl each on 3 consecutive days (total 900 μl). Blood smears were prepared at 24 hours after the last treatment (as in Experiment 2).

(c) 300 μl each, three weekly treatments, 7 days apart, on the same mouse (total 900 μl). Blood smears were prepared at 72 hours after the first, second, and third weekly treatment.

RESULTS

The group mean incidence of MN in 2000 PCE (+ standard deviation and confidence intervals at 95% level of the means) recorded in all three separate experiments are summarized in Table 13.1. The data from the mice treated with a single 300 μl of JP-8 were not significantly different from those observed in the concurrent untreated control animals; the only exception was in the blood samples examined at 72 hours following treatment with JP-8. Experiments 2 and 3 were conducted using different doses of JP-8 and several treatment schedules; the data from these studies (i.e., treatment with increasing doses of JP-8 on three consecutive days, or on a weekly basis) showed no statistically significant increases in the incidence of MN as compared to those in concurrent untreated control animals. As expected, in all experiments, significantly elevated frequencies of MN were observed in positive control animals that were injected with cyclophosphamide.

DISCUSSION

Genetic toxicology tests are important and have been routinely used to predict the carcinogenic potential of biological, chemical, and physical agents [Tennant and Zeiger, 1993; Zeiger, 2001]. The rodent MN assay has been widely applied as an in vivo assay for detecting genotoxic agents and became a standard test system for genotoxicity evaluations in regulatory agencies in several countries [Auletta et al., 1993; Health Protection Branch Genotoxicity Committee, 1993; Kirkland, 1993; Sofuni, 1993]. The genotoxic potential of JP-8 was evaluated in a series of investigations in experimental mice using various treatment schedules, including repeated

TABLE 13.1
Incidence of Micronuclei (MN) in the Blood and Bone Marrow of Mice Treated on the Skin with Various Doses of JP-8 and Treatment Regimens

Group	No. Mice Studied	Smears Prepared Hour	MN/2000 PCE Blood	MN/2000 PCE Bone Marrow
Experiment 1				
Untreated controls	10	0	4.7 ± 1.6	
		24	5.4 ± 2.1	
		48	6.4 ± 2.5	
		72	4.4 ± 3.4	5.3 ± 3.1
(a) 300 μl	10	0	4.4 ± 1.6	
		24	5.2 ± 2.2	
		48	7.3 ± 2.4	
		72	8.2 ± 4.6[a]	7.4 ± 2.5
Positive controls	10	0	4.9 ± 2.2	
		24	14.0 ± 6.8[a]	
		48	34.9 ± 10.4[a]	
		72	14.5 ± 3.3[a]	9.3 ± 2.4[a]
Experiment 2				
Untreated controls	10	24	4.4 ± 1.8 (3.1–5.7)	5.1 ± 1.5 (4.0–6.2)
(b) 3 × 50 μl	10	24	4.8 ± 1.4 (3.8–5.8)	4.2 ± 1.8 (2.9–5.5)
(b) 3 × 100 μl	10	24	4.6 ± 1.1 (3.8–5.4)	4.6 ± 0.7 (4.1–5.0)
(b) 3 × 300 μl	10	24	5.0 ± 1.3 (4.1–5.9)	4.8 ± 0.8 (4.2–5.4)
Positive controls	10	24	28 ± 4.8 (24.5–31.5)[a]	25.2 ± 8.2 (19.3–31.0)[a]
Experiment 3				
Untreated controls	10	72	4.4 ± 1.4 (3.4–5.4)	Not done
(a) 300 μl (repeat of Expt. 1)	10	72	4.9 ± 0.8 (4.3–5.4)	Not done
(b) 3 × 300 μl (repeat of Expt. 2)	10	24	4.9 ± 1.5 (4.3–5.4)	Not done
(c) 300 μl – week 1	10	72	5.2 ± 1.9 (4.0–6.7)	Not done
(c) 300 μl – week 2	10	72	4.5 ± 1.8 (3.2–5.8)	Not done

TABLE 13.1 (*Continued*)
Incidence of Micronuclei (MN) in the Blood and Bone Marrow of Mice Treated on the Skin with Various Doses of JP-8 and Treatment Regimens

Group	No. Mice Studied	Smears Prepared Hour	MN/2000 PCE	
			Blood	Bone Marrow
(c) 300 μl – week 3	10	72	4.4 ± 1.8 (3.1–5.6)	Not done
Positive controls	5	72	36.0 ± 7.3 (29.6–42.4)[a]	Not done

Note: (a): Treatment on a single day. Smears prepared at 72 hours after the treatment. (b): Treatment on 3 consecutive days. Smears prepared at 24 hours after the last treatment. (c): Treatment at weekly intervals. Smears prepared at 72 hours after each weekly treatment. Data are Mean + Standard Deviation (confidence intervals at 95% level of the means).

[a] Significant difference between cyclophosphamide and all other groups.

application because of such exposure regimen in jet fuel handlers and aircraft maintenance personnel.

The overall results indicated no significant increase in the incidence of MN in both blood and bone marrow tissues in mice treated on the skin with JP-8 as compared with concurrent untreated control animals. The only exception was in Experiment 1, where a significant increase in MN was observed in blood samples collected at 72 hours following a single treatment with 300 μl of JP-8 [Vijayalaxmi et al., 2004]. However, this result could not be confirmed in the subsequent Experiment 3 [Vijayalaxmi et al., 2006]. The absence of a statistically significant increase in MN in mice treated with three consecutive 300-μl doses of JP-8 in Experiment 2 was, in fact, confirmed in the repeat Experiment 3. More than likely, the exceptional observation made in Experiment 1 was "false positive." Such false-positive results were reported to happen even in well-designed experiments using adequate sample size. In a mouse bone marrow MN assay, 13 of the 49 chemicals evaluated (27%) gave contradictory results in repeat experiments; the final conclusion was that the data were "inconclusive" and "the results of a single test are not always definitive or reproducible" [Shelby et al., 1993]. Among the JP-8 investigations, the small but significant increase in MN in blood at 72 hours following the treatment was not considered definitive because this observation could not be confirmed in subsequent studies. The overall results cannot be "inconclusive" because in the subsequent studies, (1) higher amounts of JP-8 were applied, and (2) several different protocols were used, all of which indicated no significant induction of MN in blood and bone marrow tissues.

ACKNOWLEDGMENTS

This work is supported by the U.S. Air Force Office of Scientific Research through research grants F4962003-1-0079 (V) and F49620-1-0102 (SEU). Sincere thanks to the excellent collaborative effort of Stephen E. Ullrich, Department of Immunology, M.D. Anderson Cancer Center, Houston, Texas, and Dr. Andrew D. Kligerman, Environmental Carcinogenesis Division, U.S. Environmental Protection Agency (U.S. EPA), Research Triangle Park, North Carolina.

REFERENCES

Auletta, A.E, Dearfield, K.L., and Cimino, M.C. Mutagenicity test schemes and guidelines: US EPA office of pollution prevention and office of pesticide programs. *Environ. Mol. Mutagen.*, 21, 38–45, 1993.

Brusick, D.J., and Matheson, D.W. Mutagen and Oncogen Study on JP-8, Aerospace Medical Research Laboratory. Technical Report 78-20. Wright-Patterson Air Force Base, OH, 1978.

Grant, G.M., Shaffer, K.M., Kao, W.Y., Stenger, D.A., and Pancrazio, J.J. Investigation of in vitro toxicity of jet fuels JP-8 and jet A. *Drug Chem. Toxicol.*, 23, 279–291, 2000.

Grant, G.M., Jackman, S.M., Kolanko, C.J., and Stenger, D.A. JP-8 jet fuel-induced DNA damage in H4IIE rat hepatoma cells. *Mutat. Res.*, 490, 67–75, 2001.

Health Protection Branch Genotoxicity Committee. The assessment of mutagenicity: health protection branch mutagenicity guidelines. *Environ. Mol. Mutagen.*, 21, 15–37, 1993.

Jackman, S.M., Grant, G.M., Kolanko, C.J., Stenger, D.A., and Nath, J. DNA damage assessment by comet assay of human lymphocytes exposed to jet propulsion fuels. *Environ. Mol. Mutagen.*, 40, 18–23, 2002.

Kirkland, D.J., Genetic toxicology testing requirements: Official and unofficial views from Europe. *Environ. Mol. Mutagen.*, 21, 8–14. 1993.

Lemasters, G.K., Livingston, G.K., Lockey, J.E., Olsen, D.M., Shukla, R., New G.R., Selevan, S.G., and Yiin, J.R. Genotoxic changes after low-level solvent and fuel exposure on aircraft maintenance personnel. *Mutagenesis*, 12, 237–243, 1997.

Lemasters, G.K., Lockey, J.E., Olsen, D.M., Selevan, S.G., Tabor, M.W., Livingston, G.K., and New, G.R. Comparison of internal dose measures of solvents in breath, blood, and urine and genotoxic changes in aircraft maintenance personnel. *Drug Chem. Toxicol.*, 22, 181–200, 1999.

Shelby, M.D., Erexson, G.L., Hook, G.J., and Tice, R.R. Evaluation of three-exposure mouse bone marrow micronuclei protocol: Results with 49 chemicals. *Environ. Mol. Mutagen.*, 21, 160–179, 1993.

McKee, R.H., Amoruso, M.A., Freeman, J.J., and Przygoda, R.T. Evaluation of the genetic toxicity of middle distillate fuels. *Environ. Mol. Mutagen.*, 23, 234–238, 1994.

NRC (National Research Council). *Toxicologic Assessment of Jet-Propulsion Fuel.* Vol. 8, National Academy Press, Washington, D.C., 2003.

Pitarque, M.A., Creus, A., Marcos, R., Hughes, J.A., and Anderson, D. Examination of various biomarkers measuring genotoxic endpoints from Barcelona airport personnel. *Mutat. Res.*, 440, 195–204, 1999.

Ramos, G., Nghiem, D.X., Walterscheid, J.P., and Ullrich, S.E. Dermal application of jet fuel suppresses secondary immune reactions. *Toxicol. Appl. Pharmacol.*, 180, 136–144, 2002.

Riviere, J.E., Brooks, J.D., Monteiro-Riviere, N.A., Budsaba, K., and Smith, C.E. Dermal absorption and distribution of topically dosed jet fuels Jet-A, JP-8, and JP-8 (100). *Toxicol. Appl. Pharmacol.*, 160, 60–75, 1999.

Rogers, J.V., Gunasekhar, P.G., Garrett, C.M., Kabbur, M.B., and McDougal, J.N. Detection of oxidative species and low-molecular-weight DNA in skin following dermal exposure with JP-8 jet fuel. *J. Appl. Toxicol.*, 21, 521–525, 2001.

SAS. *SAS User's Guide: Statistics*, Vol. 2, Version 6.12, 6th ed., SAS Institute, Cary, NC, 1977.

Sofuni, T. Japanese guidelines for mutagenicity testing, *Environ. Mol. Mutagen.*, 21, 2–7, 1993.

Tennant, R.W., and Zeiger, E. Genetic toxicology: Current status of methods of carcinogen identification, *Environ. Health Perspect.*, 100, 307–315, 1993.

Tice, R.R., Erexson, G.L., and Shelby, M.D. The induction of micronucleated polychromatic erythrocytes in mice using single and multiple treatments, *Mutat. Res.*, 234, 187–193, 1990.

Ullrich, S.E. Dermal application of JP-8 jet fuel induces immune suppression. *Toxicol Sci.*, 52, 61–67, 1999.

Vijayalaxmi, Kligerman, A.D., Prihoda, T.J., and Ullrich, S.E. Cytogenetic studies in mice treated with the jet fuels, Jet-A and JP-8. *Cytogenet. Genome Res.*, 104, 371–375, 2004.

Vijayalaxmi, Kligerman, A.D., Prihoda, T.J., and Ullrich, S.E. Micronucleus studies in the peripheral blood and bone marrow of mice treated with jet fuels, JP-8 and Jet-A. *Mutat. Res.*, 608, 82–87, 2006.

Zeiger, E. Genetic toxicity tests for predicting carcinogenicity. In *Genetic Toxicology and Cancer Risk Assessment*. Choy, W.N. (Ed.). Marcel Dekker, Inc., New York. pp. 29–45, 2001.

14 Computational Analyses of JP-8 Fuel Droplet and Vapor Depositions in Human Upper Airway Models

Clement Kleinstreuer
and Zhe Zhang

CONTENTS

INTRODUCTION

As recently reviewed (see NRC, 2003; ATSDA, 2005), jet fuel performance enhancers (i.e., additives such as BHT, DiEGME, EGME, BTEX, etc.) are highly toxic and may pose a significant health risk to Air Force personnel, fuelers, and people living near military fields. To understand the possible health effects from fuel exposure in the form of inhaled droplets and vapor, it is crucial to know where and at what concentrations fuel aerosols and their vapors deposit in the human respiratory system, given a set of realistic inlet conditions.

Direct deposition tests of toxic fuel in human lung airways are impossible and usually cost-intensive experimental deposition data, using human cast replicas, lack *detailed* resolution. Inhalation studies with small animals are often questionable because of the very different lung morphologies and air-particle dynamics as well as aerosol uptake mechanisms and health impact. Thus, computational fluid-particle dynamics (CFPD) simulations, if experimentally validated, offer a noninvasive, less-expensive, and effective means of obtaining fuel vapor and droplet transport and deposition data for representative airways. The human respiratory system features very complicated structures. The oral and nasal airways are two aerosol entries into the trachea. After the trachea, the tracheobronchial airways may be approximated as a network of repeatedly bifurcating tubes with progressively decreasing dimensions (Weibel, 1963; Horsfield et al., 1971; Raabe et al., 1976; Finlay, 2001). So far, computational airflow and particle transport simulations have concentrated on selected parts of the respiratory system. CFPD studies of micro-size particle deposition in human oral airways and tracheobronchial airways, employing CT-scanned or idealized segmental models, have been extensively conducted (Balashazy et al., 1999; Zhang et al., 2005; Choi et al., 2007; Li et al., 2007a; Longest and Vinchurkar, 2007; Finlay and Martin, 2008), as reviewed by Kleinstreuer et al. (2008a, 2008b). These studies assumed that the size of inhaled particles stays constant in the respiratory tract. However, in many situations, solid or liquid particles undergo size changes because they absorb moisture or lose mass due to hygroscopity or evaporation. The effect of hygroscopity on deposition fraction in the human respiratory system has been investigated experimentally and theoretically (Ferron et al., 1988a, 1988b; Finlay and Stapleton, 1995; Broday and Georgopoulos, 2001; Zhang et al., 2004, 2006c). For example, hygroscopity can substantially increase the size and hence the airway deposition fraction of drug particles with initial dry sizes of about 0.5 to 2 μm (Ferron et al., 1989). In contrast, as highly volatized compounds, jet fuel droplets vaporize quickly and may undergo dramatic size reduction as they move through the respiratory tract.

Focusing in this chapter on JP-8 fuel droplet and vapor deposition in human upper airway models, the contrasting deposition pattern between aerosols and vapor as well as the resultant exposure risk are of interest. Considering different inspiratory flow

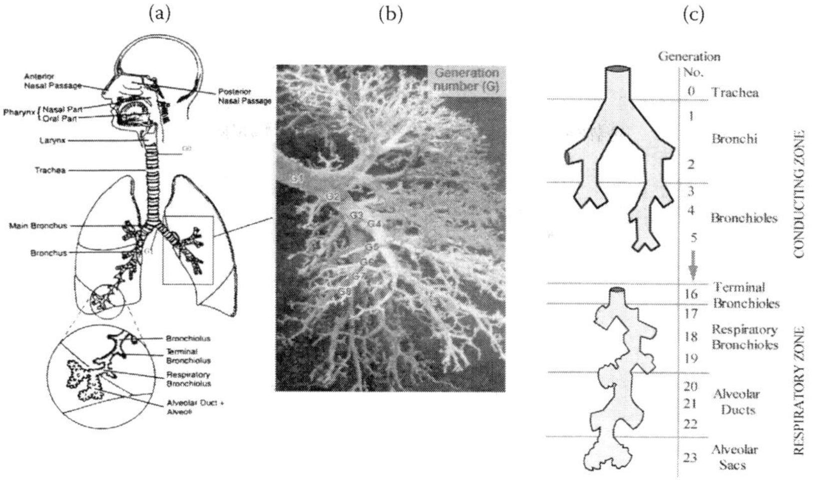

FIGURE 14.1 (See color insert following page 272) Schematics of the human respiratory system: (a) schematic of the respiratory system; (b) image of actual lung generations; and (c) lung conducting and respiratory zones.

rates, droplet trajectories, evaporation, and deposition are modeled with Newton's Second Law and scalar equations (i.e., using a decoupled Euler-Lagrange approach), whereas fuel vapor transport and deposition are described with a modified mass transfer equation (i.e., in an Eulerian framework) (Kleinstreuer, 2006).

THEORY

AIRWAY GEOMETRIES

Figure 14.1 and Figure 14.2 depict airways of the human respiratory tract and present geometric models, respectively. The trend is to employ modern imaging techniques that allow for patient-specific air-particle dynamics simulations in three to six bifurcations (Kleinstreuer et al., 2008b). The representative nasal airway geometry (see Figure 14.2a) was obtained from widely used MRI (magnetic resonance imaging) files of the nose of a healthy, 53-year-old, nonsmoking male (73 kg mass, 173 cm height), provided by The Hamner Institutes (formerly CIIT, Research Triangle Park, North Carolina). A detailed description of the nasal airway model is given by Shi et al. (2006), while the tracheobronhcial airway model (Figure 14.2c) is described by Li et al. (2007b). The dimensions of the oral airway model (see Figure 14.2b) were adapted from a human cast as reported by Cheng et al. (1997a). Variations in the actual cast include a short mouth inlet with a diameter of 2 cm, a modified soft palate, and a strong bend. The dimensions of the four-generation airway model are similar to those given by Weibel (1963) for adults with a lung volume of 3,500 ml. The 16-generation (G0 to G15) tracheobronchial (TB) model (Figure 14.2b) relies on "triple-bifurcation units" and is a computationally efficient approach to simulate local and averaged aerosol depositions in TB airways (Kleinstreuer and Zhang,

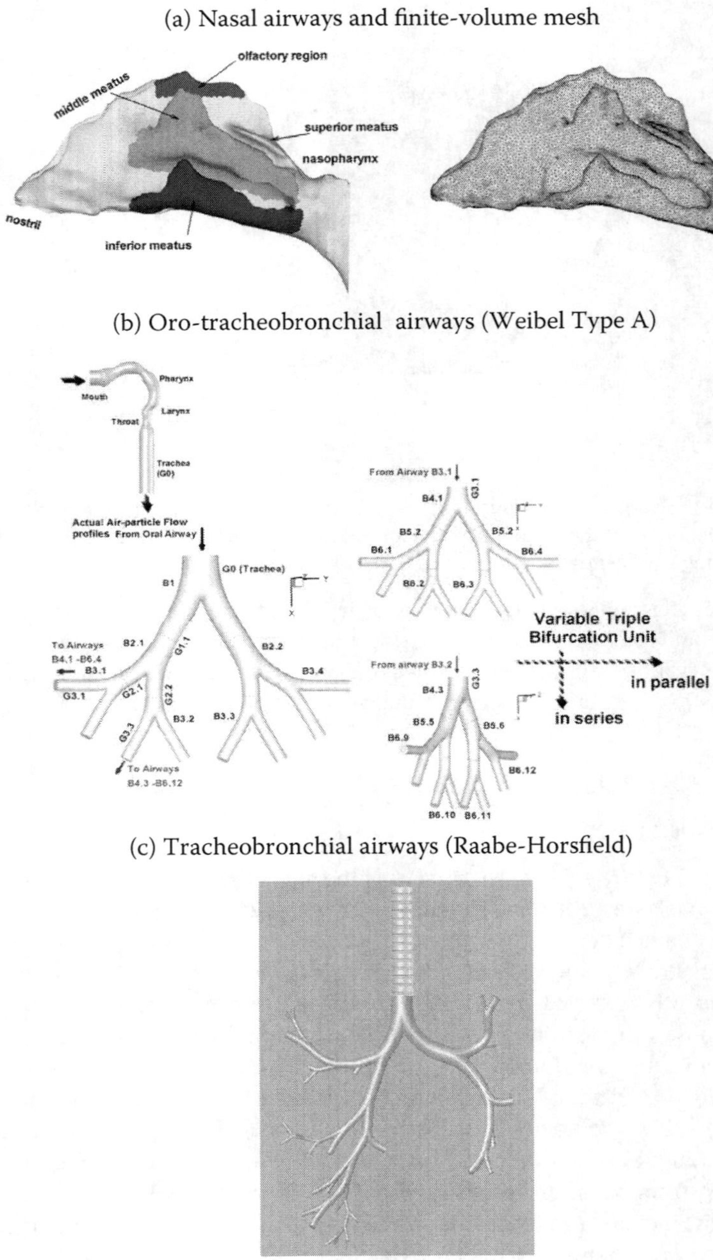

FIGURE 14.2 Human airway models: (a) nasal airways and finite-volume mesh; (b) oro-tracheobronchial airways (Weibel Type A); and (c) tracheobronchial airways (Raabe-Horsfield).

2008). The Weibel configuration is internationally applied for ease-of-use and consistent data comparisons. The airway conduit is assumed to be rigid and smooth, where the effects of cartilaginous rings in the trachea (see Figure 14.2c) have been analyzed by Li et al. (2007), Martonen et al. (1994), and Zhang and Finlay (2005), among others.

Governing Equations

Airflow and Heat Transfer

To capture the airflow structures in the laminar-to-turbulent flow regimes, that is, $0 < Re_{local} < 10{,}000$ for the maximum (transient) inhalation flow waveform, the low-Reynolds-number (LRN) k-ω model of Wilcox (1998) has been selected and adapted. It has been demonstrated to be appropriate for such internal flows (Varghese and Frankel, 2003; Zhang and Kleinstreuer, 2003a). All air transport equations, including the heat transfer equation, as well as initial and boundary conditions, are given in Zhang and Kleinstreuer (2003a, 2003b) and Kleinstreuer and Zhang (2003). The effect of wall roughness on the laminar flow profile in the nasal cavity is included in the airflow simulations by introducing the roughness-viscosity (υ_R) in a manner similar to the eddy viscosity (Shi et al., 2007).

Transport of Micro-size Droplets

With any given ambient concentration of noninteracting spherical droplets, a Lagrangian frame of reference for the trajectory computations of the evaporating droplets can be employed. In light of the large particle-to-air density ratio, dilute particle suspensions, negligible particle rotation and small thermophoretic forces, drag and gravity are considered as the dominant point force. Hence, the particle trajectory equation can be written as

$$\frac{d}{dt}(m_p u_i^p) = \frac{1}{8}\pi\rho d_p^2 C_{Dp}(u_i - u_i^p)\left|u_j - u_j^p\right| + m_p g_i \tag{14.1}$$

where ρ is the air density; u_i^P, m_p, and d_p are the velocity, mass, and diameter of the particle, respectively; g_i is the gravitational acceleration; and C_{Dp} is the drag force coefficient given by

$$C_{Dp} = C_D / C_{slip} \tag{14.2}$$

where C_{slip} is the slip correction factor (Clift et al., 1978) and

$$C_D = \begin{cases} 24(1+0.15\,\mathrm{Re}_p^{0.687})/\mathrm{Re}_p & \text{for } 0.0 < \mathrm{Re}_p \le 1000 \\ 0.44 & \text{for } 1000 < \mathrm{Re}_p \end{cases} \tag{14.3}$$

The particle Reynolds number is

$$Re_p = \rho \left| u_j - u_j^p \right| d_p / \mu \tag{14.4}$$

where μ is the fluid viscosity. In Equation (141), u_i is the instantaneous fluid velocity with $u_i = \bar{u}_i + u_i'$, where $\bar{u}_i$ is the time-averaged or bulk velocity of the fluid, and u_i' are its fluctuating components. Traditionally, turbulence is assumed to consist of a collection of randomly directed eddies; hence, an eddy-interaction model (EIM) is used to simulate the particle trajectories, and the fluctuating velocities u_i' are obtained by (Gosman and Ioannides, 1981, Matida et al., 2000):

$$u_i' = \xi_i \left(\frac{2}{3} k \right)^{1/2} \tag{14.5}$$

where k is the turbulence kinetic energy; ξ_i are random numbers with zero-mean, variance of one, and Gaussian distribution. In the EIM, each particle is allowed to interact successively with various eddies, and the random numbers are maintained constant during one eddy interaction while the corresponding turbulence intensities vary with the particle positions (Macinnes and Bracco, 1992; Matida et al., 2000). Each eddy has a lifetime t_E and length scale l_E given by

$$t_E = 1.5^{\frac{1}{2}} C_\mu^{\frac{3}{4}} \frac{k}{\varepsilon} = \frac{0.2}{\omega} \quad \text{and} \quad l_E = C_\mu^{\frac{3}{4}} \frac{k^{\frac{3}{2}}}{\varepsilon} = 0.164 \frac{k^{\frac{1}{2}}}{\omega} \tag{14.6}$$

where C_μ (= 0.09) is a turbulence constant, ε is turbulence dissipation rate, and ω is dissipation per unit turbulence kinetic energy. However, due to the assumption of turbulence isotropy, the fluctuating velocities normal to the wall calculated with Equation (14.5) may be higher than the actual values (Kim et al., 1987), which may overpredict the particle deposition in some cases (Matida et al., 2004). As proposed by Matida et al. (2004), a near-wall correction can be used to simulate the near-wall particle trajectories, that is, the component of fluctuating velocity normal to the wall u_n' can be expressed as

$$u_n' = f_v \xi \left(\frac{2}{3} k \right)^{1/2} \tag{14.7}$$

with

$$f_v = 1 - e^{-0.02 y^+} \tag{14.8}$$

where f_v is a damping function component normal to the wall considering the anisotropy of turbulence near the wall (Wang and James, 1999), and y^+ is the nondimensional wall distance, that is, $y^+ = u_\tau y / \nu$, with u_τ being the friction velocity at the nearest wall, y being the distance to the nearest wall, and ν being the local kinematic viscosity of the fluid. Usually, Equation (14.8) is used for $y^+ < 10.0$, while $f_v = 1$ elsewhere.

At the mouth inlet, presently uniform droplet distributions were prescribed. The initial droplet velocities were set equal to that of the fluid, and one-way coupling was assumed between the air and droplet flow fields because the maximum mass loading ratio (mass of droplets/mass of fluid) is below 10^{-5} in the present analysis. According to sampling data by Witten (2002), the size range of JP-8 fuel droplet exposure to airfield personnel is within 1 to 5 μm. Droplet positions, velocities, temperature, and diameter at the trachea of the oral airway model were adjusted as the inlet droplet conditions of the bifurcating airway segment (see Figure 14.2b).

Droplet Evaporation

Rapid mixing assumptions are implemented for the discrete liquid phase in this study; that is, infinite conduction and diffusion are assumed with the droplet such that liquid temperature and concentration are spatially constant but vary with time. This assumption is reasonable because of the small Biot number for micro-size droplets (i.e., $Bi = hd_p/k_d << 1$, where h is the overall heat transfer coefficient, d_p is the droplet diameter, and k_d is the thermal conductivity of the droplet). A small Biot number implies that heat conduction inside a body is much slower than at its surface, and hence internal temperature gradients are negligible. Considering convective heat and mass transfer occurs over the whole surface of the spherical droplets, the change in droplet mass as well as coupled heat transfer equation for liquid droplets can be expressed as (Berlemont et al., 1995; Benaissa et al., 2002):

$$\frac{dm_p}{dt} = -\pi d_p \rho Sh\tilde{D}C_m \ln\frac{1-Y_\infty}{1-Y_{surf}} \tag{14.9}$$

$$m_p c_g(dT/dt) = \pi d_p k_g Nu C_T(T_a - T_d) - \gamma(dm_p/dt) \tag{14.10}$$

In the above equations, m_p and d_p are the droplet mass and diameter, respectively; $\tilde{D}$ is the vapor diffusivity in the air; Sh is the Sherwood number, $Sh = (1+\mathrm{Re}_p\,Sc)^{1/3}\max[1,\mathrm{Re}_p^{0.077}]$; c_g is the gas specific heat; k_g is the gas thermal conductivity; Nu is the Nusselt number, $Nu = (1+\mathrm{Re}_p\,\mathrm{Pr})^{1/3}\max[1,\mathrm{Re}_p^{0.077}]$; T_d and T_a are the temperature of droplet and surrounding air, respectively; γ is the latent heat of vaporization (the calculation of γ for JP-8 fuel is given by Benaissa et al. (2002)); and C_m and C_T are correction factors (i.e., Knudsen number corrections) considering the non-continuum effects; specifically (Finlay, 2001),

$$C_m = \frac{1+Kn}{1+\left(\dfrac{4}{3\alpha_m}+0.377\right)Kn+\dfrac{4}{3\alpha_m}Kn^2} \tag{14.11}$$

$$C_T = \frac{1+Kn}{1+\left(\dfrac{4}{3\alpha_T}+0.377\right)Kn+\dfrac{4}{3\alpha_T}Kn^2} \tag{14.12}$$

Here, Kn is the Knudsen number, defined as $Kn = 2\lambda / d_p$ with λ being the mean free path of the gas surrounding the droplets, and α_m and α_T are the mass and thermal accommodation coefficients, respectively. Values of $\alpha_m = \alpha_T = 1$ were used in the present study (Ferron et al., 1988b).

In Equation (14.9), Y_{surf} and Y_∞ are the gas-phase mass fractions of the component on the droplet surface and far from the droplet, respectively. Specifically,

$$Y_{surf} = \frac{x_l P_{sat}(T_d)K}{\rho R_s T_d} \tag{14.13}$$

In the above expression, x_l is the liquid-phase mole fraction of the species of interest, R_s is the gas constant, T_d is the droplet temperature, and $P_{sat}(T_d)$ is the temperature-dependent saturation pressure. The saturated vapor pressure $p_{sat}(T_w)$ can be obtained using the Clausius-Clapeyron equation:

$$p_{sat}(T_w) = f_1 \exp\left(-f_2 / T_w\right) \tag{14.14}$$

For the multicomponent JP-8 fuel, the parameters M_f, f_1, and f_2 depend on the mass fraction of fuel vaporized and can be calculated using the method described by Bardon and Raw (1984) as well as Benaissa et al. (2002). In addition, $\rho_p = 0.81$ kg/cm^3 was used in the present simulation. K in Equation (14.13) is the correction factor considering the Kelvin effect for small droplets, which can be given as

$$K = \exp[4\sigma M / (R_u \rho d_p T_d)] \tag{14.15}$$

where α is the surface tension at the droplet surface (which is about 0.07 Nm^{-1} for water at 303K), M is the molar mass of the vapor molecules, and R_u is the universal gas constant.

Mass Transfer of Vapor or Nanoparticles

The convection mass transfer equation of nanoparticles, or (JP-8) fuel vapor, whose dominant radial transfer mechanisms are Brownian motion and turbulent dispersion can be given as

$$\frac{\partial Y}{\partial t} + \frac{\partial}{\partial x_j}\left(u_j Y\right) = \frac{\partial}{\partial x_j}\left[\left(\tilde{D} + \frac{\nu_T}{\sigma_Y}\right)\frac{\partial Y}{\partial x_j}\right] \tag{14.16}$$

where $\sigma_Y = 0.9$ is the turbulence Schmidt number for Y. JP-8 fuel is a multicomponent mixture that consists mainly of C_8 to C_{16} paraffinic hydrocarbons with other hydrocarbons and additives also present; however, the diffusivity in air does not vary significantly from compound to compound (Gustafson et al., 1997). Thus, the conservative assumption presently implemented was to set $\tilde{D} = 0.05$ cm²/sec for all fractions of JP-8 jet fuel at T = 293K with little loss in accuracy according to Gustafson et al. (1997). The effect of temperature change on diffusivity is expressed by the following semi-empirical correlation (Bejan, 1995):

$$\tilde{D}(T) / \tilde{D}(T = 293K) = (T / 293)^{1.75} \tag{14.17}$$

The aerosol diffusion coefficient is calculated as follows (Finlay, 2001):

$$\tilde{D} = (k_B T C_{slip}) / (3\pi\mu d_p) \tag{14.18}$$

where k_B is the Boltzmann constant (1.38×10^{-23} J/K); and C_{slip} is the Cunningham slip correction factor:

$$C_{slip} = 1 + \frac{2\lambda_m}{d_p}\left[1.142 + 0.058\exp\left(-0.999\frac{d_p}{2\lambda_m}\right)\right] \tag{14.19}$$

Assuming that the airway wall is a perfect sink for aerosols or vapors upon touch, the boundary condition on the wall is $Y_w = 0$. This assumption is reasonable for fast gas-wall reaction kinetics (Fan et al., 1996), or vapors of high solubility and reactivity, and is also suitable for estimating the maximum deposition of toxic vapor in the airways. For less soluble vapors, the wall concentration would be greater than zero so that transport in tissue and in airways must be considered simultaneously for simulating vapor uptake. Assuming that the surface of respiratory epithelium is covered by a mucus layer with uniform thickness and a lipid layer lies below the mucus lining simulating the transport barrier by the epithelial cell membrane, Keyhani et al. (1997) derived a flux condition at the airway boundary, including the vapor transport in the tissue. This boundary condition was inferred from the mass conservation and mass diffusion of vapor from the airway to the mucus layer and to the tissue, which is given as

$$D_{in}\frac{\partial Y}{\partial n} + KY = 0 \tag{14.20}$$

where D_{in} is the inlet tube diameter, n is the direction normal to the airway wall, and K is a dimensionless parameter (here we call it the absorption parameter), which is defined as

$$K = \frac{D_{in}\tilde{D}_m\xi}{\tilde{D}_a\beta\tanh(\xi H_m)}\left[1 - \frac{2\tilde{D}_m / H_m}{(\tilde{D}_m / H_m + P_l)(e^{\xi H_m} - e^{-\xi H_m})}\right] \tag{14.21}$$

where $\tilde{D}_a$, $\tilde{D}_m$ are the vapor diffusivity in the air and the liquid mucus phase, respectively; H_m is the thickness of the mucus layer; and β is the equilibrium partition coefficient for a given contaminant molecular, which can be determined by Henry's law; $\xi = \sqrt{k_r / \tilde{D}_m}$ with k_r being a single rate constant considering the chemical reactions of vapors in the mucus layer; $P_l = \gamma\tilde{D}_l / H_l$ is the lipid permeability coefficient with γ, $\tilde{D}_l$, H_l being the lipid-mucus partition coefficient, vapor diffusivity in the lipid, and the thickness of the lipid layer, respectively.

Clearly, for a highly soluble ($\beta \times 0$) or high reactive ($k_r \to \infty$) vapor, the boundary condition (Equation (14.23)) reduces to Y = 0. If vapor is insoluble ($\beta \times \infty$), the boundary condition reduces to the zero mass flux condition, that is, $\partial Y / \partial n = 0$. Ignoring removal of contaminant molecules in the mucus by chemical reaction and the resistance to transport across the lipid barrier, the absorption parameter K reduces to

$$K = \frac{D_{in}\tilde{D}_m}{\tilde{D}_a\beta H_m} \tag{14.22}$$

Deposition Parameters

The regional deposition of micro-droplets in human airways can be quantified in terms of the deposition fraction (DF) or deposition efficiency (DE) in a specific region (e.g., oral airway, first, second, and third bifurcations, etc.); they are defined as

$$DF_{particle} = \frac{\text{Number of deposited particles in a specific region}}{\text{Number of particles entering the mouth}} \tag{14.23}$$

$$DE_{particle} = \frac{\text{Number of deposited particles in a specifc region}}{\text{Number of particles entering this region}} \tag{14.24}$$

The DEs and DFs are the same for the oral airway model in this study.

In addition to the above two traditional deposition parameters DE and DF, a deposition enhancement factor (DEF) is considered to be employed to quantify local particle deposition patterns (Balashazy et al., 2003; Zhang et al., 2005). The DEF is defined as the ratio of local to average deposition densities, that is,

$$DEF = \frac{DF_i / A_i}{\sum_{i=1}^{n} DF_i \Big/ \sum_{i=1}^{n} A_i} \tag{14.25}$$

where A_i is the area of a local wall cell (i), n is the number of wall cells in one specific airway region, and DF_i is the local deposition fraction in the local wall cell (i).

Clearly, the presence of high DEF values indicates inhomogeneous deposition patterns, including "hot spots" in terms of locally maximum particle depositions.

The DF of ultrafine particles can also be calculated with the regional mass balance or the sum of local wall mass flux with the Eulerian-Eulerian modeling technique. According to the mass balance in one specific region, DF is defined as

$$DF = 1 - \left(\sum_{i=1}^{n} \int u_i Y_i dA_i \right) \Big/ \left(\int u_0 Y_0 dA_0 \right) \tag{14.26}$$

where A is the tube cross-sectional area, and u is the axial velocity. The subscripts "0" and "i" denote the properties at the inlet and the i-th daughter tube of a selected generation, respectively. In the oral airway model, the "i" denotes the trachea and $n = 1$. As for the local wall mass flux of nanoparticles, it can be determined as

$$\dot{m}_w = \rho A_i j_{wall,i} \tag{14.27}$$

where A_i is the area of local wall cell (i) and $j_{wall,i}$ is the particle flux at the local wall cell given by

$$j_{wall,i} = -\tilde{D} \frac{\partial Y}{\partial n} \bigg|_{wall,i} \tag{14.28}$$

The local particle deposition fraction, which is defined as the ratio of local wall mass flux to the inlet mass flux, can be expressed as

$$DF_{local} = (A_i j_{wall,i}) / (Q_{in} Y_{in}) \tag{14.29}$$

and the regional deposition fraction can be determined as

$$DF_{region} = \sum_{i=1}^{n} (A_i j_{wall,i}) / (Q_{in} Y_{in}) \tag{14.30}$$

where n is the number of wall cells in one specific airway region (e.g., oral airway, first airway bifurcation, etc.). The local vapor deposition patterns can also be quantified in terms of the DEF, that is,

$$DEF = j_{wall,i} \Bigg/ \left[\sum_{i=1}^{n} (A_i j_{wall,i}) \Big/ \sum_{i=1}^{n} A_i \right] \tag{14.31}$$

Again, the DEF will indicate particle deposition "hot spots" in certain regions.

Calculation of the respiratory mass transfer coefficient, h_m, is helpful in quantitatively predicting the regional uptake of inhaled nano-size particles (Cheng et al., 1997b). At the same time, the accurate values of segmental h_m can be directly incorporated in a separate physiologically based pharmacokinetics (PBPK) modeling effort. The mass balance for one airway unit under steady-state conditions is

$$N \cdot A_{wall} = \sum_{i=1}^{n} \dot{m}_i Y_i - \dot{m}_0 Y_0 \tag{14.32}$$

which yields the wall mass flux of species N, so that

$$h_m = \frac{N}{\overline{\rho Y_m} - \rho Y_{wall}} \tag{14.33}$$

For simplicity, we can only use the average values of the inlet and outlet cross-sectional mass fractions to determine the regional $\overline{\rho Y_m}$, that is,

$$\overline{\rho Y_m} = \frac{1}{2}\left[\frac{\int \rho_0 u_0 Y_0 dA_0}{\int \rho_0 u_0 dA_0} + \frac{1}{n}\sum_{i=1}^{n}\frac{\int \rho_i u_i Y_i dA_i}{\int \rho_i u_i dA_i}\right] \tag{14.34}$$

The respiratory mass transfer coefficient can also be expressed in terms of the dimensionless Sherwood number (*Sh*):

$$Sh = (h_m D)/\tilde{D} \tag{14.35}$$

The diameters (D) in the trachea and the parent tubes can be employed to determine Sh for the oral airways and bifurcation units, respectively.

Once the mass transfer coefficient h_m is known, the deposition fraction in one airway region can be calculated with the following equation (Zhang et al., 2006b):

$$DF(\%) = 1 - \exp\left[\frac{h_m \tilde{D} K}{h_m D_{in} - \tilde{D} K} \cdot \frac{\pi D_m L}{Q_{in}}\right] \times 100\% \tag{14.36}$$

where D_m and L are the mean diameter and length of the airway segment, respectively.

Numerical Method

The numerical solutions of the governing equations, including flow, heat, and mass transfer equations, were carried out with a user-enhanced commercial finite-volume-based program (i.e., CFX from ANSYS, Inc.). The micron particle transport equation was solved with an off-line F90 code with parallelized algorithms. Details of these two numerical modeling methodologies are described in Zhang et al. (2005).

Droplet deposition occurs when its center comes within a radius from the wall; that is, local surface effects such as droplet evaporation (blowing), migration, or resuspension have been currently ignored. In the present model, the number of particles, n = 10,000 to 200,000, was determined by increasing the inlet particle concentration until the deposition fraction became independent of the number of particles simulated. The computational mesh was generated with dense near-wall cells to fully contain the viscous sublayers and to resolve any geometric features present there. The mesh topology was determined by refining the mesh until grid independence of the solutions of both flow and mass fraction fields as well as particle deposition was achieved.

RESULTS

Model Validations

For accurate computational fluid-particle dynamics (CFPD) simulations, matching comparisons with data sets for airflow and particle deposition are necessary. Figures 14.3a through 14.3g summarize comparisons of experimental data sets and present computational simulations of tubular flow (Figures 14.3a, b), nano- and micron- particle depositions in nasal, oral, and upper airways (Figure 14.3c), micron particle deposition in the oral airways (Figure 14.3d), nanomaterial deposition in the oral and upper airways (Figures 14.3e, f), and stationary droplet evaporation (Figure 14.3g). Of special interest is the simulated temporal diameter variation of a JP-8 fuel droplet as conducted by Runge et al. (1998), where a JP-8 fuel droplet with an initial diameter of 639 μm was suspended in 294K air moving at a constant speed of 3 m/sec. It can be seen that the simulated vaporization law for JP-8 fuel, using $D_{effective}$ = 0.05 cm^2/sec, is in excellent agreement with experimental data. Additional literature citations supportive of the computer model accuracy may be found in Zhang et al. (2002b, 2005, 2006c); Kleinstreuer and Zhang (2003, 2008); Zhang and Kleinstreuer (2003a); and Shi et al. (2004, 2006, 2007).

In summary, the good agreement between experimental findings and theoretical predictions instills confidence that the present computer simulation model is sufficiently accurate to analyze transport and deposition of JP-8 fuel droplet and vapor in three-dimensional oral and upper bronchial airways associated with laminar-to-turbulent airflows.

Airflow Structures

Nasal Flow

A typical inspiratory airflow structure in the human nasal cavities is given in Figure 14.4, assuming steady laminar flow under resting condition (Q = 7.5 l/min). After the airflow enters a nasal cavity, the majority of flow passes through the middle-to-low portion of the main passageway between middle and inferior meatuses (see Figures 14.4d, ff). In particular, two high-speed regions are located under the middle and inferior meatuses. The narrow olfactory region and the upper part of the middle

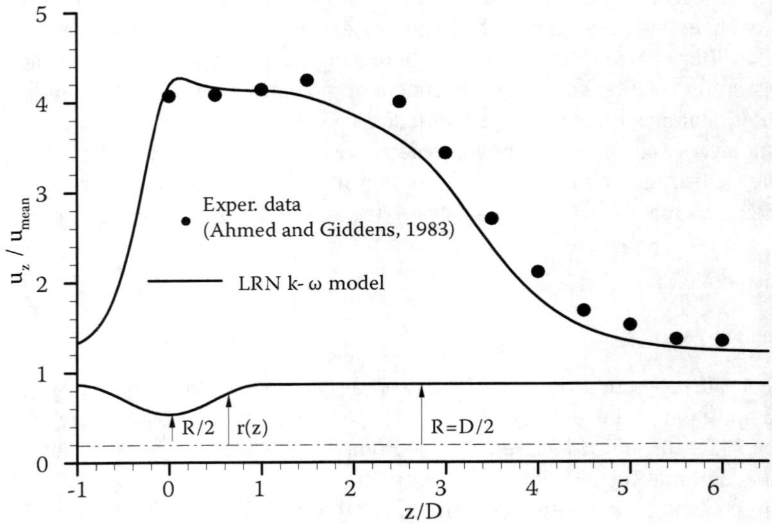

(a)

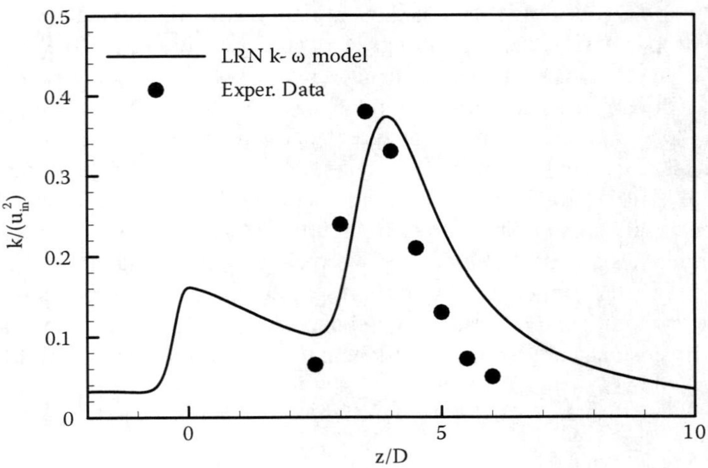

(b)

FIGURE 14.3 Examples of computational model validations: (a) comparison of the centerline velocity of turbulent flow in stenosed tube; (b) axial turbulence kinetic energy development in a stenosed tube. *Continued*

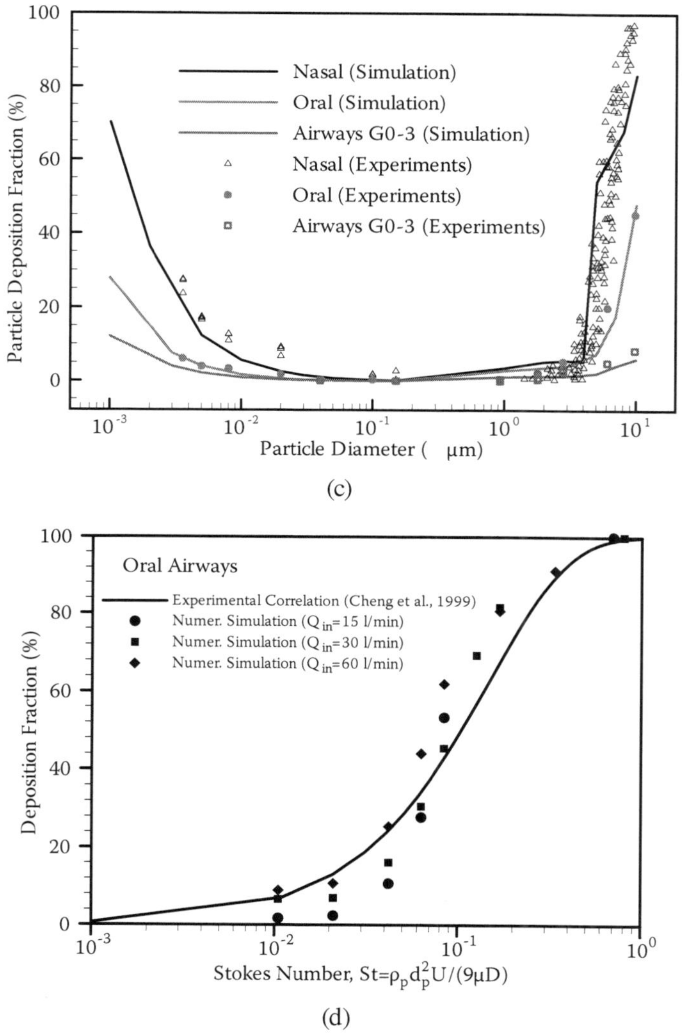

FIGURE 14.3 (*Continued*) Examples of computational model validations: (c) comparison between present modeling results and experimental data sets; (d) comparison of experimental and computational results of micron particle deposition for different inhalation flow rates in the oral airways.

Continued

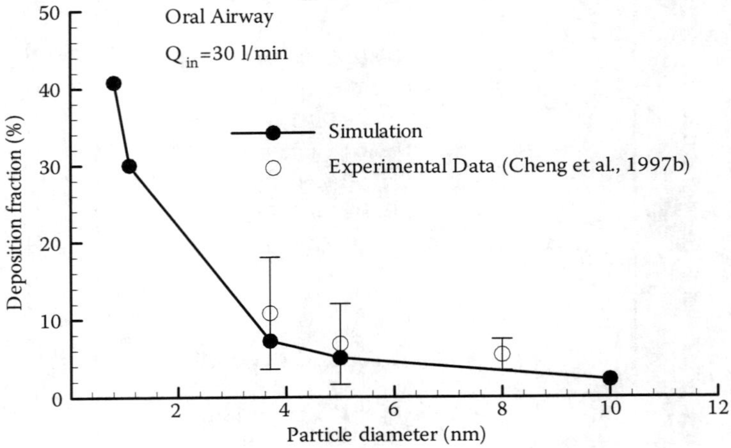

(e)

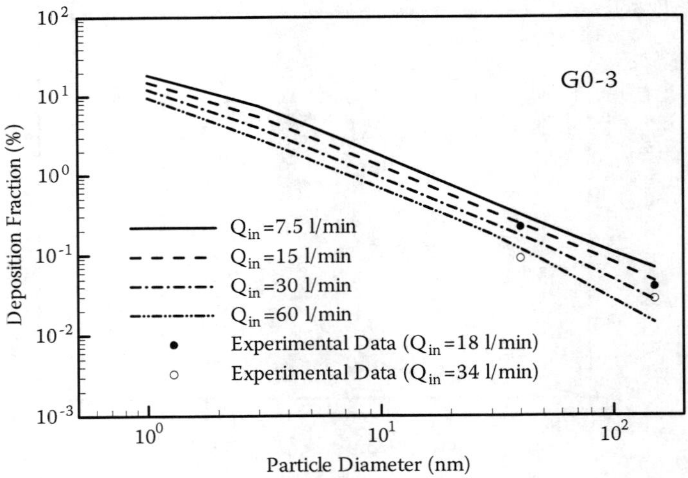

(f)

FIGURE 14.3 (*Continued*) Examples of computational model validations: (e) nanoparticle deposition in the oral airways; (f) nanoparticle deposition in generations 0 to 3 for different inhalation flow rates.

Continued

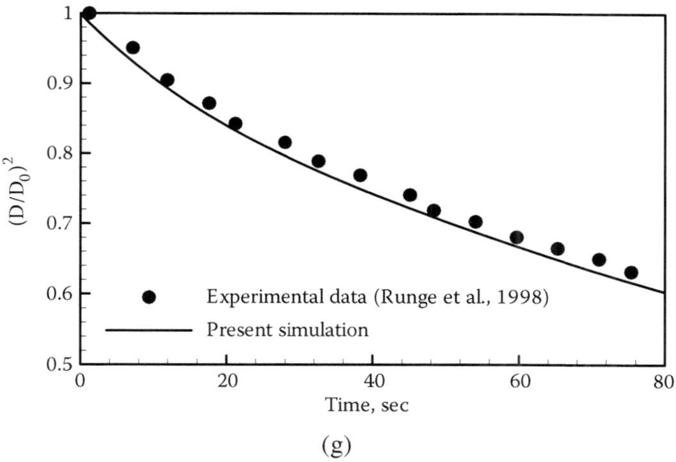

FIGURE 14.3 (*Continued*) Examples of computational model validations: (g) transient droplet evaporation.

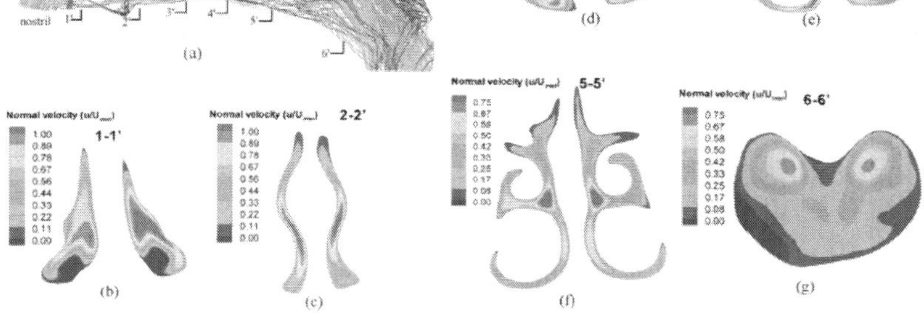

FIGURE 14.4 (See color insert following page 272) Velocity fields in human nasal cavity at constant inlet flow rate of 7.5 l/min: (a) three-dimensional streamlines and velocity contours; and (b–g) velocity fields in six selected slices.

and inferior regions receive only small amounts of air, which is believed to protect the cells for the sense of smell. Although airflow enters the nose almost vertically, the quasi-funnel shape of the vestibule redirects the airflow horizontally after the nasal valve, that is, toward the lower nasal passageway. Then, most of the inhaled air flows through the wider middle-to-low portion of the main passageway, which is free of obstacles. The secondary flow fields are quite strong in the middle part of the nasal cavities, that is, Slices 3-3′ and 5-5′, because of the locally complex geometric features and the airflow direction changes in the vestibule (Shi et al., 2008).

Oral-Tracheobronchial Flow

Figures 14.5a and 14.5b show typical mean velocity profiles in an oral airway model starting from mouth to trachea and a bifurcating airway model G0 to G3 (G0 is the trachea) with an inspiratory flow rate of Q = 30 l/min and T_{in} = 310K, respectively. The selected cross-sectional views display the axial velocity contours as well as secondary velocity vectors. Clearly, skewed velocity profiles generated by the centrifugal force can be observed in the curved portion from the oral cavity to the pharynx/larynx. Moreover, a central asymmetric jet and a recirculation zone are created in the laryngeal region because of the restriction of the (assumed stationary) vocal folds. The secondary motion is set up when the flow turns a bend from the mouth to the pharynx because of the centrifugally induced pressure gradient (see Kleinstreuer and Zhang, 2003). Turbulence may occur locally

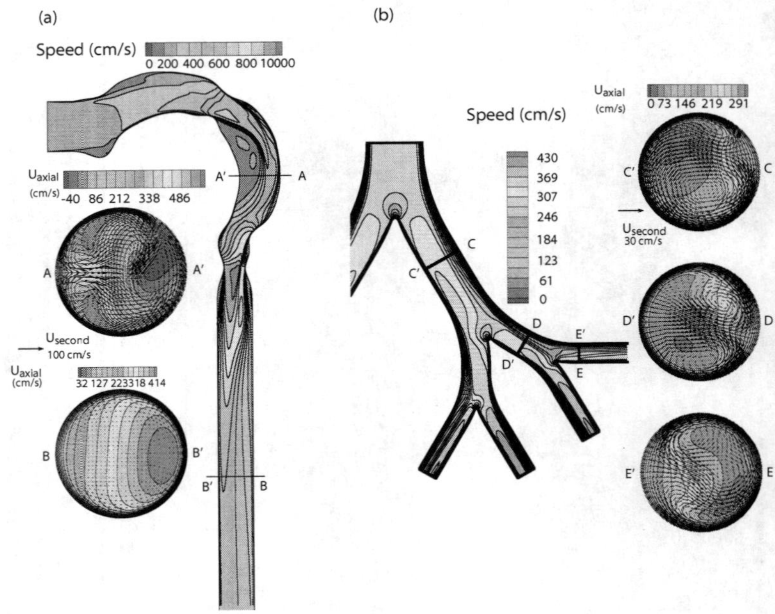

FIGURE 14.5 (See color insert following page 272) Side-plane or mid-plane velocity profiles as well as selected cross-sectional axial velocity contours and secondary velocity vectors with an inspiratory flow rate Q = 30 l/min in: (a) an oral airway model; and (b) a bifurcating airway model G0 to G3.

during medium-level inhalation (i.e., Q = 30 l/min), where the turbulent fluctuations are weak in the oral cavity; but they become strong after the constriction of the soft palate and rise rapidly after the glottis, eventually decaying more slowly and asymptotically. With further redistribution of the kinetic energy of the flow over most of the cross-section accompanied by the onset of turbulence, the velocity profiles become more blunt from the left to the right wall, and the maximum velocity zone moves to the anterior wall at the cross-section six diameters from the glottis (see Figure 14.5a). Figure 14.5b shows the mean-velocity fields in the planar triple-bifurcation airways, which are physiologically 90° turned with respect to the oral airways. The velocity profiles and turbulence quantities extracted from the trachea in the oral airway model were adjusted as the inlet conditions for the bifurcating airway model G0 to G3. The air stream in the branching airways splits at each flow divider and new boundary layers are generated at the inner walls of the daughter tubes. The velocity profiles are naturally skewed in each daughter tube, and hence each daughter tube, or generation, may experience a different flow rate. The primary characteristics of the flow fields at cross-sections C-C′, D-D′, and E-E′ are quite similar. They feature (1) skewed velocity profiles with the maximum velocity near the inner wall around the divider, and (2) two distinct secondary vortices appearing at the upper and lower side of the tube, which moves the high-speed fluid up around the top of the tube toward the outside of the bifurcation and low-speed fluid from the outside of the bifurcation along the symmetry plane toward the inside of the bifurcation.

The details of the airflow structures and particle distributions in both oral and tracheobronchial (TB) airways under cyclic breathing conditions, as well as the effects of airway geometric features, inspiratory flow rate, and thermal condition of incoming air, can be found in Zhang and Kleinstreuer (2002, 2003b, 2004), Zhang et al. (2002a), and Li et al. (2007a, 2007b, 2007c).

Vapor Deposition

As aforementioned, the degree of vapor absorption by the airway wall may influence fuel vapor transport and deposition in the airways. Because JP-8 fuel is a multicomponent mixture and the solubility of its components changes greatly (Gustafson et al., 1997), the absorption parameter K in Equation (14.22) may vary from 10^{-3} to 10^{3} for different compounds. As a case for estimating the maximum deposition of toxic vapor in the airways, the following results are based on the perfect absorption assumption of vapor in the airway surface (i.e., $Y_{wall} = 0$), except in the subsection that discusses the impacts of the absorption parameter K.

Nasal Deposition

Figure 14.6 depicts the distributions of the vapor DEF for quasi-steady flow rates of 7.5 and 20 l/min. Clearly, the majority of vapor deposition occurs in the anterior part of the nasal cavity because of the high diffusivity. The meatuses regions experience only small deposition fluxes because most particles are taken up before they can reach the deeper regions of the meatuses. High DEF values can be seen around the nasal valve region due to the locally narrowing airway. The top views of Figure 14.6,

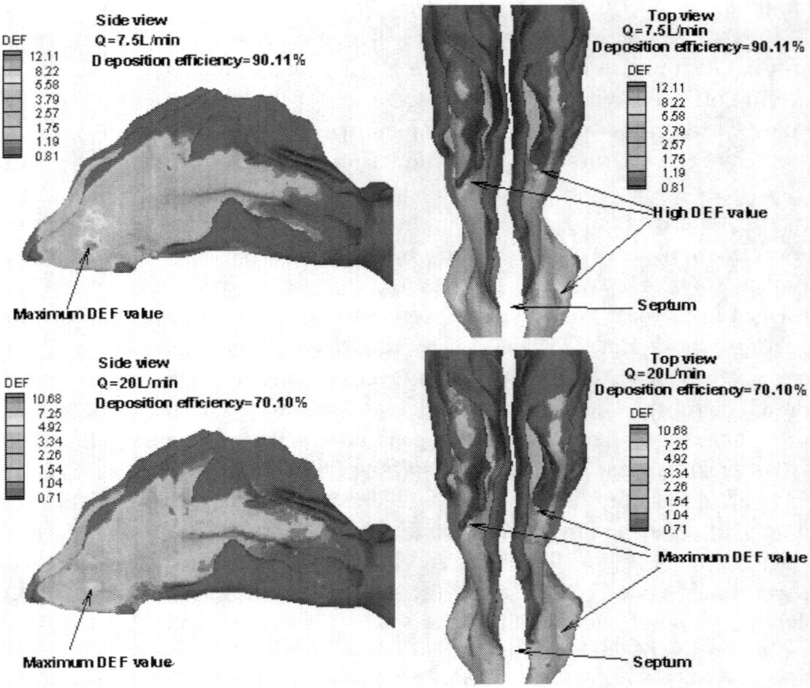

FIGURE 14.6 (See color insert following page 272) Vapor deposition enhancement factor (DEF) contours at constant inlet flow rates of 7.5 and 20 l/min.

where the middle meatus emerges, reveal the highest DEF values. Overall, 70% to 90% of the incoming vapor mass deposits if a perfect absorbing airway wall is assumed. The effects of mucus layer and airway wall absorption are discussed by Shi et al. (2008). A change in flow rate does not influence the deposition pattern significantly. However, higher flow rates will decrease the deposition efficiency due to a reduction in residence times. Vapor deposition decreases with decreasing diffusivities and with higher flow rates due to shorter residence times. Clearly, the influence of flow rate on deposition efficiency is important but less pronounced than diffusivity (Shi et al., 2008).

In terms of simulated results, the laminar vapor deposition efficiency in the human nasal cavity can be expressed, in terms of nanoparticle deposition where $d_p \leq 10$ nm, as:

$$\eta_d^{Lam} = C \cdot \left(\frac{\tilde{D}L \cdot \pi}{4 \cdot Q} \right)^{1/2} \cdot Sc^{-1/6} \tag{14.37}$$

Here, Sc is the Schmidt number, $L \approx 10$ cm is the equivalent nasal airway length, and C is a curve-fitted coefficient representing geometric complexities, given as

$$C = a + b \cdot \ln \Delta \tag{14.38}$$

where Δ is the diffusion parameter ($\Delta = \tilde{D}L/4UR^2$), $a = 0.568$ and $b = -0.69$ for $7.5 \leq Q \leq 20$ l/min and $5 \times 10^{-4} \leq \tilde{D} \leq 5.5 \times 10^{-2}$ cm^2/sec.

Oral and Tracheobronchial Deposition

Figure 14.7 shows examples of vapor deposition patterns, expressed as distributions of DEF (see Equation (14.31)) in the oral airway model and the bifurcation airway model. Clearly, the deposition patterns are somewhat inhomogeneous. In the oral airway model, the enhanced deposition may occur at the entrance, outside bend of the pharynx, and the throat because of great degrees of mixing, that is, large concentration gradients in these regions. Turning to the bifurcation airway model G0 to G3, the enhanced deposition mainly occurs at the cranial ridges and the inside walls around the cranial ridges due to the complicated airflows and large particle concentration gradients in these regions.

The deposition fractions of JP-8 fuel vapor in the oral airways under different inhalation conditions are shown in Figure 14.8. It can be seen from Figure 14.8a that higher vapor deposition corresponds to larger inlet air temperatures, especially for the high inhalation flow rate case. This can be attributed to the relatively large diffusion coefficient at relatively high temperatures (see Equation (14.17)). Generally speaking, the deposition fraction is weakly affected by the inlet air temperature for both low and high inspiratory flow rates. That is, the thermal effects could be negligible when calculating the total or segmental deposition fraction of vapors or ultrafine particles in the upper airways. The flow rate also shows a significant effect on the deposition of vapor (Figure 14.8b). The higher the flow rate, the lower the

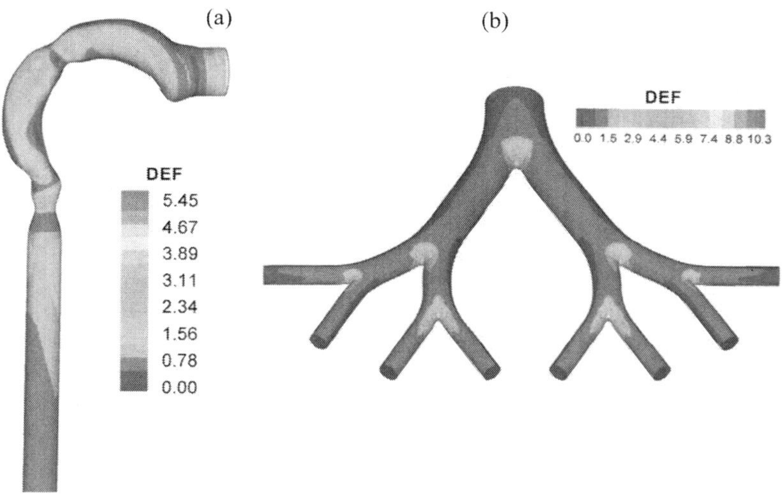

FIGURE 14.7 (See color insert following page 272) Three-dimensional distributions of deposition enhancement factor (DEF) of JP-8 fuel vapor under steady inhalation with Q = 30 l/min in (a) the oral airway model; and (b) the bifurcation airway model.

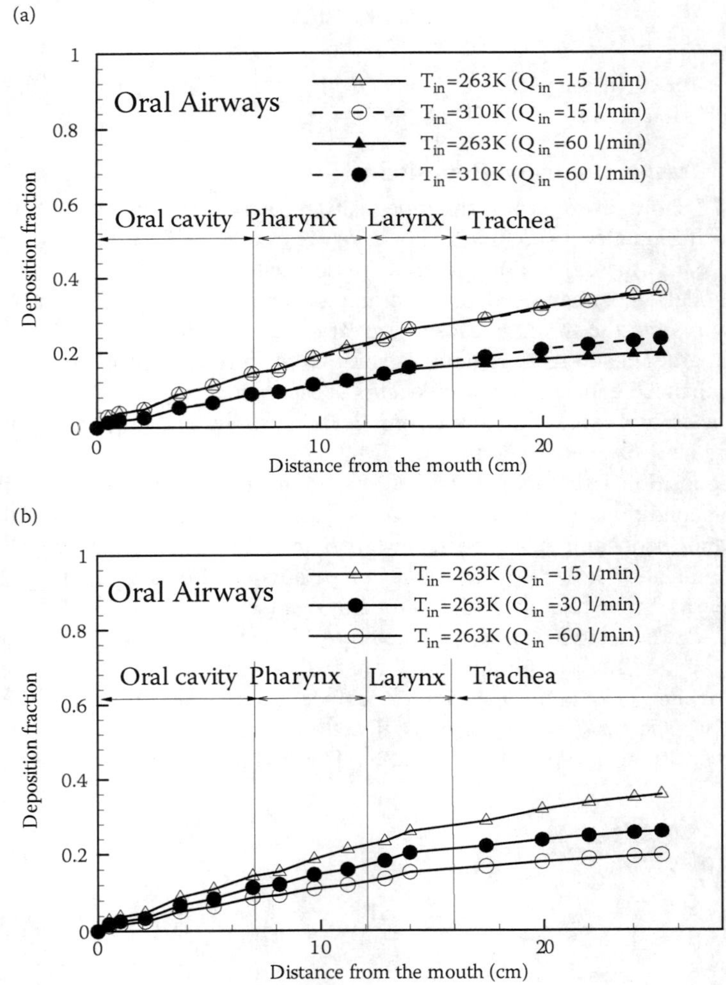

FIGURE 14.8 Deposition fractions of JP-8 fuel vapor in the oral airway.

deposition fraction. This may be because of the longer residence times for vapors with low flow rates and is consistent with the experimental observations of Li et al. (1998) for deposition of iodine vapor in a TB cast. The deposition fraction may increase relatively 75% in the oral airways when switching from exercise breathing (Q_{in} = 60 l/min)) to low-level breathing (Q_{in} = 15 l/min). The deposition fraction of JP-8 fuel vapor in the oral airway under resting conditions can be as high as 36%.

Figure 14.9 depicts the bifurcation-by-bifurcation deposition fractions of mouth-inhaled vapors in the TB airways. Clearly, vapors with perfect absorbing wall conditions deposit either in the oral airway or in the TB region; thus, they do not reach the alveolar region because of the effect of high diffusivity. Again, the higher the inhalation flow rate, the lower the vapor deposition due to the reduced residence time.

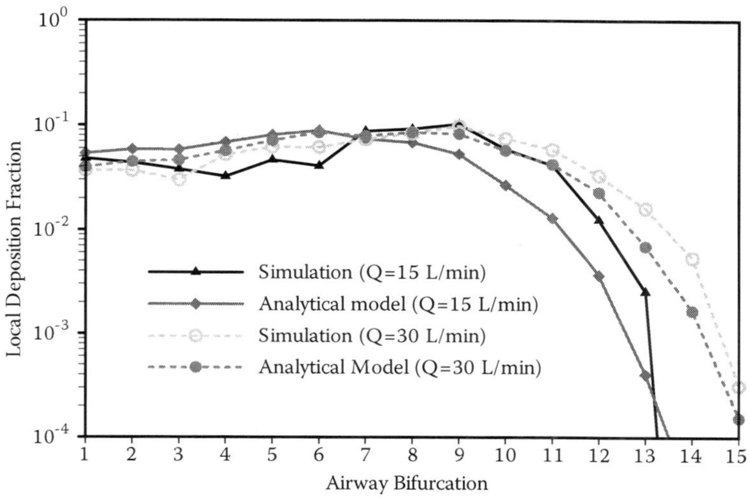

FIGURE 14.9 Bifurcation-by-bifurcation deposition fractions of vapors in the tracheobronchial airways.

Comparisons between CFPD simulations and "mathematical modeling"—that is, methodologies based on analytical or semi-empirical correlations—are also depicted in Figure 14.9. For the mathematical modeling framework, the DFs in the oral airway were assumed to be the same as those in the CFPD simulations in order to have comparable starting points. The commonly used analytical equations are employed to calculate the DEs in the present airway geometries (Kleinstreuer and Zhang, 2008). It can be observed that trend-wise these predicted DFs via mathematical modeling are similar to the simulation results. Actually, this is only true for vapor and ultrafine particle deposition (Kleinstreuer and Zhang, 2008). The measurable local differences can be attributed to geometry and upstream effects, recalling that the analytical DE equations were derived from particle transport in simple geometries (e.g., circular tubes) with idealized inlet flow profiles and particle distributions.

Figure 14.10a depicts the regional Sherwood number (Sh) for the oral airway model versus the product of Reynolds (Re) and Schmidt (Sc) numbers, both evaluated in the trachea, based on different inhalation flow rates and diffusion coefficients (i.e., nanoparticle sizes). The correlation between *Sh* and Re*Sc* group can be expressed as

$$Sh = 0.852(\mathrm{Re}\,Sc)^{0.405} \qquad 1500 < \mathrm{Re} < 7000,\ 1 < Sc < 230 \qquad (14.39)$$

where $r^2 = 0.99$. The convective mass transfer coefficients for JP-8 fuel vapor in terms of the *Sh* number has been also obtained for each individual bifurcation in airway generations G0 to G3 (Figure 14.10b). Affected by the nonhomogeneous airflow structures and mass concentration distributions, the regional *Sh* numbers are

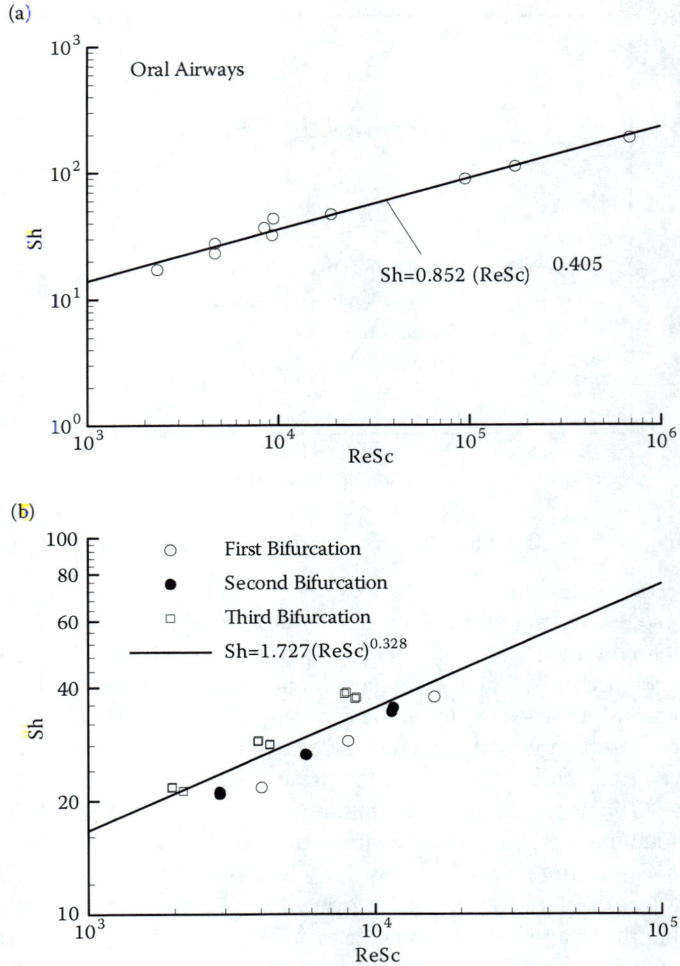

FIGURE 14.10 Regional Sherwood number (Sh) versus the product of Re and Sc for (a) human oral airways and (b) human upper bronchial tree.

slightly different for each bifurcation. However, a best-fit correlation yielded (see Figure 14.10b)

$$Sh = 1.727(\mathrm{Re}\,Sc)^{0.328} \qquad 600 < \mathrm{Re} < 6000,\ Sc \approx 3 \tag{14.40}$$

where $r^2 = 0.77$.

Impact of Airway Wall Absorption Parameter K

The absorption parameter K for different compounds of JP-8 fuel can vary from 10^{-3} to 10^3, resulting in large variations in deposition fractions. Figure 14.11a shows the

DF in the oral and bifurcation airway models as a function of K. When K is less than 1, the deposition is very low in the upper airways due to the low solubility of species in the mucus layer. The DF is greatly dependent on K for $1 < K < 1000$. If $K > 1000$, the DF is very close to that for the perfectly absorbing wall condition (i.e., $Y_{wall} = 0$). The impact of K on the species mass transfer coefficient in the airways is depicted in Figure 14.11b. Clearly, K almost has no effect on the species mass transfer coefficient for the same diffusivity.

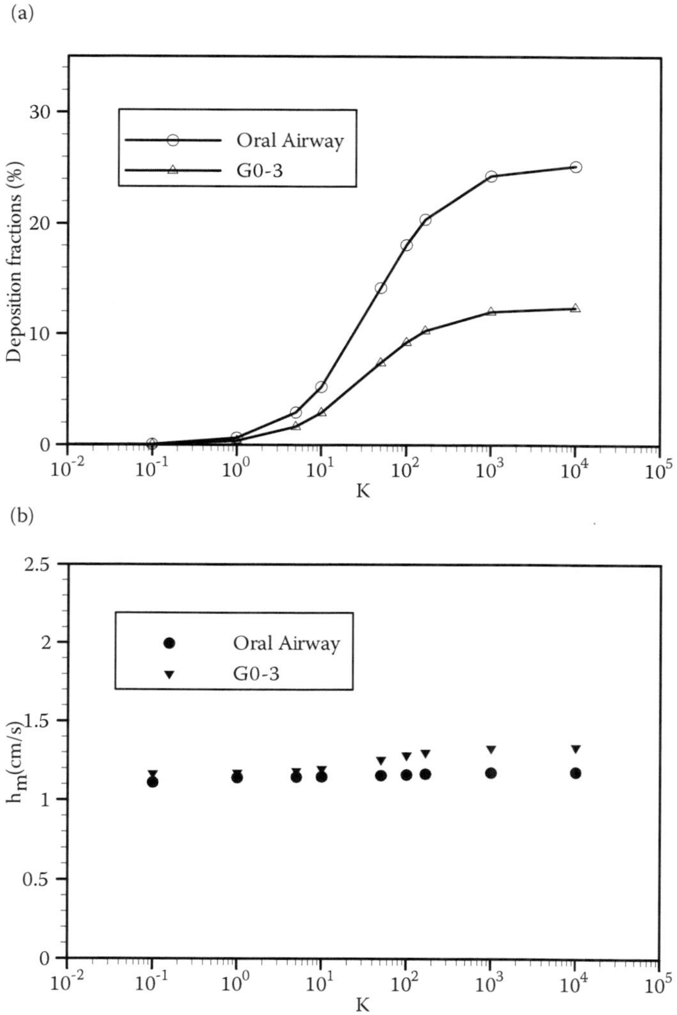

FIGURE 14.11 The effects of airway wall absorption parameter K on (a) deposition fraction and (b) mass transfer coefficients in the human upper airway model with $Q_{in} = 30$ l/min and = 0.05 cm/sec².

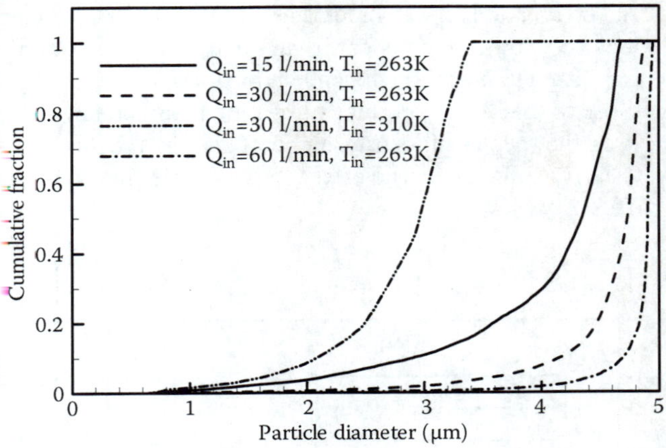

FIGURE 14.12 Cumulative distribution of JP-8 fuel droplet sizes at the inlet of the bifurcation airway model.

Droplet Dynamics

Single-Component Droplet Vaporization and Deposition

Inertial impaction is the main deposition mechanism for micron-size droplets; hence, particle size greatly affects droplet deposition. However, jet fuel is a multicomponent liquid, and the C_8 to C_{11} fractions are especially volatile. Here, one extreme case is presented: that is, it is assumed that the inlet vapor fraction of JP-8 fuel is zero and the background concentration of vapor, created by droplet evaporation, is negligible (i.e., $Y_\infty = 0$ in Equation (14.9)). As indicated by Equation (14.9), the diameter of each droplet decreases gradually in the airways due to the loss of its mass by evaporation, which is a function of the airflow momentum as well as the droplet heat and mass transfer. Affected by the highly nonuniform, occasionally turbulent airflow structures in the oral airways, the trajectories for different droplets vary; hence, the changes in droplet diameter are different because of the different local airflow fields as well as heat and mass transport. Thus, droplets enter the mouth monodisperse but turn polydisperse during vaporization. Figure 14.12 shows the cumulative distribution function of the droplet size at the outlet of the oral airways (or the inlet to the first bifurcation) for different inhalation rates and inlet air temperatures. The fraction less than a particular size can be obtained directly from the graph. The fraction of particles having diameters between two sizes can be determined by the difference of the cumulative fractions at these two sizes. The initial droplet size is a uniform distribution with d_p = 5 μm. As expected, the change in droplet size decreases with increasing flow rate due to decreasing residence times. The variation in droplet size is larger with higher inlet air temperatures because the saturation vapor pressure at the liquid/gas interface increases with temperature. The mean droplet diameters after the oral airways are d_p = 4.03, 4.57, 4.83 μm for Q_{in} = 15, 30, 60 l/min with T_{in} = 263K, respectively, and d_p = 2.80 μm for Q_{in} = 30 l/min with T_{in} = 310K. The variations in droplet size should greatly influence droplet deposition.

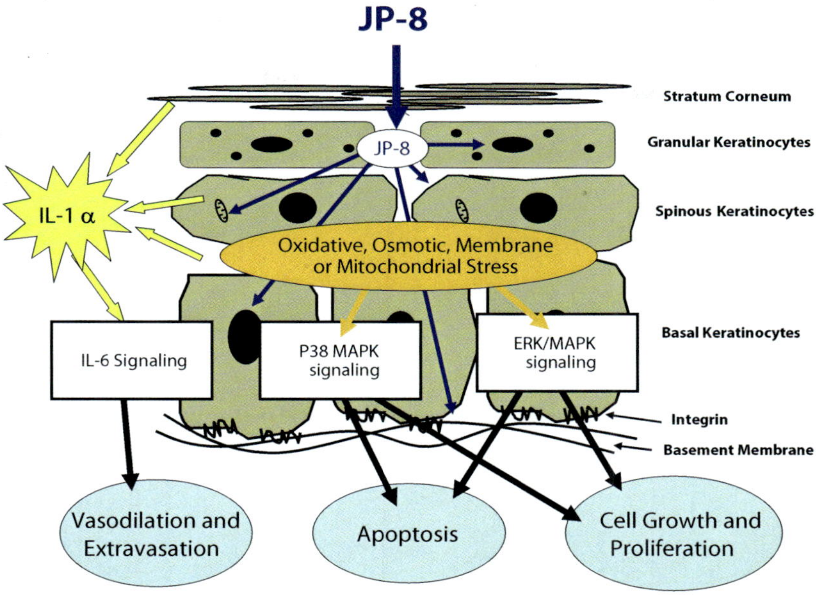

FIGURE 9.9 Schematic of the proposed mechanism of "physical" damage in the epidermis caused by JP-8 exposure. The stress response involves increases in signaling pathways that result in inflammation, apoptosis, growth, and proliferation. (Source: From McDougal, J.N., Garrett, C.M., Amato, C.M., and Berberich, S.J. *Toxicol. Sci.*, 95(2), 495–510, 2007.)

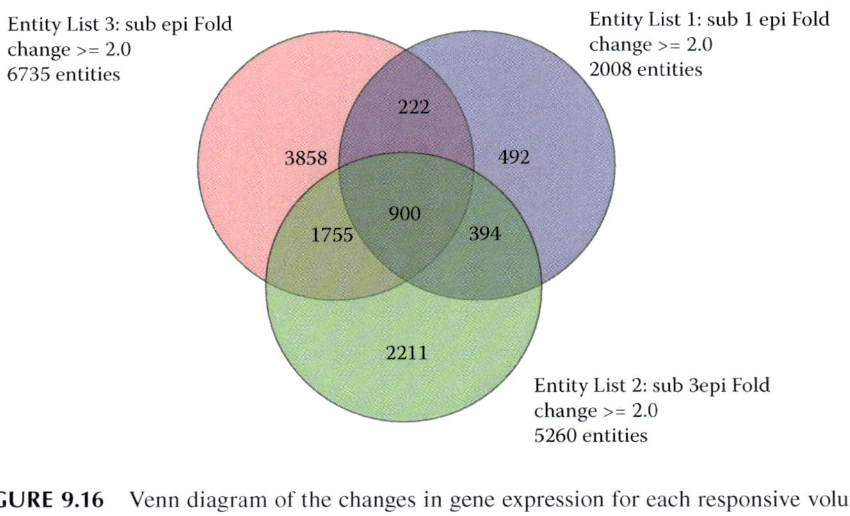

FIGURE 9.16 Venn diagram of the changes in gene expression for each responsive volunteer showing numbers of common and unique transcript changes.

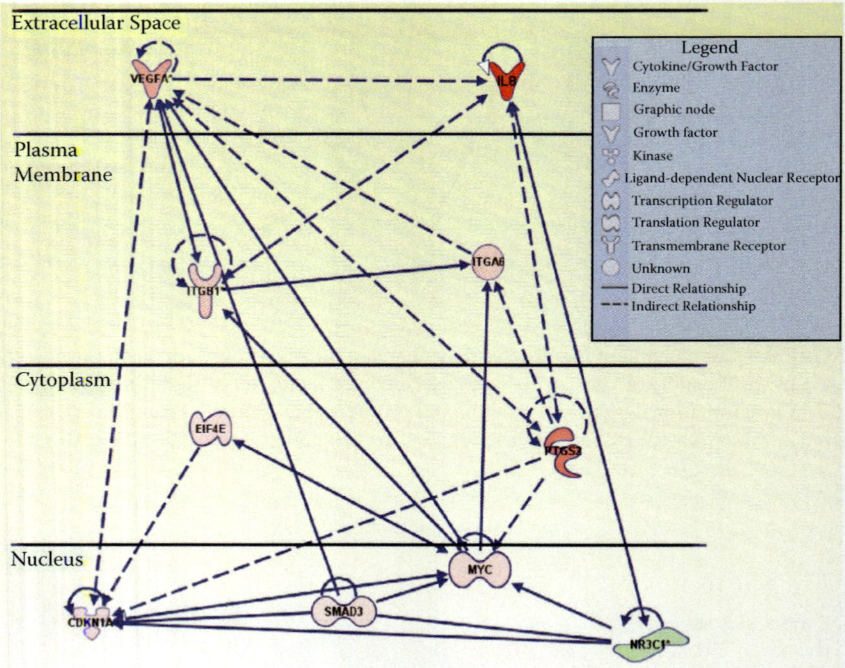

FIGURE 9.17 Genes related to proliferation of epithelial cells that were changed in each responsive individual. Genes colored red were up-regulated and genes down-regulated were colored green. The intensity of the color is related to the level of up- and down-regulation of subject 4. See text for abbreviations.

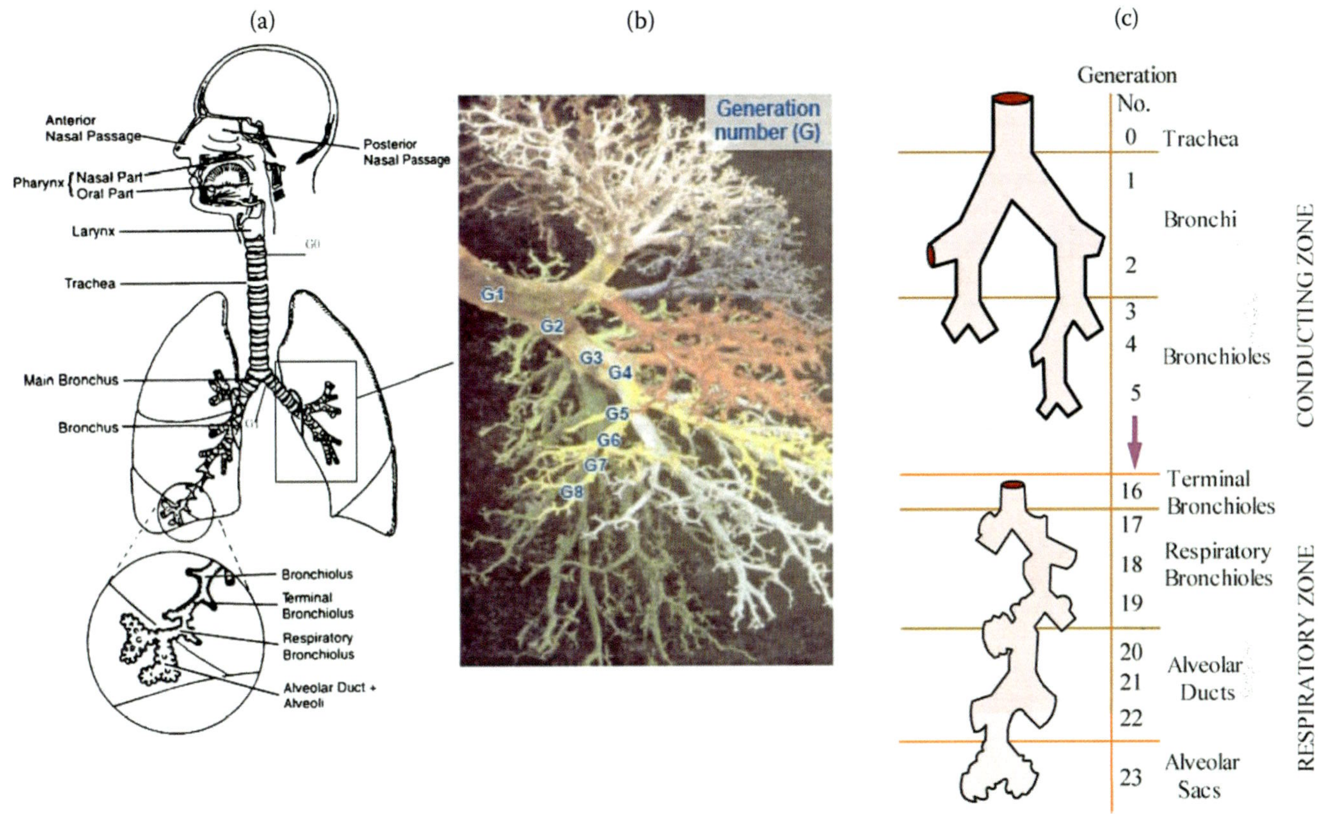

FIGURE 14.1 Schematics of the human respiratory system: (a) schematic of the respiratory system; (b) image of actual lung generations; and (c) lung conducting and respiratory zones.

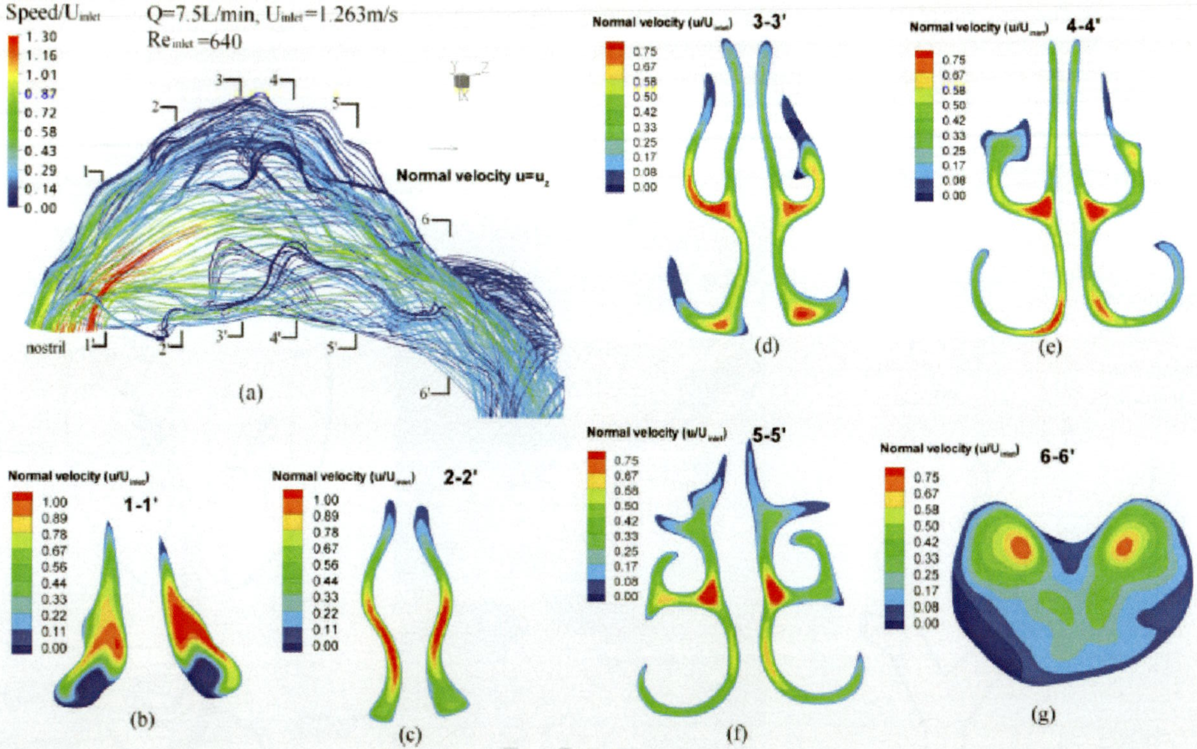

FIGURE 14.4 Velocity fields in human nasal cavity at constant inlet flow rate of 7.5 l/min: (a) three-dimensional streamlines and velocity contours; and (b–g) velocity fields in six selected slices.

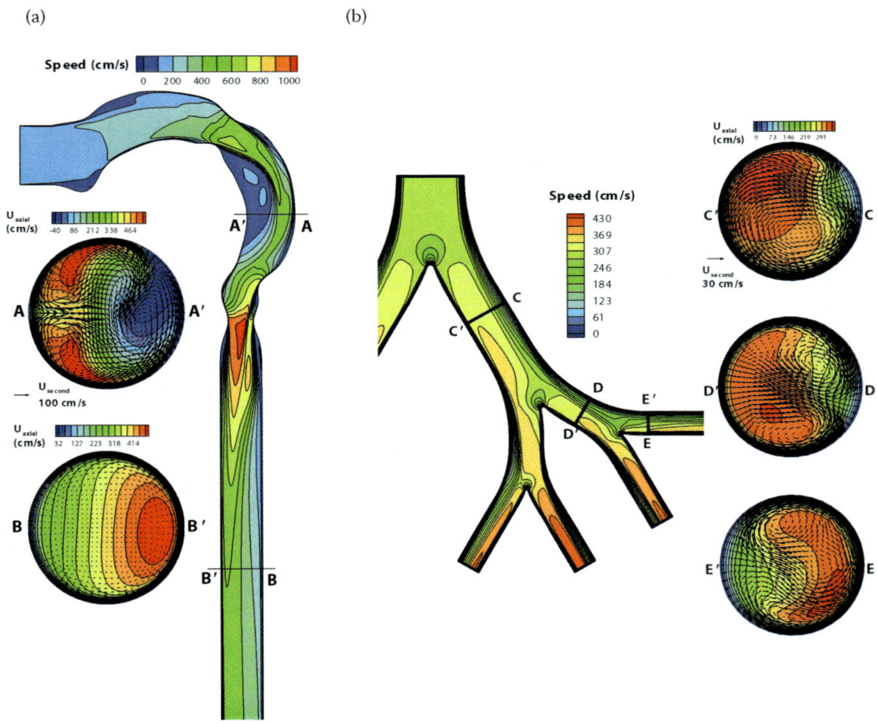

FIGURE 14.5 Side-plane or mid-plane velocity profiles as well as selected cross-sectional axial velocity contours and secondary velocity vectors with an inspiratory flow rate Q = 30 l/min in: (a) an oral airway model; and (b) a bifurcating airway model G0 to G3.

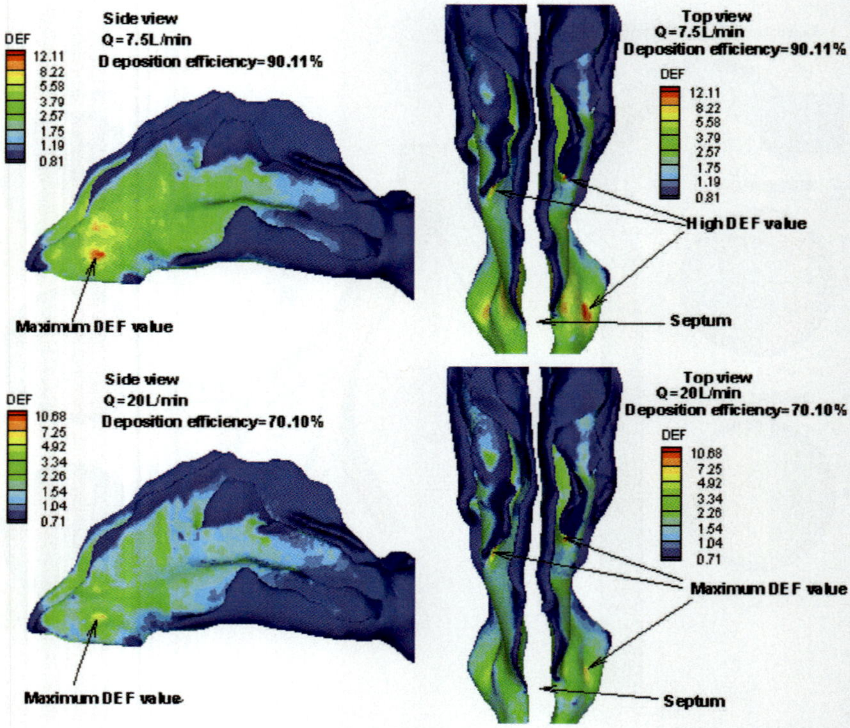

FIGURE 14.6 Vapor deposition enhancement factor (DEF) contours at constant inlet flow rates of 7.5 and 20 l/min.

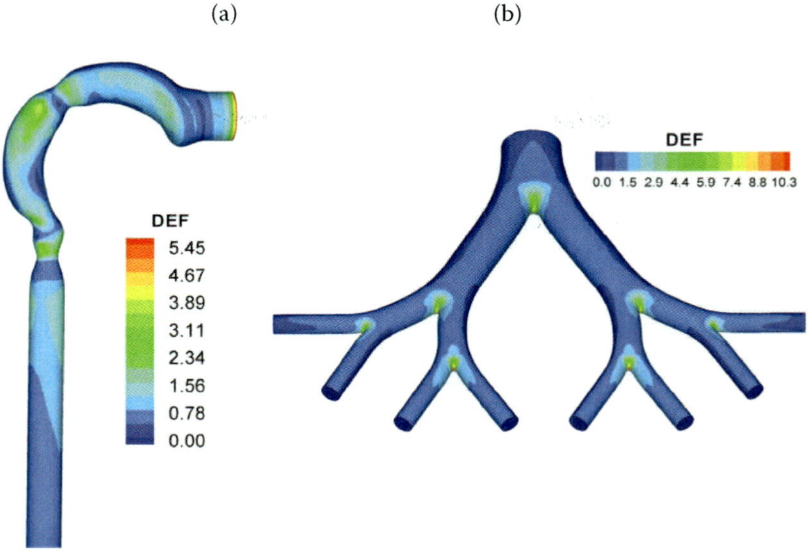

FIGURE 14.7 Three-dimensional distributions of deposition enhancement factor (DEF) of JP-8 fuel vapor under steady inhalation with Q = 30 l/min in : (a) the oral airway model; and (b) the bifurcation airway model.

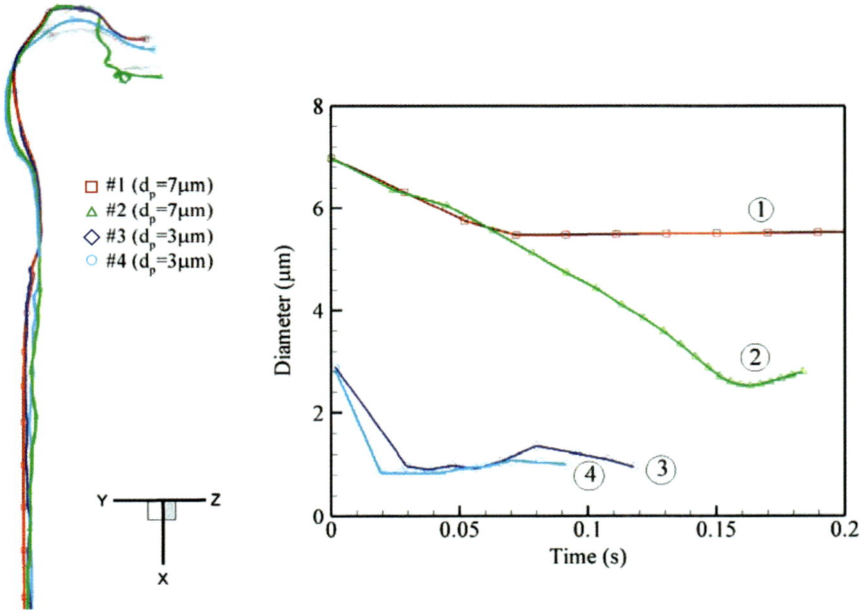

FIGURE 14.19 Trajectories (left) and diameter evolution (right) of selected isotonic saline droplets in the oral airway model with Q_{in} = 30 l/min, T_{in} = 303K, and RH_{in} = 60%.

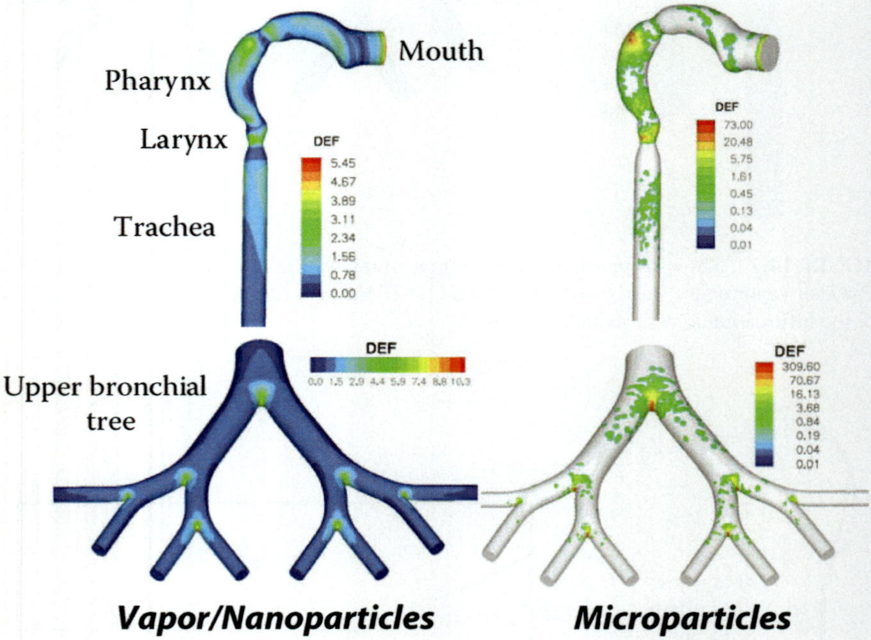

FIGURE 14.20 Local deposition patterns of vapor (nanoparticles) (left) and micron aerosols (right).

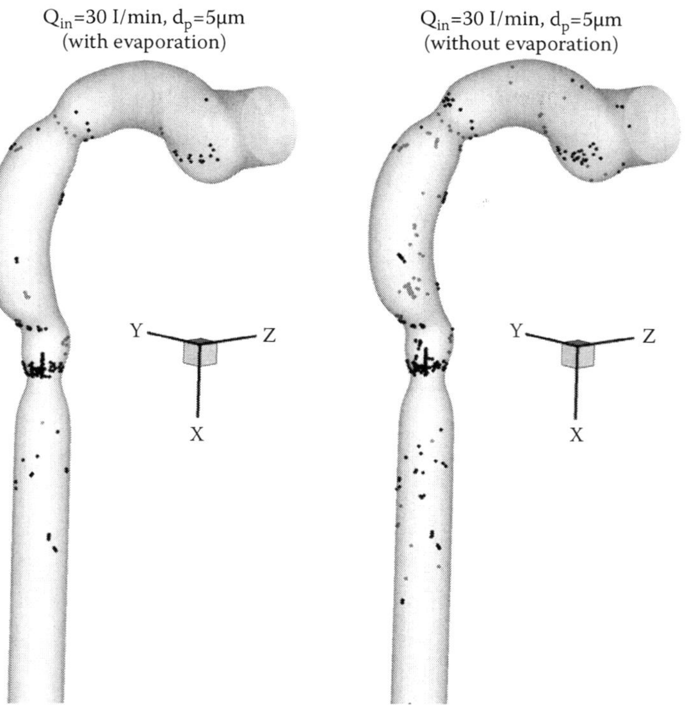

FIGURE 14.13 Droplet deposition patterns in the oral airway model.

The local deposition patterns in the oral airway model are shown in Figure 14.13 for initially $d_p = 5$ μm and $Q_{in} = 30$ l/min with and without evaporation. Due to inertial impaction, droplets mainly deposit at stagnation points for axial particle motion, such as the tongue portion in the oral cavity, the outer bend of the pharynx/larynx, and the regions just upstream of the glottis. A few droplets may also deposit outside these pronounced regions, being influenced by turbulent dispersion as well as secondary and recirculating flows. Clearly, droplet deposition decreases due to the reduced droplet diameters with evaporation.

Figure 14.14 shows the percentage of evaporated droplets in the oral airways. All droplets, initially with $d_p \leq 3$ μm, have evaporated when $Q_{in} = 15$ l/min. Thus, droplet evaporation is very important when simulating the deposition of volatile aerosols (e.g., JP-8 fuel droplets) in human airways. This also can be seen from the comparisons of predicted droplet DFs in the oral airways for different inhalation conditions (see Figure 14.15). Figure 14.15 indicates that the DFs of droplets decrease significantly if vaporization occurs. The DF increases with increasing initial particle diameter and inhalation flow rate due to enhanced inertial impaction.

Figure 14.16 depicts the overall droplet DE in the upper bronchial airways (G0 to G3), which is defined as the ratio of the number of deposited droplets to the number entering the inlet of airway segment G0 to G3. The deposition efficiency with evaporation ($T_{in} = 263$K) is still lower than that without evaporation. There is no

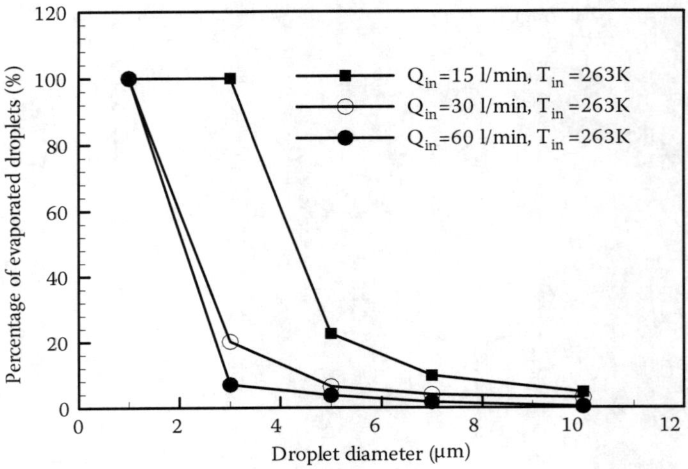

FIGURE 14.14 Variations of the percentage of evaporated droplets in the oral airway model under different inhalation conditions.

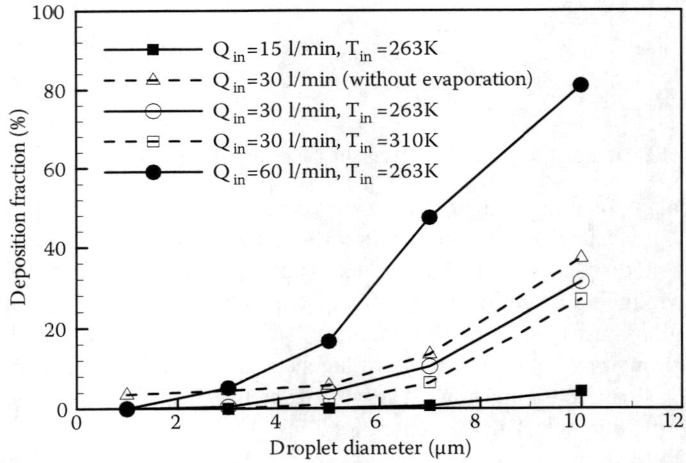

FIGURE 14.15 Deposition fractions of JP-8 fuel droplets in the oral airway model.

droplet deposition for the case of $d_p = 3$ μm with evaporation. The percent decrease in DE due to evaporation is higher for the relatively smaller size droplets (say, 3 and 5 μm). This is because large-sized particles tend to shrink much slower than smaller-sized particles. The overall DE in G0 to G3 with an ideal inlet condition (i.e., fully developed air flow with corresponding (parabolic) particle distribution) is also given in Figure 14.16 for comparison. Clearly, the inlet air velocity profile and particle distribution strongly affect the particle deposition efficiency, where idealized inlet flow conditions may overpredict actual deposition efficiencies.

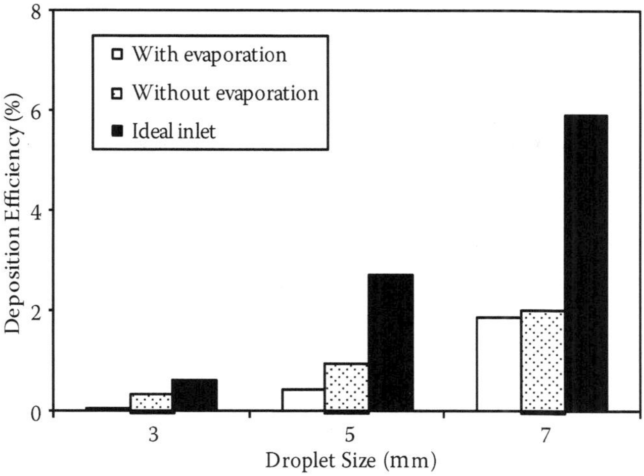

FIGURE 14.16 Deposition efficiencies of JP-8 fuel droplets in the bifurcation airway model with an inspiratory flow rate Q_{in} = 30 l/min. The inlet air temperature is 263K for the evaporation case.

Airway Geometry Effects

More realistic airway models (or patient-specific models) (e.g., asymmetric and non-planar geometries (see Figure 14.2c)), may result in some different local deposition rates when compared to the symmetric Weibel Type A model (Nowak et al., 2003, Li et al., 2007c). Figure 14.17 displays the comparisons of deposition fractions in the first bifurcation region for a solid particle (or droplet without evaporation) in the Weibel Type A model (Figure 14.2a) and asymmetric airway model (Figure 14.2c). With the realistic upstream configuration used, the Weibel Type A model provides reasonable results for relatively low and high Stokes numbers. Differences are due to geometric characteristics (e.g., diameters, length, and branching angles) and hence airflow effects on particle transport/deposition. Details concerning geometric effects on micron particle (droplet) and vapor depositions may be found in Li et al. (2007a) and Zhang et al. (2008a, 2008b).

Development of Droplet Vaporization Model and JP-8 Fuel Surrogate

An effective model for predicting multicomponent aerosol evaporation in the upper respiratory system that is capable of estimating the vaporization of individual components is needed for accurate dosimetry and toxicology analyses. In this study, the performance of evaporation models for multicomponent droplets over a range of volatilities is evaluated based on comparisons with available experimental results for conditions similar to aerosols in the upper respiratory tract (Longest and Kleinstreuer, 2005). Models considered include a semiempirical correlation approach as well as resolved-volume computational simulations of single and multicomponent aerosol evaporations to test the effects of variable gas-phase properties, surface blowing velocity, and internal droplet temperature gradients. Of the parameters assessed,

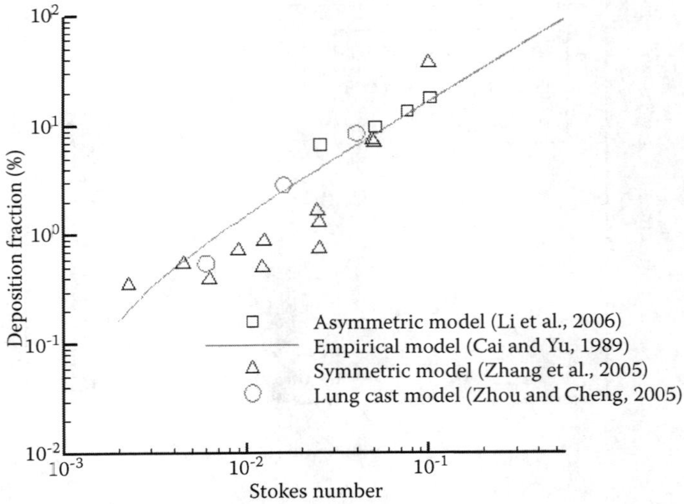

FIGURE 14.17 Comparison of experimental and computational deposition results in the first bifurcation region of different airway models.

concentration-dependent gas-phase specific heat had the largest effect on evaporation and should be taken into consideration for respiratory aerosols that contain high-volatility species (e.g., n-heptane) at significant concentrations. For heavier droplet components or conditions below body temperature, semiempirical estimates were shown to be appropriate for respiratory aerosol conditions. To reduce the number of equations and properties required for complex mixtures, a resolved-volume evaporation model (Longest and Kleinstreuer, 2005) was used to identify a 12-component surrogate representation of potentially toxic JP-8 fuel based on comparisons with experimentally reported droplet evaporation data. Due to the relatively slow evaporation rate of JP-8 aerosols, preliminary results indicate that a semiempirical evaporation model in conjunction with the identified surrogate mixture provides a computationally efficient method for computing droplet evaporation that can track individual toxic markers. However, semiempirical methodologies are in need of further development to effectively compute the evaporation of other higher-volatility aerosols for which variable gas-phase specific heat does play a significant role.

Solutions of selected evaporation models (i.e., RMM1 and RMM2) have been evaluated for a 12-component surrogate mixture suggested by Edwards and Maurice (2001) (Table 14.1) and are compared to the experimental results of Runge et al. (1998) (see Figure 14.18). The significant discrepancy between the computed and empirical results is likely due to an incorrect surrogate mixture for droplet evaporation or an inaccuracy of the assumed rapid mixing model for droplets consisting of a significant number of compounds. Inclusion of concentration gradients in the droplet will slow the initial evaporation rate by limiting transport of high-volatility compounds to the surface, in the absence of a Hill vortex formation. However, evaporation at later times will be increased due to the continued availability of the lighter components.

TABLE 14.1
Twelve-Component JP-8 Surrogate

Component	Wt.% Edwards and Maurice (2001)	Wt.% Computed with RMM2
iso-Octane	5.0	0.25
Methylcyclohexane	5.0	0.25
m-Xylene	5.0	0.25
Cyclooctane	5.0	0.25
Decane	15.0	7.0
Butylbenzene	5.0	3.0
1,2,4,5-Tetramethylbenzene	5.0	14.0
Tetralin	5.0	15.0
Docecane	20.0	23.0
1-Methylnaphthalene	5.0	8.0
Tetradecane	15.0	17.0
Hexadecane	10.0	12.0

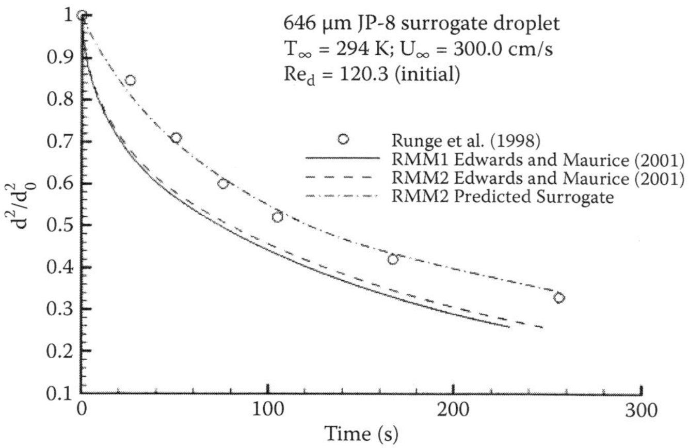

FIGURE 14.18 Computational estimates of normalized droplet surface area (d^2/d_σ^2) over time compared to the experimental results of Runge et al. (1998) for a 12-component JP-8 surrogate mixture. (*Source:* From Longest, P.W., and Kleinstreuer, C. *Aerosol Sci. Technol.*, 39, 124–138, 2005.)

It appears that the slopes of the predicted evaporation curves remain greater than or equal to the experimental data. This indicates that internal gradients may not be responsible for the discrepancy between predicted and experimental values, which is consistent with the computed results for a binary heptane–decane droplet. Based on calculations with the RMM2 approximation, a surrogate mixture that accurately predicts droplet evaporation is provided in Table 14.1, and evaporation rates are shown in Figure 14.18. For the suggested surrogate model, differences among the RMM1 and RMM2 solutions are largely negligible, which indicates that either a reduced presence of volatile hydrocarbons or a reduced evaporation rate associated with multiple components minimizes the need to account for variable gas-phase specific heat in this case. As such, the ODE-based semiempirical solution used with the suggested surrogate mixture provides a computationally efficient method for evaluating the evaporation of potentially toxic JP-8 aerosols in the respiratory tract.

Transport and Deposition of Multicomponent Droplets

The impurities of multicomponent droplets may reduce the surface vapor pressure and hence decrease the droplet evaporation or cause droplet growth. As an example, the evaporation/hygroscopicity, transport, and deposition of *isotonic and hypertonic saline* droplets have been simulated for the human upper airways and analyzed for different steady inhalation conditions (Zhang et al., 2006a, 2006c). The insight gained from these studies will be valuable when analyzing actual JP-8 fuel aerosols. Figure 14.19 depicts the evolution of both trajectories and diameters in the oral airways, considering four different isotonic saline droplets (NaCl ≈ 0.9%) at Q_{in} = 30 l/min, RH_{in} = 60%, and T_{in} = 303K. Clearly, the diameter of each droplet, starting at

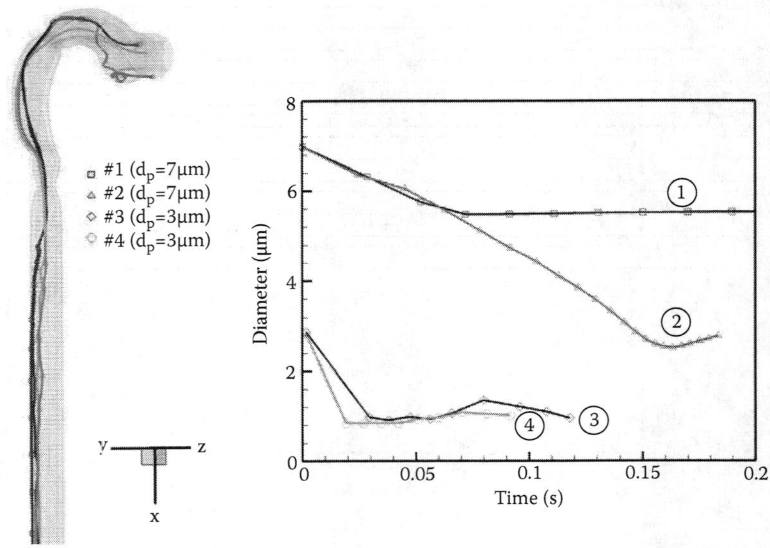

FIGURE 14.19 (See color insert following page 272) Trajectories (left) and diameter evolution (right) of selected isotonic saline droplets in the oral airway model with Q_{in} = 30 l/min, T_{in} = 303K, and RH_{in} = 60%.

the mouth inlet, decreases gradually due to mass loss by evaporation, and then may increase due to water condensation by hygroscopicity, driven by the airflow field and convective heat/mass transfer. Affected by the highly nonuniform, occasionally turbulent airflow structures in the oral airways, the trajectories for different droplets vary; hence, the changes in droplet diameters are different because of the different local airflow as well as heat and mass transfer. Even when starting with the same diameters (i.e., 3 and 7 μm) but different release positions, they undergo contrasting changes depending upon local (moist) environment. Specifically, the diameter of droplet #1 gradually decreases to about 5.5 μm from the mouth inlet to the glottis due to vaporization; however, it may increase slightly in the trachea when it moves near the airway wall where RH_∞ is very high. Droplet #2 may decrease from 7 μm to 2.6 μm before it starts to grow after the trachea. Small-size droplets (e.g.,, $d_p = 3$ μm, i.e., droplets #3 and #4) may continuously shrink from the inlet until reaching an equilibrium size where they are neither losing nor absorbing water, and then they start to grow or shrink as they move through a high- or low-humidity environment.

The increase in initial solute concentration for multicomponent droplets (e.g., hypertonic saline droplets) may reduce solution (e.g., water) evaporation and increase its condensation (i.e., hygroscopic effects) at droplet surfaces so that the droplet deposition can increase due to increasing particle diameter and density (Zhang et al., 2006a).

Deposition Comparisons between Vapor (Nanoparticles) and Micron Particles (Droplets)

Typical deposition patterns of both vapor (nanoparticles) and micron-droplets (particles) are shown in Figure 14.20 in terms of DEF distributions in oral and bifurcation airway models. Micron-size aerosol deposition during inhalation is mainly due to impaction, secondary flow convection, and turbulent dispersion. Thus, they largely deposit at stagnation points for axial aerosol motion, such as the tongue portion in the oral cavity, the outer bend of the pharynx/larynx, and the regions just upstream of the glottis and the straight tracheal tube, as well as the regions of carinal ridges in the bronchial tree. As for vapors, the enhanced deposition mainly occurs at the carinal ridges and the inside walls around the carinal ridges again due to the complicated airflows and large concentration gradients in these regions. Although the enhanced deposition sites (i.e., those with high DEF-values) in the bifurcating airways are similar for vapors and micron droplets, the maximum DEF values for vapor [nanoparticles with $d_p < 10$ nm] are two to three orders of magnitude smaller than for micron aerosols. Specifically, the DEF_{max} for micron aerosols is on the order of 10^2 to 10^3, while DEF_{max} for vapors (nanoparticles) is on the order of 1 (Zhang et al., 2005). A more uniform distribution of deposited vapors may relate to a different toxicity effect when compared to fine aerosols made of the same materials. Specifically, not only the larger surface areas relative to the particle mass but also, and more importantly, the larger surface areas with a *near-uniform* deposition can generate a higher probability of interaction with cell membranes. Hence, it may result in a greater capacity to absorb and transport toxic substances into tissue, blood, and the whole body, and the enhanced possibility of systematic diseases, such as cardiovascular diseases (Hoet et

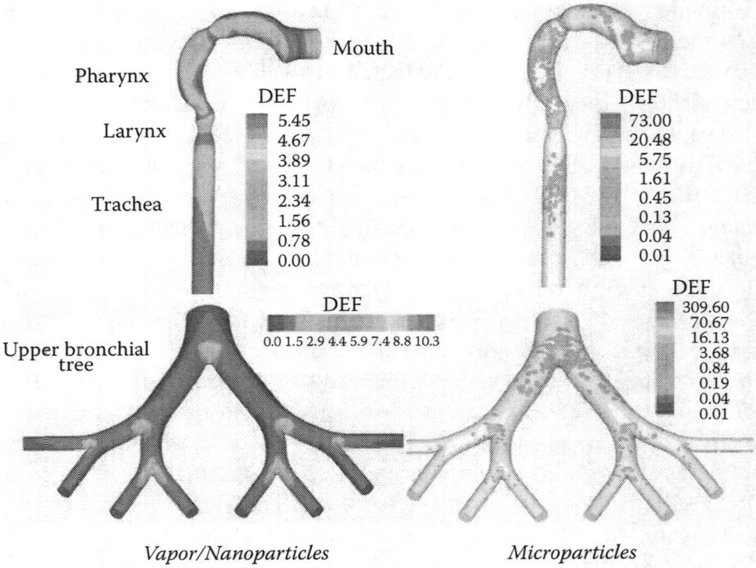

FIGURE 14.20 (See color insert following page 272) Local deposition patterns of vapor (nanoparticles) (left) and micron aerosols (right).

al., 2004; Oberdörster et al., 2005). In contrast, the extremely high local DEF values for micron aerosols indicates the possibility of local pathological changes in bronchial airways, such as the formation of lung tumors (Balashazy et al., 2003).

CONCLUSIONS

Jet fuel performance enhancers—that is, additives such as BHT, DiEGME, EGME, BTEX, etc.—are highly toxic and may pose a significant health risk to Air Force personnel, fuelers, and people living near military fields. To understand the possible health effects from fuel exposure in the form of inhaled droplets and vapor, it is crucial to know where and at what concentrations fuel aerosols and their vapors deposit in the human respiratory system, given a set of realistic inlet conditions. Direct deposition tests of toxic fuel in human lung airways are impossible and usually cost-intensive experimental deposition data, using human cast replicas, lack *detailed* resolution. Inhalation studies with small animals are often questionable because of the very different lung morphologies and air-particle dynamics as well as aerosol uptake mechanisms and health impact. Thus, CFPD simulations, if experimentally validated, offer a noninvasive, less-expensive, and effective means of obtaining fuel vapor and droplet transport and deposition data for representative airways.

An experimentally validated computer model has been developed to simulate micron-droplet and vapor (nanoparticle) transport and deposition in nasal and lung airway models. The inhaled aerosols represent main characteristics of JP-8 fuel droplets and vapor. The results can be summarized as follows:

1. Unique flow characteristics, such as flow separation, secondary flows, and turbulent fluctuations, can occur in representative human nasal cavities and lung airways, which may influence the deposition of inhaled vapor or droplets.
2. Assuming a perfect absorbing airway wall, approximately 70% to 90% of the incoming fuel vapor mass deposits in the nasal cavity while the deposition fractions in the oral airways are about 20% to 36% in case of oral inhalation. The inhalation flow rate has a significant effect on fuel vapor deposition; that is, the higher the flow rate, the lower the deposition fraction.
3. The respiratory uptake of vapors of different fuel components is greatly influenced by airway wall absorption in terms of the dimensionless absorption parameter K. The deposition is very low in the upper airways when K is less than 1, and the deposition fraction is very close to that assuming a perfectly absorbing wall when $K > 1000$. Thus, the deposition fraction is greatly dependent on materials with $1 < K < 1000$. The deposition is greater for vapors with higher diffusivity and solubility. The parameter K has almost no effect on the mass transfer coefficients of species with the same diffusivities. Thus, the mass transfer coefficients of vapors in the upper airway can be correlated as a function of flow and diffusion parameters.
4. JP-8 fuel droplet evaporation greatly affects deposition concentrations in human airways. Droplet deposition fractions due to evaporation decrease with elevated ambient temperatures and lower inspiratory flow rates. The impurities of droplets may reduce the surface vapor pressure and hence decrease the droplet evaporation or cause droplet growth. A semi-empirical evaporation model in conjunction with the identified surrogate mixture provides a computationally efficient method for computing fuel droplet evaporation that can track individual toxic markers.
5. As with micron-droplets, vapor (or nanoparticle) deposition occurs at greater concentrations around the carinal ridges when compared to the straight segments in the bronchial airways; however, deposition distributions for nanoparticles are much more uniform along the airway branches than for microparticles. This may imply different toxicities and hence health effects for micron aerosols and vapors.

ACKNOWLEDGMENTS

This effort was sponsored by the Air Force Office of Scientific Research, Air Force Material Command, USAF, under grant number FA9550-07-1-0461 (Dr. Walt Kozumbo, Program Manager).

REFERENCES

Agency for Toxic Substances and Disease Registry (ATSDR). (2005). Toxicological Profile for Naphthalene, 1-Methylnaphthalene, and 2-Methylnaphthalene. U.S. Department of Health and Human Services.

Ahmed, S.A., and Giddens, D.P. (1983). Velocity measurements in steady flow through axisymmetric stenoses at moderate Reynolds number. *J. Biomechanics*, 16, 505–516.

Balashazy, I., Hofmann, W., and Heistracher, T. (1999). Computation of local enhancement factors for the quantification of particle deposition patterns in airway bifurcations. *J. Aerosol Sci.,* 30, 185–203.

Balashazy, I., Hofmann, W., and Heistracher, T. (2003). Local particle deposition patterns may play a key role in the development of lung cancer. *J. Appl. Physiol.,* 94, 1719–1725.

Bardon, M.F., and Rao, V.K. (1984). Calculation of gasoline volatility. *J. Inst. Energy,* 57, 343–348.

Bejan, A. (1995). *Convection Heat Transfer* New York, New York: John Wiley & Sons, Inc.

Benaissa, A., Gauthier, J.E.D., Bardoni, M.F., and Laviolette, M. (2002). Modelling evaporation of multicomponent fuel droplets under ambient temperature conditions. *J. Inst. Energy,* 75, 19–26.

Berlemont, A., Grancher, M.S., and Gouesbet, G. (1995). Heat and mass-transfer coupling between vaporizing droplets and turbulence using a Lagrangian approach. *Int. J. Heat Mass Transfer,* 38, 3023–3034.

Broday, D.M., and Georgopoulos, P.G. (2001). Growth and deposition of hygroscopic particulate matter in the human lungs. *Aerosol Sci. Technol.,* 34, 144-–159.

Cai, F.S., and Yu, C.P. (1988). Inertial and interceptional deposition of spherical particles and fibers in a bifurcating airway. *J. Aerosol Sci.,* 6, 679–688.

Cheng, K.H., Cheng, Y.S., Yeh, H.C., and Swift, D.L. (1997a). An experimental method for measuring aerosol deposition efficiency in the human oral airway. *Am. Ind. Hyg. Assoc. J.,* 58, 207–213.

Cheng, K.H., Cheng, Y.S., Yeh, H.C., and Swift, D.L. (1997b). Measurements of airway dimensions and calculation of mass transfer characteristics of the human oral passage. *J. Biomechan. Eng.-Trans. ASME,* 119, 476–482.

Cheng, Y.S., Zhou, Y., and Chen, B.T. (1999). Particle deposition in a cast of human oral airways. *Aerosol Sci. Technol.*, 31, 286–300.

Choi, L.T., Tu, J.Y., Li, H.F., and Thien, F. (2007). Flow and particle deposition patterns in a realistic human double bifurcation airway model. *Inhalat. Toxicol.,* 19, 117–131.

Clift, R., Grace, J.R., and Weber, M.E. (1978). *Bubbles, Drops, and Particles,* New York: Academic Press.

Edwards, T., and Maurice, L.Q. (2001). Surrogate mixtures to represent complex aviation and rocket fuels. *J. Propulsion Power,* 17, 461–466.

Fan, B.J., Cheng, Y.S., and Yeh, H.C. (1996). Gas collection efficiency and entrance flow effect of an annular diffusion denuder. *Aerosol Sci. Technol.,* 25, 113–120.

Ferron, G.A., Haider, B., and Kreyling, W.G. (1988a). Inhalation of salt aerosol-particles .1. Estimation of the temperature and relative-humidity of the air in the human upper airways. *J. Aerosol Sci.,* 19, 343–363.

Ferron, G.A., Kreyling, W.G., and Haider, B. (1988b). Inhalation of salt aerosol-particles. 2. Growth and deposition in the human respiratory-tract. *J. Aerosol Sci.,* 19, 611–631.

Ferron, G.A., Oberdörster, G., and Henneberg, R. (1989). Estimation of the deposition of aerosolized drugs in the human respiratory tract due to hygroscopic growth. *J. Aerosol Med.,* 2, 271–284.

Finlay, W.H. (2001). *The Mechanics of Inhaled Pharmaceutical Aerosols: An Introduction,* London, U.K.: Academic Press.

Finlay, W.H., and Martin, A.R. (2008). Recent advances in predictive understanding of respiratory tract deposition. *J. Aerosol Med. Pulmon. Drug Delivery,* 21, 189–205.

Finlay, W.H., and Stapleton, K.W. (1995). The effect on regional lung deposition of coupled heat and mass-transfer between hygroscopic droplets and their surrounding phase. *J. Aerosol Sci.,* 26, 655–670.

Gosman, A.D. and Ioannides, E. (1981). Aspects of computer simulation of liquid-fueled combustors. *J. Energy,* 7, 482–490.

Gustafson, J.B., Tell, J.G., and Orem, D. (1997). *Selection of Representative TPH Fractions Based on Fate and Transport Consideration,* Amherst, MA: Amherst Scientific Publishers.

Hoet, P., Brueske-Hohlfeld, I., and Salata, O. (2004). Nanoparticles – known and unknown health risks. *J. Nanobiotechnol.,* 2, 12.

Horsfield, K., Dart, G., Olson, D.E., Filley, G.F., and Cumming, G. (1971). Models of the human bronchial tree. *J. Appl. Physiol.,* 31.

Keyhani, K., Scherer, P.W., and Mozell, M.M. (1997). A numerical model of nasal odorant transport for the analysis of human olfaction. *J. Theoret. Biol.,* 186, 279–301.

Kim, J., Moin, P., and Moser, R. (1987). Turbulence statistics in fully-developed channel flow at low Reynolds-number. *J. Fluid Mechan.,* 177, 133–166.

Kleinstreuer, C. (2006). *Biofluid Dynamics: Principles and Selected Applications,* Boca Raton, FL: CRC Press.

Kleinstreuer, C., and Zhang, Z. (2003). Laminar-to-turbulent fluid-particle flows in a human airway model. *Int. J. Multiphase Flow,* 29, 271–289.

Kleinstreuer, C., and Zhang, Z. (2009). An adjustable triple-bifurcation unit (TBU) model for air-particle flow simulations in human tracheobronchial airways. *J. Biomechan. Eng. – Trans. ASME,* in press.

Kleinstreuer, C., Zhang, Z., and Donohue, J.F. (2008a). Targeted drug-aerosol delivery in the human respiratory system. *Annu. Rev. Biomed. Eng.,* 10, 195–220.

Kleinstreuer, C., Zhang, Z., and Li, Z. (2008b). Modeling airflow and particle transport/deposition in pulmonary airways. *Respir. Physiol. Neurobiol.,* in press.

Li, W., Xiong, J.Q., and Cohen, B.S. (1998). The deposition of unattached radon progeny in a tracheobronchial cast as measured with iodine vapor. *Aerosol Sci. Technol.,* 28, 502–510.

Li, Z., Kleinstreuer, C., and Zhang, Z. (2007a). Particle deposition in the human tracheobronchial airways due to transient inspiratory flow patterns. *J. Aerosol Sci.,* 38, 625–644.

Li, Z., Kleinstreuer, C., and Zhang, Z. (2007b). Simulation of airflow fields and microparticle deposition in realistic human lung airway models. I. Airflow patterns. *Eur. J. Mechanics – B/Fluids,* 26, 632–649.

Li, Z., Kleinstreuer, C., and Zhang, Z. (2007c). Simulation of airflow fields and microparticle deposition in realistic human lung airway models. II. Particle transport and deposition. *Eur. J. Mechanics – B/Fluids* 26, 650–668.

Longest, P.W., and Kleinstreuer, C. (2005). Computational models for simulating multicomponent aerosol evaporation in the upper respiratory airways. *Aerosol Sci. Technol.,* 39, 124–138.

Longest, P.W., and Vinchurkar, S. (2007). Effects of mesh style and grid convergence on particle deposition in bifurcating airway models with comparisons to experimental data. *Med. Eng. Phys.,* 29, 350–366.

MacInnes, J.M., and Bracco, F.V. (1992). Stochastic particle dispersion modeling and the tracer-particle limit. *Phys. Fluids A,* 4, 2809–2824.

Martonen, T.B., Yang, Y., and Xue, Z.Q. (1994). Influences of cartilaginous rings on tracheobronchial fluid-dynamics. *Inhalat. Toxicol.,* 6.

Matida, E.A., Finlay, W.H., Lange, C.F., and Grgic, B. (2004). Improved numerical simulation of aerosol deposition in an idealized mouth-throat. *J. Aerosol Sci.,* 35, 1–19.

Matida, E.A., Nishino, K., and Torii, K. (2000). Statistical simulation of particle deposition on the wall from turbulent dispersed pipe flow. *Int. J. Heat Fluid Flow,* 21, 389–402.

National Research Council (NRC). Board on Environmental Studies of Toxicology (BEST). 2003. *Toxicologic Assessment of Jet-Propulsion Fuel-8.* Washington, D.C.: National Academies Press.

Nowak, N., Kakade, P.P., and Annapragada, A.V. (2003). Computational fluid dynamics simulation of airflow and aerosol deposition in human lungs. *Ann. Biomed. Eng.,* 31, 374–390.

Oberdörster, G., Oberdörster, E., and Oberdörster, J. (2005). Nanotoxicology: An emerging discipline evolving from studies of ultrafine particles. *Environ. Health Perspect.,* 113, 823–839.

Raabe, O.G., Yeh, H.C., Schum, G.M., and Phalen, R.F. (1976). Tracheobronchial Geometry: Human, Dog, Rat, Hamster, LF-53. Lovelace Foundation Report. Albuquerque, NM.

Runge, T., Teske, M., and Polymeropoulos, C.E. (1998). Low-temperature vaporization of JP-4 and JP-8 fuel droplets. *Atomization and Sprays,* 8, 25–44.

Shi, H., Kleinstreuer, C., and Zhang, Z. (2006). Laminar airflow and nanoparticle or vapor deposition in a human nasal cavity model. *J. Biomechan. Eng. - Trans. ASME,* 128, 697–706.

Shi, H., Kleinstreuer, C., and Zhang, Z. (2007). Modeling of inertial particle transport and deposition in human nasal cavities with wall roughness *J. Aerosol Sci.,* 38, 398–419.

Shi, H., Kleinstreuer, C., and Zhang, Z. (2008). Dilute suspension flow with nanoparticle deposition in a representative nasal airway model. *Phys. Fluids,* 20(1).

Shi, H., Kleinstreuer, C., Zhang, Z., and Kim, C.S. (2004). Nanoparticle transport and deposition in bifurcating tubes with different inlet conditions. *Phys. Fluids,* 16, 2199–2213.

Varghese, S.S., and Frankel, S.H. (2003). Numerical modeling of pulsatile turbulent flow in stenotic vessels. *J. Biomechan. Eng. – Trans. ASME,* 125, 445–460.

Wang, Y., and James, P.W. (1999). On the effect of anisotropy on the turbulent dispersion and deposition of small particles. *Int. J. Multiphase Flow,* 25, 551–558.

Weibel, E.R. (1963). *Morphometry of the Human Lung,* New York: Academic Press.

Wilcox, D.C. (1998). *Turbulence Modeling for CFD,* La Canada, CA: DCW Industries, Inc.

Witten, M. (2002). JP-8 fuel aerosol exposure/inhalation data, personal communication.

Zhang, Y., and Finlay, W.H. (2005). Measurement of the effect of cartilaginous rings on particle deposition in a proximal lung bifurcation model. *Aerosol Sci. Technol.,* 39, 394–399.

Zhang, Z., and Kleinstreuer, C. (2002). Transient airflow structures and particle transport in a sequentially branching lung airway model. *Phys. Fluids,* 14, 862–880.

Zhang, Z., and Kleinstreuer, C. (2003a). Low-Reynolds-number turbulent flows in locally constricted conduits: A comparison study. *AIAA J.,* 41, 831–840.

Zhang, Z., and Kleinstreuer, C. (2003b). Species heat and mass transfer in a human upper airway model. *Int. J. Heat Mass Transfer,* 46, 4755–4768.

Zhang, Z., and Kleinstreuer, C. (2004). Airflow structures and nano-particle deposition in a human upper airway model. *J. Computat. Phys.,* 198, 178–210.

Zhang, Z., Kleinstreuer, C., Donohue, J.F., and Kim, C.S. (2005). Comparison of micro- and nano-size particle depositions in a human upper airway model. *J. Aerosol Sci.,* 36, 211–233.

Zhang, Z., Kleinstreuer, C., and Kim, C.S. (2002a). Cyclic micron-size particle inhalation and deposition in a triple bifurcation lung airway model. *J. Aerosol Sci.,* 33, 257–281.

Zhang, Z., Kleinstreuer, C., and Kim, C.S. (2002b). Gas-solid two-phase flow in a triple bifurcation lung airway model. *Int. J. Multiphase Flow,* 28, 1021–1046.

Zhang, Z., Kleinstreuer, C., and Kim, C.S. (2006a). Isotonic and hypertonic saline droplet deposition in a human upper airway model. *J. Aerosol Med. - Deposition Clearance and Effects in the Lung,* 19, 184–198.

Zhang, Z., Kleinstreuer, C., and Kim, C.S. (2006b). Transport and uptake of MTBE and ethanol vapors in a human upper airway model. *Inhalat. Toxicol.,* 18, 169–184.

Zhang, Z., Kleinstreuer, C., and Kim, C.S. (2006c). Water vapor transport and its effects on the deposition of hygroscopic droplets in a human upper airway model. *Aerosol Sci. Technol.,* 40, 1–16.

Zhang, Z., Kleinstreuer, C., and Kim, C.S. (2008a). Airflow and nano-particle deposition in a 16-generation tracheobronchial airway model. *Ann. Biomed, Eng.,* submitted.

Zhang, Z., Kleinstreuer, C., and Kim, C.S. (2008b). Comparison of analytical and CFD models with regard to micron particle deposition in a human 16-generation tracheobronchial airway model. *J. Aerosol Sci.,* in press.

Zhang, Z., Kleinstreuer, C., Kim, C.S., and Cheng, Y.S. (2004). Vaporizing microdroplet inhalation, transport, and deposition in a human upper airway model. *Aerosol Sci. Technol.,* 38, 36–49.

Zhou, Y., and Cheng, Y.S. (2005). Particle deposition in a cast of human tracheobronchial airways. *Aerosol Sci. and Technol.,* 39, 492–500.

15 Human Exposure to Jet Propellant-8

Raymond H. Tu and Terence H. Risby

CONTENTS

INTRODUCTION

Human exposure to Jet Propellant-8 (JP-8) has increased dramatically over the past 20 years when JP-8 became the preferred fuel for military aircraft and land vehicles in the United States and NATO countries (Ritchie et al., 2001a). JP-8 is the single largest source (9.58 billion liters in 2000) of chemical exposure to military personnel in the United States (Henz, 1998, from Egeghy et al., 2003; Ritchie et al., 2003). Worldwide consumption of kerosene-based jet fuels has grown from approximately 227 billion liters in 1992 (Defense Fuels Supply Center, 1997) to 240 billion liters in 1998 (Armbrust Aviation Group, 1998). The rise in the use of kerosene-based fuels worldwide has expanded human exposure to these fuels from 1.3 million U.S. workers in 1992 (Defense Fuels Supply Center, 1997) to 2.0 million workers in 2003 (Ritchie et al., 2003). As worldwide demand for JP-8 and its commercial airline equivalent, Jet A or Jet A-1, increases annually, the number of workers exposed to kerosene-based fuels is expected to continually multiply.

OCCUPATIONAL EXPOSURE TO JP-8

Field studies on military bases have identified activities that result in human JP-8 exposure: spills during the transportation and storage of fuel, fueling, general maintenance and operation of aircrafts and vehicles, cold engine starts, performance testing, and cleaning and degreasing of parts with fuel (Kim et al., 2007; Subcommittee on Jet-Propulsion Fuel 8, 2003). Military personnel who work directly with the fuel undoubtedly have the greatest potential for JP-8 exposure, including pilots, flight crews, and maintenance workers (Zeiger and Smith, 1998). Fuel cell maintenance personnel have the highest exposure to JP-8 because in order to perform repair work, they must enter the aircraft fuel tanks where fuel vapors accumulate, and residual JP-8 covers the bottom of the tank. Before maintenance can be performed, the polyurethane foam, which is saturated with JP-8 that fills the fuel cell, must be removed (Carlton and Smith, 2000). This foam is in fuel cells to improve the safety and stability of the aircraft. Air Force fuel cell maintenance personnel work in groups of three and consist of an entrant (enters the fuel tank, removes polyurethane foam from the tank, and performs maintenance work inside the tank); an attendant (assists entrant from outside the fuel tank and hands foam to the runner); and a runner (moves foam to a temporary storage location and provides tools to the attendant) (Serdar et al., 2004).

Protective equipment offers some defense against exposure to JP-8. Entrants wear respirators when working inside the tank; however, the attendant and runner who provide support to the entrant outside the tank are not required to wear respirators (Rhodes et al., 2003). Respirators reduce the amount of JP-8 vapor entrants are exposed to during fuel cell work, whereas the attendants and runners who do not wear respirators have significant inhalation exposure to JP-8. All three fuel cell maintenance personnel wear cotton clothing to protect against static charge; however, cotton provides very little protection against dermal exposure because it becomes quickly saturated with JP-8 and prolongs dermal exposure (Mattorano et al., 2004). To reduce skin exposure, military personnel are supplied with multiple pairs of cotton sweatshirts, sweatpants, coveralls, head-covers, and booties, and are encouraged to change into clean clothes whenever their clothes are contaminated with JP-8. Recently, the U.S. Air Force (USAF) approved protective coveralls manufactured from a tri-layer fabric consisting of a nylon filament outer fabric, a breathable GORE-TEX® membrane, and an antistatic nylon knit backer to replace the cotton coveralls. This new material provides dermal protection to jet fuel exposure while maintaining antistatic properties. Another protective measure that fuel cell maintenance workers take to reduce dermal exposure is wearing heavy-duty, chemical-resistant gloves when they handle and carry foam saturated with JP-8. In some instances, they wear disposable gloves under these gloves (double gloving) to further minimize their exposure to jet fuel when the chemical-resistant gloves are removed (Tu et al., 2004). Disposable booties have been shown to significantly decrease naphthalene exposure, a surrogate for JP-8 exposure, for subjects whose foot or leg was used for exposure monitoring (Chao et al., 2005).

Other categories of military personnel who work directly with fuel include crew chiefs, mechanics, and fuel specialists. Crew chiefs perform inspections on the aircraft before and after flight operations and may be exposed to JP-8 aerosols during low ambient temperature engine starts. Mechanics acquire JP-8 exposure when working with engine or other fuel-soaked parts that need repair. Fuel specialists are responsible for receiving JP-8 on base, refueling aircraft on the flightline, and checking the fuel for contamination and proper additive concentrations. Exposure to fuel specialists can occur when handling the fuel or during fuel spills that occasionally occur on the flightline (Tu et al., 2004). Even individuals on military bases who do not work directly with JP-8, such as hospital staff, military police, and office workers, have incidental exposure to jet fuel because it is present in the ambient air or from contact with individuals exposed to JP-8 either on their skin or on their clothing. All military personnel on base, regardless of their occupation, had measurable levels of JP-8 exposure (Pleil et al., 2000; Tu et al., 2004).

JP-8 EXPOSURE ASSESSMENT

Assessing JP-8 exposure is difficult because it is a mixture of aliphatic hydrocarbons (81%) and aromatic compounds (19%). The formulation of JP-8 is performance based so the actual composition of a particular JP-8 fuel will depend on the feedstock (Mattorano et al., 2004; Serdar et al., 2004). Consequently, a marker compound (a component found in all jet fuel mixtures) is often identified and used to represent exposure to the complete mixture. Naphthalene and less frequently benzene have been used as marker compounds for JP-8 exposure. Benzene is only present in JP-8 at concentrations below 0.02% (DoD specifications); but as a human carcinogen, it is important from a public health perspective that human exposure to benzene be identified and monitored. Using benzene as a marker of exposure is problematic because benzene is present in cigarette smoke and gasoline. Activities such as smoking, automobile refueling, or engine repair will contribute to body burden of benzene (Egeghy et al., 2003). Furthermore, due to the high volatility of benzene, benzene exposure primarily occurs via inhalation and would therefore not provide a good measure of estimating dermal absorption of JP-8 (Serdar et al., 2003). On the other hand, using naphthalene as a surrogate for JP-8 exposure is more appropriate for several reasons. It is one of the 13 components that comprise 29% of the base fuel and is present at concentrations greater than 1% v/v. It is a minor constituent of cigarette smoke and gasoline exhaust, so environmental factors and personal activities would not confound the results. Unlike benzene, naphthalene is readily absorbed into the blood by both inhalation and dermal contact, and provides a more complete assessment of JP-8 exposure (Mattorano et al., 2004; Serdar et al, 2003). Instead of one marker of exposure, other investigators have taken another approach to estimating JP-8 exposure by selecting a number of aliphatic and aromatic hydrocarbons and using these compounds as a "JP-8 fingerprint" to represent JP-8 exposure (Pleil et al., 2000; Tu et al., 2004).

Several methods have been developed to estimate human exposure to JP-8 and they utilize different sampling media, such as ambient air, skin samples, exhaled breath, blood, and urine. Marker compounds like naphthalene and benzene, JP-8 fingerprint compounds (aliphatic: nonane, decane, undecane, and dodecane; aromatic: benzene, toluene, ethylbenzene, *m*-xylene, *o*-xylene, and *p*-xylene), or their metabolites have been measured in various media. The advantages and limitations of each approach are discussed further.

AMBIENT AIR MEASUREMENTS

Estimating JP-8 exposure on the basis of ambient air is usually performed with personal monitoring devices. Personal monitoring devices are attached to the collars of workers for the duration of their shift and passively collect air samples (via gaseous diffusion). The accuracy of these devices depends on the worker's surroundings. If the workers remain in their area of work (e.g., a hangar or mechanic shop), then personal monitoring devices would provide a reasonable estimate of potential inhalation exposure to JP-8. However, if workers left their workplaces to take a break, get a snack, or smoke a cigarette, then the exposure measured by the personal monitoring devices would include additional exposure contributions from other sources during that work shift.

Let's examine exposure of a tank entrant on the fuel cell maintenance team. The day before fuel cell maintenance is performed, the tanks are drained and vented for several hours to remove as much JP-8 fuel and vapor as possible, but residual fuel always remains. The next day, the entrant enters the confined space of the tank and is immediately exposed to JP-8 fuel on the floor and to JP-8 vapors that have accumulated inside the tank. However, because the entrant wears a respirator to minimize inhalation exposure inside the fuel tank, the personal monitoring devices would overestimate JP-8 exposure via inhalation. The actual exposure of the entrant to JP-8 during fuel cell maintenance work occurs primarily through dermal absorption. Personal monitoring devices are limited and insufficient in this situation because they only estimate potential inhalation exposure, and cannot estimate dermal exposure. To assess an entrant's JP-8 exposure as well as the exposure in other military personnel, novel methods to measure an individual's internal dose following JP-8 exposure need to be developed.

Dermal Estimation of Exposure

Current methods to quantify chemical contaminants on the skin include the use of passive exposure patches, clothing, skin swabs, liquid rinses, and tracers. While these methods estimate the potential amount deposited on the skin, like personal monitoring devices, they fail to provide a measure of the internal dose of the compound (Chao and Nylander-French, 2004). Once a chemical is placed on the skin, it diffuses into the stratum corneum (SC) or evaporates from the skin surface. After absorption, the chemical may be stored in the hydrophobic layer of the SC, it may be stored in the hydrophilic layer of the viable epidermis, or it may penetrate the vascularized layer of the dermis and become bioavailable to the systemic circulation. Moreover, some of the absorbed chemical may also diffuse back to the skin surface and evaporate (Kim et al., 2006a).

The tape-stripping technique measures the amount of chemical on the skin, giving an indication of the dermal kinetics of a chemical by absorption and evaporation. Using successive tape-strip samples, it is possible to infer the depth of absorption as well as some indication of the rate of absorption (Surber et al., 1999; Reddy et al., 2002; Loffler et al., 2004). A simple, noninvasive tape-stripping technique coupled with gas chromatography-mass spectrometry (GC-MS) using naphthalene as a surrogate of JP-8 exposure was developed to measure the amount of JP-8 in the skin. Before this technique could be applied to field measurements, it was validated under laboratory conditions. Human volunteers with no previous known exposure to JP-8 were given 25 μl of JP-8 to non-occluded skin at two sites on the ventral surface of

TABLE 15.1
The Average Mass of Stratum Corneum Keratin (μg/cm²) Removed by Successive Tape-Strips at Different JP-8 Exposure Times

Tape-Strip	Unexposed	10 minutes	15 minutes	20 minutes	25 minutes
1st	154 (75.3)[a]	135 (60.7)[a]	90.8 (56.9)[b]	138 (43.7)[a]	123 (70.1)[c]
2nd	107 (39.6)[c]	140 (56.0)[a]	98.0 (53.9)	108 (45.8)[d]	104 (60.3)[d]
3rd	82.0 (41.7)[d]	82.2 (24.5)[d]	81.1 (40.7)	88.2 (35.8)[d]	87.9 (39.5)
4th	69.8 (27.4)	74.9 (22.7)	72.7 (35.8)	78.9 (36.8)	69.5 (36.7)
5th	52.8 (17.3)	56.7 (18.3)	65.4 (24.3)	56.3 (17.7)	61.1 (29.5)
All	465 (163)	489 (139)	408 (190)	469 (124)	445 (199)

Source: Adapted from Chao, Y.-C.E., and Nylander-French, L.A. *Ann. Occup. Hyg.*, 48(1): 65–73, 2004.

Note: Numbers in parentheses are SD.

[a] Significantly different from tape-strips 3 through 5; all $p < 0.011$.

[b] Significantly different from the first tape-strip at unexposed and 20-min sites; $p = 0.030$ and $p = 0.035$, respectively.

[c] Significantly different from tape-strips 4 and 5; all $p < 0.028$.

[d] Significantly different from tape-strip 5; all $p < 0.043$.

each arm in an application chamber. The four sites of exposure corresponded to 10, 15, 20, and 25 minutes of JP-8 exposure and a control tape-strip sample was taken from an unexposed site on each arm. After the desired exposure time had elapsed, adhesive tape was applied to the exposed site and removed. Four successive tape-strips were applied to the same site (Chao and Nylander-French, 2004). Several factors that could affect the tape-strip data included the amount of SC collected with each tape, evaporation of the fuel from the skin, and chamber loss indicated by the amount of JP-8 residue on the chamber (Mattorano et al., 2004).

To account for the variability in the amount of SC removed with each tape-strip, Chao and Nylander-French (2004) calculated the average mass of keratin protein removed by five successive tape-strips at four different JP-8 exposure times (Table 15.1). With each successive tape-strip, the average mass of keratin removed decreased but there was no significant difference between the total mass of keratin removed using the five tape-strips or the average mass of keratin removed with each tape-strip at different exposure times when comparing unexposed sites and JP-8-exposed sites. Furthermore, no effects of covariates such as gender, age, race, and skin type were observed on the total mass of keratin removed. Therefore, the variability in the amount of SC that was removed with each tape-strip was negligible because the average total mass of keratin removed was consistent among all subjects and exposure groups. The average total mass of naphthalene removed at each exposure period was also similar by gender, age, race, and skin type.

As noted earlier, when JP-8 was applied to skin, the JP-8 surrogate naphthalene could penetrate into the skin, remain on the walls of the application chamber, or evaporate. For each exposure time, the average amount of naphthalene removed with the first tape-strip was over 50 orders of magnitude higher than in the successive tape-

TABLE 15.2
The Average Mass of Naphthalene (ng/cm²) Removed by Five Successive Tape-Strips at Different JP-8 Exposure Times

Tape-Strip	10 minutes	15 minutes	20 minutes	25 minutes
1st	2,510 (674)[a]	2,410 (875)	1,870 (678)	1,570 (1120)
2nd	30.9 (13.2)	31.6 (15.7)	33.9 (11.5)	29.9 (15.2)
3rd	8.45 (2.61)	10.1 (3.76)	9.84 (2.72)	9.78 (3.45)
4th	5.66 (1.13)	6.38 (2.72)	6.48 (1.21)	6.12 (1.59)
5th	4.59 (1.05)	5.36 (2.03)	5.42 (1.21)	5.47 (1.32)
All	2,560 (681)[b]	2,470 (888)	1,920 (685)	1,620 (1130)

Source: Adapted from Chao, Y.-C.E., and Nylander-French, L.A. *Ann. Occup. Hyg.*, 48(1): 65–73, 2004.

Note: Numbers in parentheses are SD.

[a] Significantly different from the first tape-strip at 20- and 25-minute sites; $p = 0.037$ and $p = 0.028$, respectively.

[b] Significantly different from all tape-strips at 20- and 25-minute sites; $p = 0.040$ and $p = 0.029$, respectively.

strips, suggesting that the majority of naphthalene still resided on the skin surface after 25 minutes. Interestingly, when comparing the first tape-strip over time, the average amount of naphthalene removed decreased from 2510 ± 674 ng/cm², 2410 ± 875 ng/cm², 1870 ± 678 ng/cm², to 1570 ± 1120 ng/cm² for 10, 15, 20, and 25 minutes, respectively (Table 15.2). This indicates that as time passed, there was less naphthalene on the outer layer of the skin because it had either evaporated or been absorbed by the skin. Investigation of the second tape-strip data demonstrated that the amount of naphthalene removed marginally increased with exposure time, supporting the hypothesis that more naphthalene was being absorbed and penetrated into the skin as time progressed. This was also apparent in the third, fourth, and fifth tape-strip data.

The percent recovery of naphthalene from the tape-strips decreased from 60.2% at 10 minutes to 38.1% at 25 minutes. Perhaps this decrease in naphthalene recovery in the tape-strips could be attributed to naphthalene evaporation. The percent recovery of naphthalene from evaporation increased as expected with exposure time from 1.45% at 10 minutes to 7.21% at 25 minutes, but does not completely account for the naphthalene that was not recovered. Another potential explanation may be naphthalene residue on the chamber, but the percent recovery of naphthalene from naphthalene residue on the chamber decreased with increasing exposure time from 15.1% at 10 minutes to 6.41% at 25 minutes. Because the total naphthalene recovered (tape-strips, evaporation, and chamber residue) fell from 77% for 10-minute exposure to 52% for 25-minute exposure (Table 15.3), this indicates that naphthalene rapidly penetrates the skin following JP-8 exposure in humans (Chao and Nylander-French, 2004).

Naphthalene concentrations in the SC were normalized by dividing the mass of naphthalene removed by each tape-strip by the mass of the keratin removed by the matching tape-strip and produced log-normally distributed results. The log-normal

TABLE 15.3
Results of the Best Estimates for the Total Amount of Naphthalene Recovered (ng), and Recovery Efficiency (%) at Different JP-8 Exposure Times

	10 minutes	15 minutes	20 minutes	25 minutes
Tape-strips	2.56e04 (6.81e03)	2.47e04 (8.88e03)	1.92e04 (6.85e03)	1.62e04 (1.13e04)
Recovery efficiency	60.2 (16.0)	58.0 (20.9)	45.2 (16.1)	38.1 (26.7)
Total chamber loss	7.02e03 (3.10e03)	6.99e03 (2.46e03)	7.08e03 (2.52e03)	5.79e03 (2.50e03)
Recovery efficiency	16.5 (7.30)	16.4 (5.80)	16.7 (5.92)	13.6 (5.88)
Evaporation	616 (207)	1.24e03 (436)	2.16e03 (786)	3.07e03 (1.58e03)
Recovery efficiency	1.45 (0.49)	2.93 (1.02)	5.08 (1.85)	7.21 (3.72)
Chamber residue	6.40e03 (3.09e03)	5.75e03 (2.51e03)	4.93e03 (2.36e03)	2.72e03 (2.13e03)
Recovery efficiency	15.1 (7.30)	13.5 (5.91)	11.6 (5.56)	6.41 (5.02)
Total naphthalene	3.26e04 (8.03e03)	3.16e04 (1.04e04)	2.93e04 (8.16e03)	2.20e04 (1.25e04)
Total recovery efficiency	76.7 (18.9)	74.4 (24.5)	61.8 (19.2)	51.8 (29.5)

Source: Adapted from Chao, Y.-C.E., and Nylander-French, L.A. *Ann. Occup. Hyg.*, 48(1): 65–73, 2004.
Note: Numbers in parentheses are SD.

distribution of naphthalene concentrations and the consistent amounts of both keratin and naphthalene removed by sequential tape-strips independent of potential confounders demonstrate the ability of this tape-strip technique in quantifying dermal exposure to JP-8 when naphthalene is used as a marker for exposure to jet fuel (Chao and Nylander-French, 2004).

As the tape-strip technique measuring naphthalene proved a reliable and useful method to estimate JP-8 exposure in a laboratory setting, it was used to assess human JP-8 exposure in an occupational exposure setting. Chao et al. (2005) studied a population of U.S. Air Force fuel cell maintenance workers and included entrants, attendants, runners, and field workers. They investigated if JP-8 exposure measured by the tape-stripping technique could be correlated with the exposures anticipated with performing certain job duties. Based on the job duty descriptions noted earlier, entrants were classified as the high-exposure group, attendants and runners were placed in the medium-exposure group, and field workers who performed maintenance with occasional contact with JP-8 were categorized as the low-exposure group. Subjects identified the three regions of the body that had the greatest exposure to JP-8 during the workshift, and acceptable sampling regions included the forehead, neck, shoulders, arms, hands, legs, knees, feet, or buttocks. Three successive tape-strip samples were collected from the three most-exposed regions after the workshift, and the amount of naphthalene removed was adjusted for the surface area of that particular region to account for individual differences in exposed surface areas. Whole-body dermal exposure to naphthalene was determined by summing the estimated regional dermal naphthalene concentrations of the three sampled regions and assuming no exposure occurred at the other unsampled regions.

Areas of the body that reportedly had the most frequent JP-8 exposure were the hands, arms, and neck. Although the hands, arms, and neck were cited as the most common exposure areas, the lower regions of the body (knee, calf, and leg) generally had higher dermal exposures to naphthalene. The highest regional dermal exposure occurred on the buttocks and was attributed to a worker sitting in a puddle of JP-8 during work (Table 15.4). When subjects were divided into high-, medium-, and low-exposure groups, dermal exposure to naphthalene was significantly different among the three exposure groups. The high-exposure group had an almost 10-fold higher geometric mean exposure level than the medium-exposure group and nearly a 13-fold higher geometric mean exposure level than the low-exposure group. The geometric mean exposure levels for the high-, medium-, and low-exposure groups were 4,180 ng/m^2 ± 9.35 (GSD), 485 ng/m^2 ± 3.87 (GSD), and 343 ng/m^2 ± 3.80 (GSD), respectively. It is important to note that these values were conservative estimates because the researchers assumed that other regions besides the three sampled regions were unexposed, which is highly unlikely. Nevertheless, no significant difference in dermal

TABLE 15.4
Dermal Exposure to Naphthalene (ng/m^2) by the Sampled Body Region

Sampled Region	n	Estimated Body Surface Area (%)	Naphthalene
Right lower arm	99	3	659 (5.21)
Left lower arm	81	3	608 (5.41)
Hands	76	2.5	265 (3.42)
Neck	55	2	982 (7.10)
Right foot	3	3.5	9.23e^3 (3.63)
Right knee	2	3.5	781 (13.7)
Right calf	3	3.5	1.58e^4 (24.0)
Right lower leg	3	7	5.01e^3 (41.3)
Right upper leg	1	9.5	9.87e^4
Left foot	15	3.5	9.70e^3 (6.96)
Left knee	4	3.5	4.32e^3 (11.4)
Left calf	2	3.5	1.00e^4(1.77)
Left lower leg	1	7.	1.79e^5
Left upper leg	1	9.5	3.27e^6
Back	7	13	1.82e^4 (15.6)
Left or right shoulder	5	3.5	3.43e^3 (12.6)
Buttock	3	5	1.62e^6 (3.60)
Forehead	1	1.2	48.9
Armpit	1	4	4.19 e^3

Source: Adapted from Chao, Y.-C.E., Gibson, R.L., and Nylander-French, L.A. *Ann. Occup. Hyg.*, 49(7): 639–645, 2005.
Note: Numbers in parentheses are SD.

exposure to naphthalene was observed between the medium- and low-exposure groups (Chao et al., 2005). Additionally, the dermal exposure to naphthalene among individuals in an exposure group was consistent and nonsignificant, suggesting that a worker's job role was a good predictor of their level of JP-8 exposure.

Understanding all the factors that might contribute to JP-8 exposure is a critical component of risk assessment in order to establish or revise operational methods or technical improvements that reduce human exposure. In the USAF occupational study, four significant predictors were found to correlate with dermal exposure estimates in a linear regression model. These four predictors were skin irritation, use of booties, working inside the tank, and duration of JP-8 exposure (Chao et al., 2005). Immediate symptoms of JP-8 exposure were skin drying, burning, or irritation; therefore, the onset of these indicators should alert workers that they have been dermally exposed to JP-8 and should take steps to minimize exposure, such as changing their JP-8-soaked clothing. Military personnel whose feet or legs were sampled for JP-8 exposure and who wore booties were shown to have significantly lower dermal exposure on the basis of naphthalene (Chao et al., 2005). Booties were shown to provide an extra barrier of protection and provided a simple and effective method to reduce dermal absorption of naphthalene. As expected, working inside the fuel tank was a major contributor in predicting JP-8 exposure because entrants had significantly higher JP-8 dermal exposure compared to other workers as a result of unavoidable contact with fuel. The last predictor that correlated with dermal exposure was the duration of JP-8 exposure, or the time between the start of work and dermal sampling. As concluded by Chao and Nylander-French (2004), increased exposure time decreased the amount of naphthalene detected in the upper layers of the SC due to enhanced dermal absorption and penetration of naphthalene and, to a lesser extent, evaporative loss from the skin (Chao et al., 2005).

Although naphthalene may serve as a suitable marker of JP-8 exposure, other compounds that represent a larger portion of JP-8 may provide a better estimate of JP-8 exposure. Kim et al. (2006a) chose six aromatic and aliphatic hydrocarbons that are the most abundant hydrocarbons in JP-8 based upon mass as marker compounds for JP-8. These six compounds were naphthalene, 1-methylnaphthalene, 2-methylnaphthalene, decane, undecane, and dodecane (McDougal and Rogers, 2004). Ten adult subjects (five males and five nonpregnant females) with no occupational exposure to jet fuel participated in the study and were exposed dermally to pure JP-8 (1 ml) on one forearm at two sites using sealed exposure chambers. At the end of the 30-minute exposure period, the two exposed skin sites were wiped with gauze pads and tape-stripped as many as ten times (Table 15.5). All six compounds penetrated the skin and all but decane (measurable only up to the fourth tape-strip) were quantifiable for all ten tape-strips. No decane was detected in the tape-strips 5 through 10. These results suggested that decane, which is the most volatile hydrocarbon, was absorbed quickly from the surface of the SC. Alternatively, decane could have been lost by evaporation. Another observation was that the first tape-strip removed the greatest amount of the marker compounds, followed by smaller amounts with each successive tape-strip (Kim et al., 2006a). These results were also observed in another JP-8 dermal exposure study using tape-strips and may partially explain the decreased mass of SC removed with increasing number of tape-strips (Chao and

TABLE 15.5
Mass per Area (μg/cm²) of Aromatic and Aliphatic Hydrocarbons in Tape-Strips

Tape-Strip	Naphthalene	1-Methylnaphthalene	2-Methylnaphthalene	Decane	Undecane	Dodecane
1	216 (127)	304.4 (102)	283 (91.4)	237 (213)	4,033 (2230)	7,277 (1970)
2	17.1 (10.3)	26.3 (20.8)	25.4 (19.6)	29.7 (29.9)	291 (237)	679 (587)
3	7.6 (3.7)	8.9 (6.7)	8.7 (7.0)	7.3 (6.7)	116 (43.3)	250 (192)
4	5.3 (5.0)	4.0 (2.6)	4.0 (2.4)	1.4 (4.6)	71.1 (26.9)	111 (54.2)
5	2.6 (1.0)	2.2 (1.1)	2.1 (1.1)	ND	47.3 (17.1)	67.4 (26.6)
6	3.0 (3.4)	1.7 (0.8)	1.7 (0.8)	ND	39.5 (11.5)	54.2 (16.4)
7	2.5 (3.4)	1.1 (0.8)	1.2 (0.7)	ND	29.8 (9.3)	40.8 (9.6)
8	1.3 (0.7)	0.9 (0.6)	0.9 (0.6)	ND	24.8 (6.7)	33.2 (8.3)
9	2.0 (3.3)	0.6 (0.6)	0.7 (0.5)	ND	18.1 (4.4)	26.7 (5.5)
10	1.8 (3.2)	0.5 (0.4)	0.5 (0.3)	ND	15.1 (3.5)	20.5 (3.7)

Source: Adapted from Kim, D., Andersen, M.E., and Nylander-French, L.A. *Toxicol. Lett.*, 165: 11–21, 2006.
Note: ND = none detected. Numbers in parentheses are SD.

Nylander-French, 2004). However, because the mass of naphthalene, 1-methylnaphthalene, 2-methylnaphthalene, undecane, and dodecane on the tenth strip was less than 1% of the mass measured on the first tape-strip and because the mass of SC removed using tape-strips does not decrease as appreciably, it was concluded that these aromatic and aliphatic compounds were absorbed, penetrated the skin, and did not reside in the SC very long (Kim et al., 2006a).

It is clear, based upon studies to date, that JP-8 dermal exposure leads to the absorption and penetration of jet fuel components into the skin. The next step in understanding human exposure to JP-8 is to model what occurs following absorption. The fate of chemicals following human exposure involves contributions from absorption, distribution, metabolism, and elimination. A five-compartment dermatotoxicokinetic model—(1) surface, (2) stratum corneum (SC), (3) viable epidermis, (4) blood, and (5) storage—has been used to predict blood concentrations of JP-8 components based on dermal exposure measurements made in an occupational exposure setting (Chao et al., (2005). When these measurements were incorporated into Monte Carlo simulations of dermal exposure and cumulative internal dose, no overlap among the naphthalene concentrations was observed in the blood among low-, medium-, and high-exposure groups. This dermatotoxicokinetic model was able to associate the mass of JP-8 components in the SC collected from tape-strip measurements with concentrations of these compounds in the systemic circulation and may be useful in developing additional toxicokinetic models for multi-route exposures to JP-8 (Kim et al., 2006b).

Inhalation Estimation of Exposure

Inhalation is the other major exposure route to jet fuel. As discussed earlier, personal monitoring devices provide information about gaseous levels of JP-8 but not about dermal exposure levels. Skin sampling, on the other hand, provides information about dermal exposure but not about inhalation exposure. The presence of exogenous compounds (such as JP-8 fingerprint compounds) in exhaled breath is a clear indication of JP-8 exposure and provides an integrated estimate of exposure from both dermal and inhalation routes. Moreover, by measuring specific compounds, it may be possible to assess the relative contributions from each exposure route (Pleil et al., 2000; Tu et al., 2004; Kim et al., 2007). To assess daily exposure to JP-8, breath samples are collected before the workshift starts (pre-exposure) and after the workshift finishes (post-exposure). Chromatograms of exhaled breath from GC-MS analysis between pre-exposure and post-exposure samples clearly show an increase in chemical exposure during the course of the day. However, breath measurements can include contributions from exposures before work, including previous-day exposures depending on the half-life of the molecule, time of exposure, and an individual's physiology. Therefore, pre-exposure samples show potential cumulative exposure while the difference between pre- and post-exposure samples show incremental exposure during the work shift (Pleil et al., 2000).

Inhalation exposure studies have also used biomarkers of JP-8 to indicate exposure. A study conducted by Egeghy et al. (2003) quantified benzene and naphthalene in the breath of USAF personnel to estimate exposure (Tables 15.6 and 15.7). USAF personnel were categorized into three exposure groups: high (fuel cell maintenance

TABLE 15.6
Concentrations of Benzene (μg/m³) in Environmental Air and in Breath

Variable	Exposure Category	Number (n)	n < LOQ[a]	Median
Benzene in air	High	114	0	252
	Moderate	38	0	7.4
	Low	140	1	3.1
Benzene in pre-exposure breath	High	111	19	4.6
	Moderate	44	3	5.8
	Low	151	23	4.7
Benzene in post-exposure breath	High	114	2	11.4
	Moderate	41	3	9.0
	Low	143	11	4.6

Source: Egeghy, P.P., Hauf-Cabalo, L., Gibson, R., and Rappaport, S.M. *Occup. Environ. Med.*, 60: 959–976, 2003.

[a] LOQ = Limit of quantification.

TABLE 15.7
Concentrations of Naphthalene (μg/m³) in Environmental Air and in Breath

Variable	Exposure Category	Number (n)	n < LOQ[a]	Median
Naphthalene in air	High	113	0	485
	Moderate	38	3	10.3
	Low	139	30	1.9
Naphthalene in pre-exposure breath	High	112	77	<0.5
	Moderate	43	20	0.6
	Low	149	89	<0.5
Naphthalene in post-exposure breath	High	111	7	1.83
	Moderate	40	12	0.93
	Low	143	51	0.73

Source: Adapted from Egeghy, P.P., Hauf-Cabalo, L., Gibson, R., and Rappaport, S.M. *Occup. Environ. Med.*, 60: 959–976, 2003.

[a] LOQ = Limit of quantification.

workers who were described as entrants), moderate (regular contact with the fuel via fuel handling, distribution, recovery, and testing), and low (no contact with JP-8). Median pre-exposure breath concentrations of benzene in the low-, moderate-, and high-exposure groups were 4.7, 5.8, and 4.6 μg/m^3, respectively, and median post-exposure breath concentrations of benzene in the low-, moderate-, and high-exposure groups were 4.6, 9.0, and 11.4 μg/m^3, respectively. Median pre-exposure breath concentrations of naphthalene in the low-, moderate-, and high-exposure groups were <0.5, 0.6, <0.5 μg/m^3, respectively; and median post-exposure breath concentrations of naphthalene in the low-, moderate-, and high-exposure groups were 0.73, 0.93, and 1.83 μg/m^3, respectively. Median post-exposure breath samples for both benzene and naphthalene were significantly higher when comparing the high-exposure group to the moderate-exposure group and the low-exposure group, but no significant difference was found between the moderate- and low-exposure groups. These data show that workers assigned to the high-exposure group had more exposure to benzene and naphthalene but their use as biomarkers of JP-8 exposure needs additional validation.

Pre-exposure breath concentrations and cigarette smoking were both significant covariates that predicted benzene concentrations in post-exposure breath of fuel maintenance workers indicating that sources other than JP-8 made significant contributions to the levels of benzene present in the post-exposure breath samples (Egeghy et al., 2003). Although its suitability as a biomarker for JP-8 exposure is inappropriate, the determination of benzene exposure in workers is still warranted from a risk assessment perspective. Two fuel distribution workers (5% of subjects in the moderate-exposure category) and 21 fuel maintenance workers (15% of subjects in the high-exposure category) had benzene concentrations in breath that exceeded the 2002 TLV of 1.6 mg/m^3 (Egeghy et al., 2003). This demonstrates that individuals who work with JP-8 are at risk for developing benzene-induced toxic effects such as bone marrow damage, aplastic anemia, and leukemia (IARC, 1982).

Pre-exposure breath samples and cigarette smoking were not significant predictors of naphthalene concentrations in post-exposure breath samples of fuel maintenance workers. Significant predictors of naphthalene concentrations in post-exposure breath samples of fuel maintenance workers included skin irritation and the method of tank entry (Egeghy et al., 2003). Exposure to naphthalene from JP-8 exposure was also correlated with skin irritation in a dermal study in USAF fuel cell maintenance workers (Chao et al., 2005). This data, taken together, suggest that naphthalene, which produces skin irritation in fuel cell maintenance workers, may serve as a good predictor of human exposure to JP-8. Tank entry was another significant predictor of JP-8 exposure because tank entry from the bottom of the aircraft was associated with higher breath levels of naphthalene. This probably occurred because opening the latch from underneath the aircraft resulted in elevated levels of JP-8 exposure and, consequently, higher concentrations of naphthalene in exhaled breath of fuel cell maintenance workers (Egeghy et al., 2003). Because indicators of JP-8 exposure (skin irritation and method of tank entry) were significant predictors of naphthalene exposure without contributions from environmental sources such as cigarette smoke, naphthalene in exhaled breath may serve as a useful biomarker of human exposure to JP-8.

Although naphthalene has been used as a marker compound to estimate JP-8 exposure, the inclusion of additional marker compounds could provide a more comprehensive estimate of JP-8 exposure. This can be accomplished through breath analysis because the major components of JP-8 can be separated from molecules normally found in breath and quantified by GC-MS. The summation of C_9 to C_{12} n-alkanes (nonane, decane, undecane, and dodecane) and the sum of single-ring aromatic compounds (benzene, toluene, ethylbenzene, *m,p*-xylene, *o*-xylene, and styrene) provide a comprehensive JP-8 fingerprint to estimate JP-8 exposure. Pleil et al. (2000) used this approach to measure JP-8 exposure in USAF personnel by collecting ambient air samples around the base and from workers with different job types on base (Table 15.8). The ambient air samples were taken from indoor areas such as shops, break rooms, or office areas, from aircraft cold-start operations on the flightline for exhaust exposure, from the vicinity of aircraft undergoing fuel cell maintenance, and from inside the fuel cell itself. These ambient air samples were used for comparison with the actual exposures found in personnel located at these sampling areas. Indoor air samples were compared with breath samples collected from incidental workers who did not work directly with jet fuel, air samples were compared with breath samples collected from ground crew on the flightline such as crew chiefs, air samples from the vicinity of aircraft undergoing fuel cell maintenance were compared with breath samples collected from attendants, and samples from inside the fuel cell were compared with breath samples collected from entrants (Table 15.9). Breath measurements of JP-8 fingerprint compounds from a control group in Los Angeles (LA) who should not have exposure to JP-8 were used to compare the exposure of biomarker compounds with individuals living in an urban city. An additional categorization was made to distinguish smokers from nonsmokers to determine the contribution of benzene exposure from JP-8 with respect to benzene exposure from cigarettes.

As anticipated, the mean concentrations of the C_9 to C_{12} n-alkanes and the single-ring aromatic compounds increased in the ambient air from indoor air levels, to flightline during cold-start, to fuel tank maintenance around aircraft, and peaked inside the "empty" fuel tanks. This trend was mirrored in the breath of respective workers. JP-8 biomarkers were highest in the breath of fuel cell entrants, followed by breath of attendants, and breath of crew chiefs exposed to exhaust on the flightline, although appreciably lower than their fuel cell maintenance worker counterparts, was still five times higher than the indoor exposure group. Furthermore, the indoor exposure group had a slight, yet statistically significant, increase in JP-8 biomarkers compared to LA controls except for dodecane (Pleil et al., 2000). These results demonstrated that all individuals on base had exposure to JP-8 regardless of their job duty.

When the pre- and post-exposure breath sample concentrations of fuel workers, attendants, and entrants were examined in detail, all subjects had significant increases in the sum of C_9 to C_{12} n-alkanes and aromatic compounds at the end of the workshift (Table 15.10). Surprisingly, there were decreases in benzene levels in breath from the pre-exposure to post-exposure measurement, indicating that JP-8 exposure was not a significant source of benzene exposure. As pre-exposure benzene levels were higher in fuel workers, this demonstrated that fuel maintenance workers have a chronic accumulated exposure to benzene. If benzene exposure was

TABLE 15.8
Ambient Air Concentrations of JP-8 Components (ppbv)

Compounds	Control (RTP)	Control (LA)	Indoor Air AFB Shops	Aircraft Cold-Start Exhaust Exposure	Fuel Tank Maintenance *around* Aircraft	Fuel Tank Maintenance *inside* Fuel Tanks
Benzene	0.37 (0.05)	1.19 (0.08)	1.05 (0.33)	13.04 (4.80)	17.6 (7.52)	2,990 (1,110)
Toluene	0.44 (0.08)	3.22 (0.31)	2.51 (0.91)	8.87 (2.73)	53.2 (22.2)	16,000 (5,900)
Ethylbenzene	0.08 (0.01)	0.43 (0.04)	0.40 (0.12)	3.13 (1.41)	74.9 (45.4)	9,590 (3,470)
m,p-Xylene	0.26 (0.05)	1.52 (0.15)	1.01 (0.23)	5.13 (2.40)	112 (57.3)	14,200 (3,550)
o-Xylene	0.10 (0.02)	0.62 (0.05)	0.69 (0.21)	4.83 (2.30)	196 (110)	6,750 (1,850)
Octane	0.03 (0.00)	0.58 (0.10)	0.18 (0.08)	3.13 (1.36)	65.7 (25.5)	5,980 (2,090)
Nonane	0.05 (0.00)	0.17 (0.02)	1.19 (0.87)	9.72 (4.50)	1,820 (1,380)	34,100 (11,500)
Decane	0.03 (0.01)	0.16 (0.04)	2.70 (1.98)	9.35 (4.63)	613 (370)	31,300 (10,600)
Undecane	0.03 (0.00)	0.31 (0.10)	2.54 (1.37)	6.71 (3.60)	159 (63.9)	31,000 (12,200)
Dodecane	—	—	7.60 (4.41)	3.65 (1.25)	69.8 (19.6)	7,470 (2,270)

Source: Adapted from Pleil, J.D., Smith, L.B., and Zelnick, S.D. (2000). *Environ. Health Perspect.,* 108: 183–192, 2000.
Note: Numbers in parentheses are SEM.

TABLE 15.9
Concentrations (ppbv) of JP-8 Components in Breath

Compounds	Controls (n = 19)	Incidental Work (n = 28)	All Fuel Work (n = 49)	All Exhaust Work (n = 49)
Benzene	0.60 (0.08)	3.47 (0.67)	3.03 (0.30)	2.25 (0.22)
Toluene	1.02 (0.17)	6.85 (1.45)	6.13 (0.70)	5.36 (0.82)
Ethylbenzene	0.09 (0.01)	0.39 (0.06)	2.11 (0.39)	0.96 (0.20)
m,p-Xylene	0.15 (0.02)	0.63 (0.07)	3.11 (0.49)	1.81 (0.40)
o-Xylene	0.10 (0.02)	0.36 (0.05)	4.00 (0.77)	1.47 (0.35)
Octane	0.08 (0.03)	0.19 (0.03)	2.77 (0.56)	2.10 (0.38)
Nonane	0.17 (0.05)	0.22 (0.03)	36.1 (6.83)	1.01 (0.19)
Decane	0.12 (0.03)	0.19 (0.03)	41.4 (5.63)	0.65 (0.15)
Undecane	0.16 (0.03)	0.24 (0.05)	15.6 (2.42)	0.93 (0.19)
Dodecane	3.33 (1.36)	0.30 (0.06)	8.86 (2.04)	0.92 (0.15)

Source: Adapted from Pleil, J.D., Smith, L.B., and Zelnick, S.D. *Environ. Health Perspect.*, 108: 183–192, 2000.

Note: Numbers in parentheses are SEM.

TABLE 15.10
Concentrations (ppbv) of Hydrocarbons in the Breath of Fuel Maintenance Workers by Job Classification

Compounds	Before Work (all samples) (n = 40)	After Work Tank Entry (n = 15)	After Work Attendants (n = 30)
Benzene	3.42 (0.52)	1.91 (0.54)	3.09 (0.41)
Octane	0.75 (0.12)	3.65 (1.23)	5.00 (1.34)
Nonane	4.16 (1.05)	45.0 (20.34)	72.7 (13.72)
Decane	6.79 (1.49)	41.7 (12.75)	84.3 (9.87)
Undecane	4.40 (0.86)	42.2 (10.65)	16.8 (1.35)
Dodecane	2.93 (0.63)	29.8 (9.56)	6.23 (1.35)
Sum, aromatics[a]	12.00 (1.51)	22.0 (4.61)	27.6 (4.99)
Sum, JP-8 fingerprint[b]	18.3 (2.94)	159 (42.5)	180 (23.1)

Source: Adapted from Pleil, J.D., Smith, L.B., and Zelnick, S.D. *Environ. Health Perspect.*, 108: 183–192, 2000.

Note: Numbers in parentheses are SEM.

[a] Sum of benzene, toluene, ethylbenzene, *m,p*-xylene, *o*-xylene, and styrene.

[b] Sum of nonane, decane, undecane, and dodecane.

TABLE 15.11
Concentrations of Hydrocarbons (ppbv) in the Breath of Fuel Maintenance Workers by Smoking Status

Compounds/Groups	Smokers Before Work (n = 18)	Smokers After Work (n = 12)	Nonsmokers Before Work (n = 22)	Nonsmokers After Work (n = 33)
Benzene	6.08 (0.68)	6.04 (0.23)	1.25 (0.34)	1.49 (0.17)
Benzene[a]	9.24 (0.86)[b]	6.32 (0.28)[b]	0.84 (0.25)[c]	1.70 (0.27)[c]
Octane	0.92 (0.20)	3.07 (0.45)	0.62 (0.13)	5.09 (1.32)
Nonane	6.98 (2.04)	65.1 (22.83)	1.84 (0.61)	62.9 (13.4)
Decane	8.31 (1.70)	80.4 (15.58)	5.54 (2.32)	66.4 (9.89)
Undecane	5.10 (1.17)	21.6 (4.05)	3.82 (1.25)	26.6 (5.28)
Dodecane	4.45 (1.20)	12.9 (6.64)	1.70 (0.48)	14.5 (4.40)
Sum, aromatics[d]	19.0 (2.31)	26.2 (4.38)	6.29 (0.84)	25.6 (4.76)
Sum, JP-8 fingerprint[e]	24.8 (4.12)	180 (45.9)	12.91 (3.86)	170.4 (23.2)

Source: Adapted from Pleil, J.D., Smith, L.B., and Zelnick, S.D. *Environ. Health Perspect.*, 108: 183–192, 2000.

Note: Numbers in parentheses are SEM.

[a] From paired samples.

[b] n = 6.

[c] n = 22.

[d] Sum of benzene, toluene, ethylbenzene, *m,p*-xylene, *o*-xylene, and styrene.

[e] Sum of nonane, decane, undecane, and dodecane.

due to smoking behavior, then fuel cell maintenance work decreased their benzene exposure because it prevented them from smoking during the day and reduced their overall benzene body burden (Table 15.11, Pleil et al., 2000).

Another interesting finding occurred when the breath of the entrants and attendants were compared and contrasted. Although there was no significant difference in the summation of nonane, decane, undecane, and dodecane between entrants and attendants, the distributions of n-alkanes were different. Breath samples collected from the entrants contained less of the higher-vapor-pressure hydrocarbons (nonane and decane) and more of the less-volatile hydrocarbons (undecane and dodecane). Conversely, breath samples collected from the attendants contained more nonane and decane and less of undecane and dodecane than entrants. Entrants wear respirators and are, therefore, less likely to be exposed to the higher-vapor-pressure compounds. Entrants predominantly receive dermal exposure to JP-8, which would favor higher concentrations of undecane and dodecane in breath as these compounds would reside on the skin longer and be absorbed. Observations of entrants corroborate this deduction, as entrants are more likely to contact liquid fuel than attendants. Attendants do not wear respirators and do not typically receive as much dermal exposure as entrants, so the attendant's primary exposure route is via inhalation and is consistent with higher levels of volatile compounds (nonane and decane) in breath (Pleil et al., 2000).

TABLE 15.12
Concentrations of Hydrocarbons (ppbv) in Breath of Exhaust Workers by Aircraft Location

	Indoor		Outdoor	
Compounds	**Before Work (n = 6) (Mean)**	**After Work (n = 7) (Mean)**	**Before Work (n = 12) (Mean)**	**After Work (n = 24) (Mean)**
Benzene	2.42 (0.17)	2.08 (0.19)	1.65 (0.29)	2.55 (0.42)
Hexane	2.86 (0.48)	0.85 (0.19)	2.07 (0.54)	2.59 (0.31)
Heptane	5.66 (0.58)	1.85 (0.34)	1.80 (0.68)	1.93 (0.27)
Octane	13.8 (1.10)	4.70 (0.97)	0.95 (0.65)	2.10 (0.38)
Nonane	20.2 (1.69)	5.49 (1.20)	0.32 (0.13)	1.01 (0.19)
Decane	15.7 (1.25)	2.82 (0.51)	0.24 (0.06)	0.65 (0.15)
Undecane	9.42 (0.86)	1.41 (0.19)	0.31 (0.11)	0.93 (0.19)
Dodecane	7.67 (2.74)	0.80 (0.10)	0.54 (0.17)	0.92 (0.15)
Sum, aromatics[a]	36.2 (7.68)	13.7 (2.11)	6.91 (1.01)	8.45 (1.32)
Sum, non-JP-8[b]	8.52 (0.98)	2.70 (0.37)	3.87 (1.14)	4.52 (0.56)
Sum, JP-8 fingerprint[c]	53.0 (4.89)	10.53 (1.83)	1.38 (0.35)	3.50 (0.65)

Source: Adapted from Pleil, J.D., Smith, L.B., and Zelnick, S.D. *Environ. Health Perspect.*, 108: 183–192, 2000.

Note: Numbers in parentheses are SEM.

[a] Sum of benzene, toluene, ethylbenzene, *m*,p-xylene, and styrene.

[b] Sum of hexane and heptane.

[c] Sum of nonane, decane, undecane, and dodecane.

During the course of this study, Pleil et al. (2000) discovered an important factor that affects JP-8 exposure for exhaust workers. Subjects who were inside a hangar near an aircraft during startups had approximately five times more aromatics and 40 times more n-alkanes in their pre-exposure breath samples than their colleagues who worked outdoors (Table 15.12). Post-exposure breath samples of workers who were inside the hanger during engine starts contained higher levels of aromatics and n-alkanes compared to workers who were outdoors during engine starts. It seemed that once the hangar doors were opened and the aircraft engines started, the combination of fresh outdoor air and aircraft exhaust produced a much lower overall exposure for all fuel-related compounds. This study demonstrated how breath analysis could be used to assess the risks associated with JP-8 exposure and to identify operating procedures on base that could be adopted to help reduce human exposure to jet fuel.

Another study using breath analysis was conducted to investigate whether safety precautions taken by military personnel could minimize occupational exposure to JP-8. Figure 15.1 shows the breath of a JP-8-exposed USAF individual before and after work. Tu et al. (2004) conducted their study at an Air National Guard base (AGB) and stratified their study subjects into the following categories: fuel cell

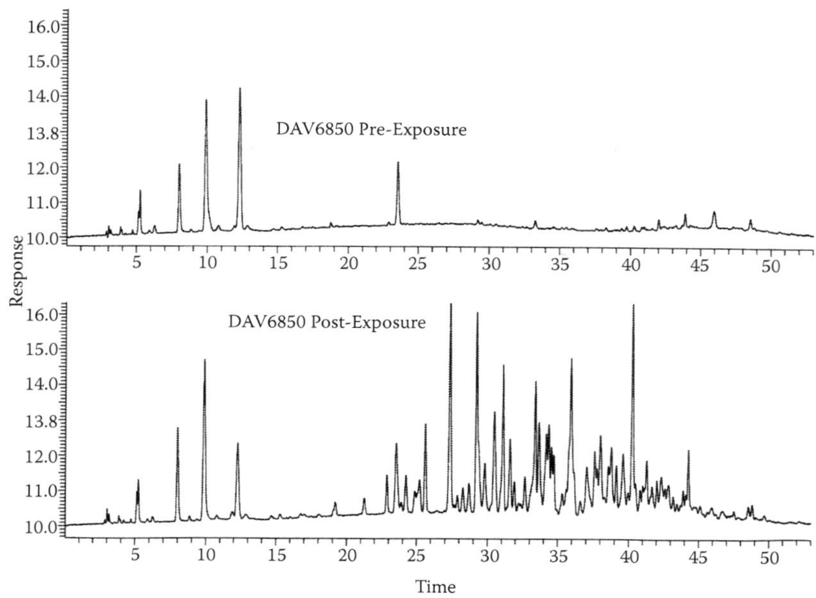

FIGURE 15.1 Breath GC chromatogram profile of an individual at a U.S. Air Force base before (upper) and after (lower) JP-8 exposure through inhalation and dermal routes. (*Source*: Tu, R.H., Mitchell, C.S., Kay, G.G., and Risby, T.H. *Human exposure to the Jet Fuel, JP-8*. Aviat Space Environ Med, 75, 49–59, 2004. With permission.)

workers (entrants, attendants, and runners), crew chiefs, mechanics, fuel specialists, and incidental workers. Because the fuel cell workers in this study were not differentiated into entrants, attendants, and runners, it is impossible to compare the results between this study and the study conducted by Pleil et al. (2000). Moreover, the personnel at the AGB were significantly older than the Air Force personnel studied by others. Another difference between these two studies was that Pleil et al. (2000) used a JP-8 fingerprint calculated by the summation of selected hydrocarbons in JP-8 to estimate JP-8 exposure, whereas Tu et al. (2004) took a more macroscopic approach. This latter study was based upon GC-MS analysis of complete breath chromatograms consisting of breath biomarkers of normal and abnormal physiology and of JP-8. Total JP-8 exposure was quantified by the summation of the concentrations of all compounds with elution times that corresponded to JP-8 components. JP-8 exposure was also fractionated on the basis of volatility in order to estimate routes of exposure, that is, JP-8 volatiles and JP-8 nonvolatiles representing inhalation and dermal exposures, respectively. Approximately 85% of the mass of JP-8 vapor consists of hydrocarbons in the range of C_8 and C_{11}. Compounds that eluted before undecane (including undecane) were considered JP-8 volatiles, and compounds eluting after undecane were considered JP-8 nonvolatiles. Additionally, concentrations of aliphatic hydrocarbons (nonane, decane, undecane, and dodecane) and aromatic hydrocarbons (benzene, toluene, ethylbenzene, *m*-xylene, *o*-xylene, and *p*-xylene) were also determined and comparisons of smokers versus nonsmokers were made.

Total JP-8 concentrations in the exhaled breath of subjects from the AGB ranged from 0 to 7.6 mg/m^3 before work, and ranged from 0.2 to 11.5 mg/m^3 after work. Human JP-8 exposure increased slightly over the course of the workday (10-hour workday). Military personnel who worked directly with fuel (i.e., the fuel cell workers and fuel specialists in this study) had the highest concentrations of total JP-8 in post-exposure breath samples (Figure 15.2). Fuel cell workers had the greatest JP-8 exposure, followed by fuel specialists. Moreover, the major route of JP-8 exposure differed for fuel cell workers and fuel specialists (Figures 15.3 and 15.4). Fuel cell workers acquired the most inhalation exposure (JP-8 volatiles fraction), whereas fuel specialists acquired the most dermal exposure (JP-8 nonvolatiles fraction). Fuel cell workers in this study were entrants, attendants, and runners. At the AGB, fuel cell workers were observed switching positions during the workday. If an individual began the day as an entrant and wore a respirator to decrease his JP-8 inhalation exposure, then his morning inhalation exposure may be low; but after work, that same individual would obtain inhalation exposure by working as an attendant or runner. Therefore, it is likely that all fuel cell workers obtained inhalation exposure at one point during their workshift. A surprising outcome was that fuel cell workers did not have more dermal exposure as measured by JP-8 nonvolatiles in their breath samples, considering the potential for dermal exposure when working inside the fuel cell or handling fuel-soaked polyurethane foam. The explanation for this outcome was evident when we examined their normal safety practices on the questionnaires the subjects completed, and was confirmed as we observed fuel cell workers on their daily routine. On this AGB, military personnel took additional precautions when handling foam or when their clothing became drenched with JP-8. Fuel cell workers handling foam wore disposable nitrile gloves under their solvent-resistant gloves (double gloving) to add an extra layer of protection for their hands and would put on clean clothes if their cotton coveralls became contaminated with JP-8. Fuel specialists, on the other hand, work directly with fuel, which predisposes them to dermal exposure, particularly during fuel spills or mishandling. Fuel specialists wore solvent-resistant gloves but did not wear any protective clothing. These personnel may have reduced their dermal exposure to JP-8 by adopting similar safety precautions as adopted by fuel cell workers. Breath collected from crew chiefs, mechanics, and incidental workers showed evidence of comparable post-exposures to JP-8. Crew chiefs in this study worked on the flightline and by working outside, instead of inside a hangar, reduced their potential exposure to JP-8. Fuel specialists and mechanics had significantly more JP-8 nonvolatiles in exhaled breath than the crew chiefs. This supports the proposition that JP-8 nonvolatiles can be used to quantify JP-8 dermal exposure because fuel specialists and mechanics are more likely to receive dermal exposure to JP-8 than crew chiefs. Incidental subjects had evidence of significantly less exposure to total JP-8, JP-8 volatiles, and JP-8 nonvolatiles on the basis of their pre- to post-breath samples. It was unlikely that incidental subjects carried JP-8 over from the previous day because their exposure diminished during the course of the day. Furthermore, incidental subjects with elevated pre-exposure values indicated on their questionnaire that they refueled their vehicle on the morning of testing. Nevertheless, at the end of the workday, all subjects on base had measurable levels of JP-8 compounds in their breath, indicating that all personnel on base had exposure

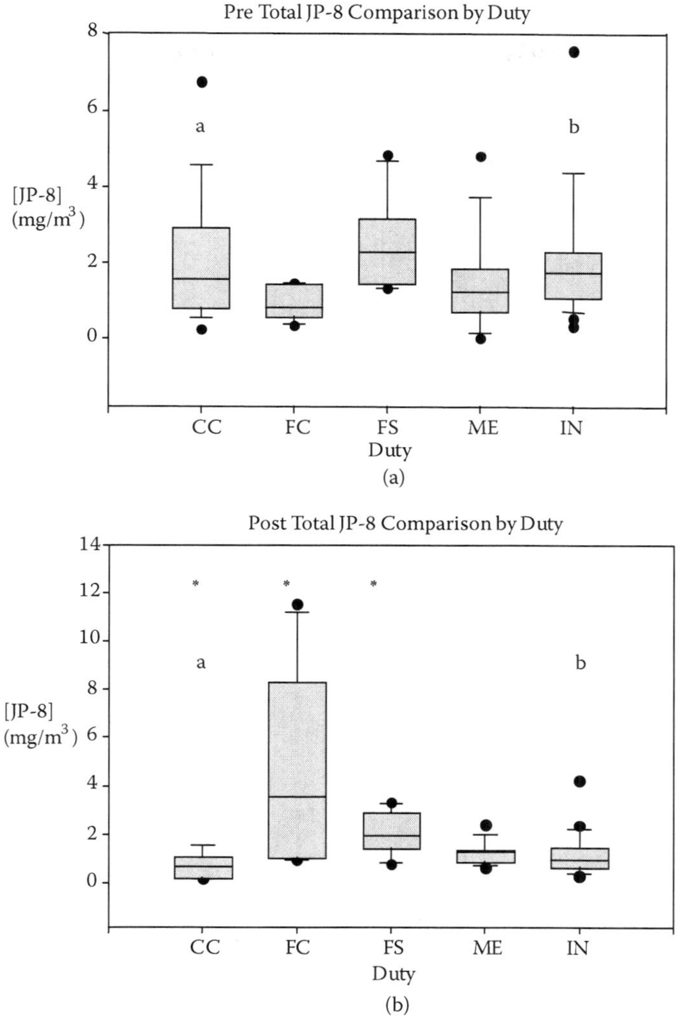

FIGURE 15.2 Total pre (upper) and post (lower) JP-8 comparison by duty: "a" denotes significance between pre and post samples for crew chiefs at $p < 0.02$; "b" denotes significance between pre and post samples for incidental workers at $p < 0.001$; and * denotes significance for FC and FS versus CC at $p < 0.004$. (CC, crew chief; FC, fuel cell worker; FS, fuel specialist; ME, mechanic; IN, incidental worker.) (*Source*: Tu, R.H., Mitchell, C.S., Kay, G.G., and Risby, T.H. *Human exposure to the Jet Fuel, JP-8.* Aviat Space Environ Med, 75, 49–59, 2004. With permission.)

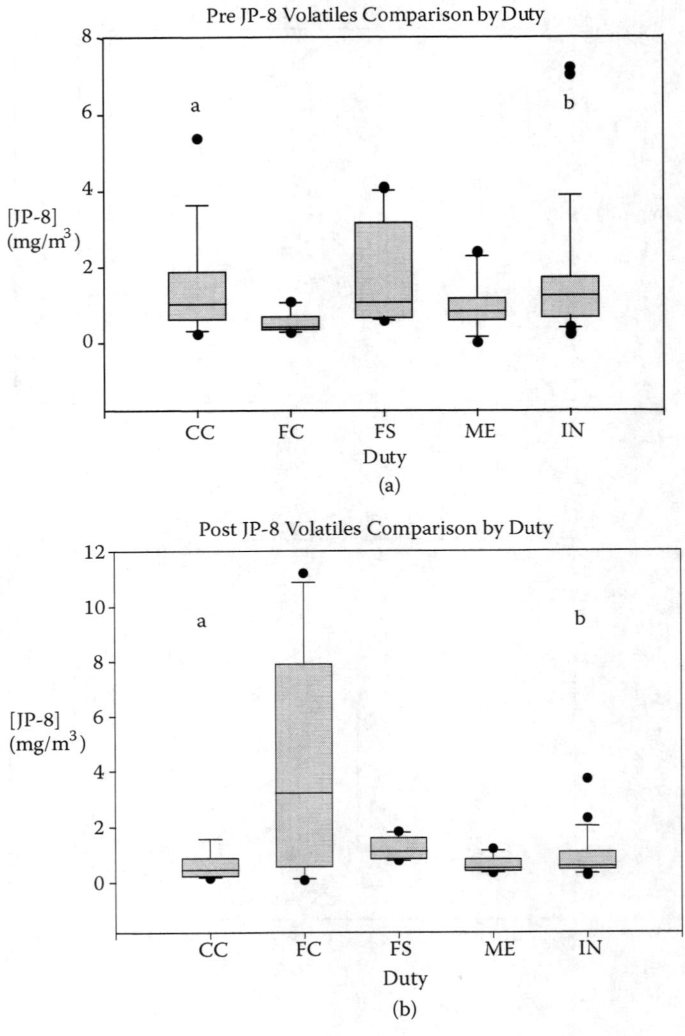

FIGURE 15.3 Pre (upper) and post (lower) JP-8 volatiles comparison by duty: "a" denotes significance between pre and post samples for crew chiefs at $p < 0.04$; and "b" denotes significance between pre and post samples for incidental workers at $p < 0.004$. (CC, crew chief; FC, fuel cell worker; FS, fuel specialist; ME, mechanic; IN, incidental worker.) (*Source*: Tu, R.H., Mitchell, C.S., Kay, G.G., and Risby, T.H. *Human exposure to the Jet Fuel, JP-8.* Aviat Space Environ Med, 75, 49–59, 2004. With permission.)

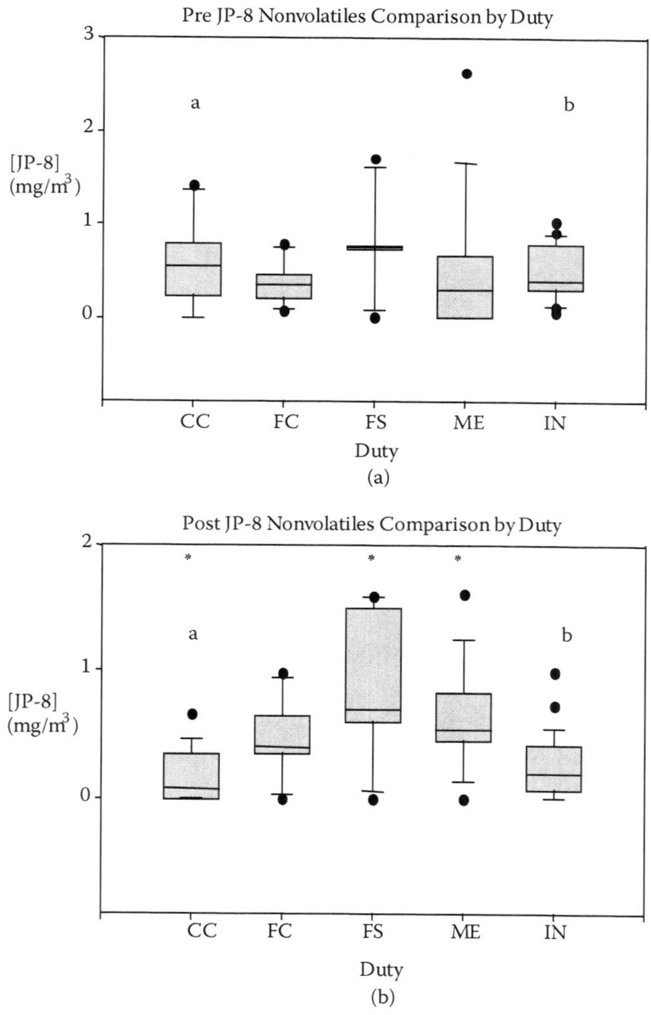

FIGURE 15.4 Pre (upper) and post (lower) JP-8 nonvolatiles comparison by duty: "a" denotes significance between pre and post samples for crew chiefs at $p < 0.01$; "b" denotes significance between pre and post samples for incidental workers at $p < 0.009$; and * denotes significance for FS and ME versus CC at $p < 0.002$. (CC, crew chief; FC, fuel cell worker; FS, fuel specialist; ME, mechanic; IN, incidental worker.) (*Source*: Tu, R.H., Mitchell, C.S., Kay, G.G., and Risby, T.H. *Human exposure to the Jet Fuel, JP-8.* Aviat Space Environ Med, 75, 49–59, 2004. With permission.)

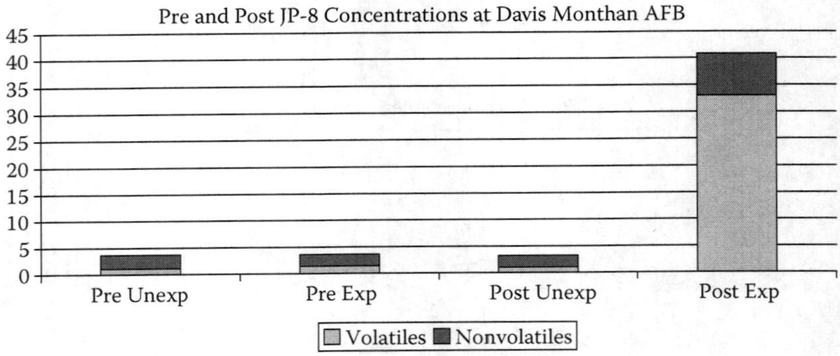

FIGURE 15.5 Concentrations of pre and post JP-8 volatiles and nonvolatiles (mg/m^3) in the breath of incidental U.S. Air Force personnel (Unexp, n = 4) and FC US Air Force personnel (Exp, n = 18) at Davis Monthan AFB.

to the JP-8. The method of collection and analysis of breath samples used by Tu et al. (2004) was used to investigate USAF personnel at Davis Monthan AFB (Risby et al., 2000). Subjects were separated by work as incidental personnel (Unexposed, n = 4) and fuel cell personnel (Exposed, n = 18), and the data were stratified on the basis of the volatility of the JP-8 components. These results are summarized in Figure 15.5 and show that breath reflects exposure. Moreover, the exposure at this AFB was significantly greater than the AGB. Figure 15.6 shows that exposure appears to be dependent upon the AFB studied.

Because smoking was such an important confounder, smokers were compared to nonsmokers. Comparisons of aliphatic and aromatic hydrocarbons between smokers and nonsmokers generated the expected results (Figure 15.7). There were no differences in the concentration of aliphatic hydrocarbons (as cigarette smoke does

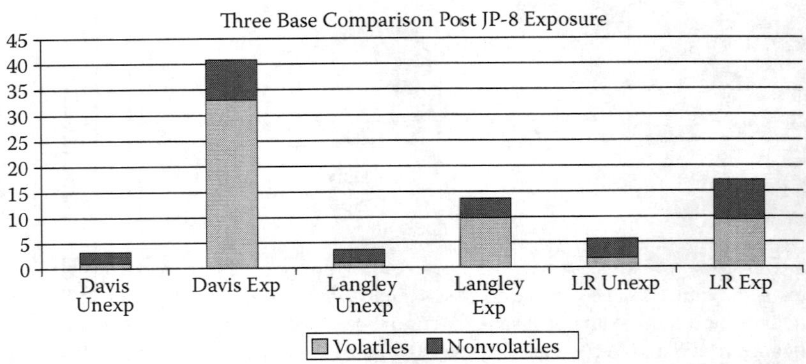

FIGURE 15.6 Three Base comparison of the concentrations of post JP-8 volatiles and nonvolatiles (mg/m^3) in the breath of incidental U.S. Air Force personnel (Unexp) and FC US Air Force personnel (Exp) at Davis Monthan AFB (Davis, Unexp, n = 4 and Exp, n = 18), Langley AFB (Unexp, n = 25 and Exp, n = 20) and Little Rock AFB (LR, Unexp, n = 12 and Exp, n = 30).

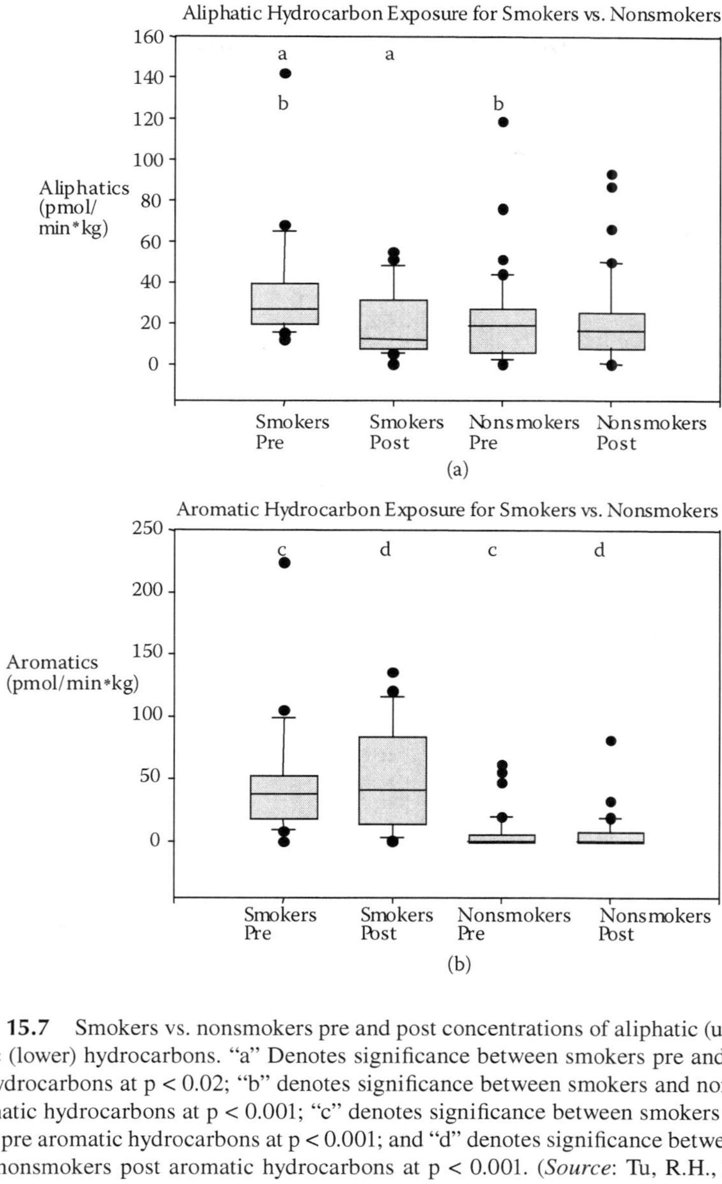

FIGURE 15.7 Smokers vs. nonsmokers pre and post concentrations of aliphatic (upper) and aromatic (lower) hydrocarbons. "a" Denotes significance between smokers pre and post aliphatic hydrocarbons at $p < 0.02$; "b" denotes significance between smokers and nonsmokers pre aliphatic hydrocarbons at $p < 0.001$; "c" denotes significance between smokers and nonsmokers pre aromatic hydrocarbons at $p < 0.001$; and "d" denotes significance between smokers and nonsmokers post aromatic hydrocarbons at $p < 0.001$. (*Source*: Tu, R.H., Mitchell, C.S., Kay, G.G., and Risby, T.H. *Human exposure to the Jet Fuel, JP-8.* Aviat Space Environ Med, 75, 49–59, 2004. With permission.)

not contain significant amounts of C_9 to C_{12} alkanes) between the exhaled breath of smokers and nonsmokers. On the other hand, smokers had significantly higher concentrations of aromatic compounds in their breath compared to the breath of nonsmokers, a finding that can be more readily attributed to cigarette smoking than to JP-8 exposure. Fuel cell workers had the greatest aromatic hydrocarbon exposure, followed by incidental subjects. Incidental subjects were able to smoke more frequently than the other duty groups because military personnel working directly with fuel were not allowed to smoke near fuel sources. As a result, incidental subjects received considerable exposure to aromatic hydrocarbons. There were some subjects who worked with fuel that had elevated breath aromatic hydrocarbons. For these subjects, high pre-breath aromatic hydrocarbons were presumably due to cigarette smoking, whereas high concentrations of post-breath aromatic hydrocarbons were attributed to JP-8 exposure. Because some smokers also had higher JP-8 exposure, it was likely that when smokers go outside to smoke, they receive additional JP-8 exposure. This explains the higher concentrations of JP-8 volatiles in smokers than in nonsmokers. Smoking had a greater effect on the amount of aromatic compounds in breath than did the duty-group. Therefore, subjects working with fuel were unable to smoke at will and actually reduced their exposure to aromatic hydrocarbons during the day. The presence of JP-8 compounds in exhaled breath is a noninvasive, validated method for estimating human exposure to JP-8, and demonstrates a chemical's absorption into the body via the skin or inhalation as well as its elimination from the body.

Although breath analysis provides an estimate of the internal dose of a compound, the concentration of a compound in blood provides the amount of a substance that was systemically available after inhalation or after absorption and dermal penetration. Permeability coefficients for aromatic compounds indicated that aromatic compounds penetrated the skin faster than aliphatic compounds, but because the concentration of aliphatic compounds in JP-8 was much greater than the concentration of aromatic compounds in JP-8, the calculated flux of aliphatic compounds was greater than the calculated flux of aromatic compounds (Table 15.13; Kim et al., 2006a). Determination of the time dependence of the blood concentrations of JP-8 compounds can be used to develop dermatotoxicokinetic models to estimate JP-8 exposure in occupational exposure settings (Kim et al., 2006b). These models will provide a better prediction of the behavior of these compounds in the body and be valuable tools when assessing exposure and risks associated with human exposure to JP-8. Blood concentrations of chemicals can provide useful information on elimination kinetics, the rate at which chemicals disappear from the blood. However, due to the invasiveness of blood collection and additional steps needed to conduct studies involving blood collection in humans, urinary biomarkers of JP-8 compounds, following their metabolism and elimination from the body, have been used to estimate JP-8 exposure.

Serdar et al. (2003) measured the concentrations of naphthalene and its metabolites (1-naphthol and 2-naphthol) in urine as markers of human exposure to JP-8 (Table 15.14). Naphthalene has been well characterized and shown to be a surrogate of JP-8 exposure. Concentrations of naphthalene, 1-naphthol, and 2-naphthol were determined in the urine collected from USAF personnel before work and after work.

TABLE 15.13
Apparent Permeability Coefficients (cm/hr) of Aromatic and Aliphatic Hydrocarbons for Each Study Volunteer

Volunteer	Naphthalene	1-Methylnaphthalene	2-Methylnaphthalene	Decane	Undecane	Dodecane
1	1.6e–04	1.3e–05	5.3e–05	8.5e–06	2.1e–07	1.3e–06
2	3.4e–05	2.8e–05	3.1e–05	8.6e–06	3.6e–07	1.5e–06
3	3.2e–05	3.1e–05	3.0e–05	7.6e–06	2.3e–07	3.0e–06
4	3.7e–05	2.7e–05	3.2e–05	1.2e–05	2.5e–07	2.0e–06
5	5.4e–05	2.8e–05	2.9e–05	3.3e–06	6.7e–07	1.3e–06
6	5.7e–05	3.3e–05	3.0e–05	8.0e–06	4.8e–07	1.5e–06
7	4.1e–05	3.1e–05	3.0e–05	7.1e–06	3.3e–07	1.5e–06
8	4.1e–05	2.9e–05	3.1e–05	5.6e–06	4.1e–07	1.3e–06
9	3.3e–05	3.1e–05	3.0e–05	2.2e–06	7.6e–07	1.1e–06
10	4.2e–05	3.5e––05	2.8e–05	2.0e–05	8.3e–07	1.2e–06
Mean	5.3e–05 (3.8e–05)	2.9e–05 (5.9e–06)	3.2e–05 (7.4e–06)	6.5e–06 (3.3e–06)	4.5e–07 (2.3e–07)	1.6e–06 (5.6e–07)

Source: Adapted from Kim, D., Andersen, M.E., and Nylander-French, L.A. *Toxicol. Lett.*, 165: 11–21, 2006.

TABLE 15.14
Urinary Measurements of U.S. Air Force Personnel Exposed to JP-8

	Pre-exposure		Post-exposure	
Urinary Analyte	**Non-smoker (n = 36)**	**Smoker (n = 26)**	**Non-smoker (n = 66)**	**Smoker (n = 51)**
Benzene	0.709 (1.90)	1.72 (2.58)	1.81 (2.80)	3.71 (3.40)
Naphthalene	0.016 (6.51)	0.033 (2.80)	0.316 (11.2)	0.457 (4.93)
1-Naphthol	2.04 (3.24)	7.76 (3.10)	15.9 (3.21)	26.3 (2.32)
2-Naphthol	2.06 (2.27)	9.49 (2.54)	22.6 (3.46)	37.3 (2.48)

Source: Adapted from Serdar, B., Egeghy, P.P., Waidyanatha, S., et al. *Environ. Health Perspect.*, 111(14): 1760–1764, 2003.

Note: Numbers in parentheses are GSD.

Subjects were categorized into three exposure groups: high (fuel cell maintenance workers), moderate (regular contact with fuel who were not fuel cell maintenance workers), and low (job titles did not require exposure to jet fuel). Post-shift urinary measurements for naphthalene, 1-naphthol, and 2-naphthol were significantly higher than pre-shift urinary measurements for these analytes in the high-exposure group, but such differences were not observed in the low- or moderate-exposure groups. Concentrations of all urinary compounds in the high-exposure group were 3- to 28-fold higher than those of the low-exposure group and 2- to 12-fold higher than those of the moderate-exposure group. Naphthalene, 1-naphthol, and 2-naphthol urinary concentrations in the moderate-exposure group were twofold higher than the low-exposure group. The levels of urinary markers appeared to correlate to the extent to which subjects in the selected jobs were thought to have contact with JP-8 (Serdar et al., 2003). Not only were urinary naphthol levels able to distinguish between categories of JP-8 exposure, but they could also reflect specific task-related differences in exposure. In fuel cell maintenance workers, a significant predictor of urinary 1-naphthol was the removal of jet fuel puddles, whereas significant predictors of urinary 2-naphthol were working inside the fuel tank and wearing protective booties. Although it is unclear why such predictors differed between 1- and 2-naphthol, working inside the fuel tank and wearing booties both related to dermal exposure to JP-8, so 2-naphthol in urine may provide a biomarker of dermal exposure to JP-8 (Serdar et al., 2004; Chao et al., 2006).

Human Health Effects of Exposure

Studies to date have improved quantification of JP-8 exposure and also the understanding of the uptake, disposition, and metabolism of JP-8 compounds. However, comparable studies assessing potential health effects associated with JP-8 exposure in humans are limited. The inadequate toxicity database that exists for JP-8 indicates that jet fuel exposure produces nausea, dizziness, headache, skin irritation, and

TABLE 15.15
Breathing Zone and Breath Levels (μg/m^3) of Naphthalene by Exposure Group

	High Exposure (n = 45)	Low/No Exposure (n = 78)
Air monitoring (mean)	583 (269)	2.47 (1.73)
Pre-breath (mean)	0.75 (0.91)	0.71 (0.49)
Post-breath (mean)	3.80 (2.17)	0.8 (0.8)

Source: Adapted from Rhodes, A.G., LeMasters, G.K., Lockey, J.E., et al. *J. Occup. Environ. Med.*, 45(1): 79–86, 2003.
Note: Numbers in parentheses are SD.

respiratory distress (ATSDR, 1998; Ritchie et al., 2001b). As fuel cell maintenance workers had the greatest potential for JP-8 exposure, it was anticipated that these personnel were most likely to present with adverse health effects of JP-8 exposure. An immunotoxicity study was conducted by comparing immune cell counts in the peripheral blood of fuel cell maintenance workers (high-exposure group) and military personnel with minimal or no exposure to jet fuel in the course of their routine work (low-exposure group). To verify that the fuel cell maintenance workers had received more JP-8 exposure than the low-exposure group, breath samples were collected before and after the workshift using naphthalene as a surrogate of JP-8 exposure (Table 15.15). Mean concentrations of pre-shift naphthalene were comparable between the low- and high-exposure groups, but mean concentrations of post-shift naphthalene were almost fivefold higher in the high-exposure group. There were no significant differences in the lymphocyte subpopulations (T cells, B cells, and natural killer cells; Table 15.16) between the two exposure groups, however, there were significant increases in the number of white blood cells, neutrophils, and monocytes in the peripheral blood of fuel cell maintenance workers (high-exposure group; Table 15.17). Although JP-8 exposure altered the number of certain types of immune cells in human peripheral blood, it has yet to be determined whether JP-8 affects the functional ability of these cells to produce lymphokines and cytokines and modulate the immune system (Rhodes et al., 2003).

Hydrocarbons have been suspected of producing adverse reproductive effects in humans. Because JP-8 is a complex mixture of aliphatic and aromatic hydrocarbons, it is necessary to assess the potential reproductive toxicity of JP-8. To accomplish this, reproductive endocrine effects in USAF female personnel with occupational exposure to fuel (primarily JP-8) were examined. Occupational exposure to fuel was determined using breath analysis by either the summation of aliphatic n-alkanes (hexane, heptane, octane, nonane, decane, and undecane) or the summation of aromatic compounds (benzene, toluene, ethylbenzene, *o*-xylene, and *m,p*-xylene). Female reproductive dysfunction was evaluated using specific urinary endocrine markers that are predictive of the probability of conception in women with a given ovulatory menstrual cycle: lower preovulatory luteinizing hormone

TABLE 15.16
Lymphocyte Subpopulation Counts and Percentages by Exposure Group

	High Exposure (n = 36)		Low Exposure (n = 57)	
	Count (cells/μl)	Percent	Count (cells/μl)	Percent
Total lymphocytes (mean)	2,041 (524)		2,065 (624)	
T cells (mean)	1,520 (423)	74.19 (5.75)	1,509 (490)	72.88 (5.89)
T suppressor cells (mean)	550 (178)	27.22 (5.95)	545 (260)	26.09 (6.69)
T helper cells (mean)	924 (283)	45.30 (6.58)	914 (284)	44.72 (5.65)
Nature killer cells (mean)	182 (96)	9.08 (4.43)	191 (95)	9.33 (3.67)
B cells (mean)	316 (119)	15.4 (4.2)	344 (166)	16.63 (5.45)

Source: Adapted from Rhodes, A.G., LeMasters, G.K., Lockey, J.E., et al. *J. Occup. Environ. Med.*, 45(1): 79–86, 2003.

Note: Numbers in parentheses are SD.

TABLE 15.17
White Blood Cell Differential Counts and Percentages by Exposure Group

	High Exposure (n = 45)		Low Exposure (n = 78)	
	Count (cells/μl)	Percent	Count (cells/μl)	Percent
White blood cells (mean)	6,515 (1,402)		5,755 (1,309)	
Neutrophils (mean)	3,960 (1,267)	59.65 (9.17)	3,328 (1,030)	57.16 (8.90)
Lymphocytes (mean)	1,827 (482)	28.74 (7.06)	1,799 (587)	31.67 (8.13)
Monocytes (mean)	518 (193)	8.06 (2.55)	440 (155)	7.76 (2.28)
Eosinophils (mean)	196 (165)	3.06 (2.07)	113 (125)	2.93 (2.15)
Basophils (mean)	16 (37)	0.49 (0.39)	12 (32)	0.48 (0.41)

Source: Adapted from Rhodes, A.G., LeMasters, G.K., Lockey, J.E., et al. *J. Occup. Environ. Med.*, 45(1): 79–86, 2003.

Note: Numbers in parentheses are SD.

(LH), mid-luteal phase pregnanediol 3-glucuronide (Pd3G), mid-luteal phase estrone-3-glucuronide ($E_1$3G), and higher follicle phase (Pd3G). Of these markers, only pre-ovulatory LH was affected. Women who had elevated exposure to aliphatic hydrocarbons had decreased levels of urinary pre-ovulatory LH. These results suggest that compounds in fuels such as JP-8 may act as reproductive endocrine disruptors (Reutman et al., 2002).

Toxic neurological effects including fatigue, impairment of hand-eye coordination, memory defects, euphoria, depression, sleep disturbances, memory impairment, and postural imbalance, have been observed in humans following JP-8 exposure (Knave et al., 1976, 1978, 1979; Porter, 1990; Struwe et al., 1983; Smith et al., 1997). Research on Persian Gulf War veterans has shown an acceleration of divided attention task performance, a key ability in airplane piloting, over the course of repeated low-level JP-8 exposures (Bell et al., 2005). A computer-administered, standardized test battery called CogScreen-Aeromedical Edition (AE) (CogScreen, Washington, D.C.) has been used to investigate neurocognitive changes in military personnel who have been exposed to JP-8 (Tu et al., 2004). CogScreen was developed for the Federal Aviation Administration, under contract to the Department of Transportation, and is used by defense research laboratories and pharmaceutical companies to detect subtle changes in cognitive functioning that are predictive of changes in "real-world" functioning (Horst and Kay, 1991; Kay, 1992; Mapou et al., 1993; Callister et al., 1996; Kay et al., 1997; Taylor et al., 2000). Performance on CogScreen subtests, which measured response speed, accuracy, memory, rule recognition, and visual tracking, were compared between Air National Guard personnel with low-level exposure to JP-8 and a non-JP-8 exposed, age- and education-matched control sample. Of the 42 CogScreen subtests, JP-8 exposed subjects performed significantly poorer on almost half of them (i.e., 20). Analysis of the CogScreen data suggests that most of the differences between JP-8-exposed subjects and control subjects were on measures of response speed, in that JP-8 exposed subjects needed more time to process information or had slower reaction times. CogScreen also tested an individual's ability to perform complicated tasks that was measured by dual-tasking exercises. Subjects exposed to JP-8 performed significantly poorer on three out of the five dual-tasking subtests. These changes in dual-tasking exercises were due to differences in response time and accuracy. The slower response time for JP-8-exposed subjects suggests decreased attention or impaired visual-motor tracking capability. Poorer accuracy suggests declines in working memory or difficulties with concentration and following directions. Impairments in attention or concentration could have safety implications if workers had more frequent or longer lapses in attention. In general, JP-8-exposed subjects showed poorer performance on measures of memory and attention, especially on the most challenging tasks of attention, compared to the unexposed subjects (Tu et al., 2004).

CONCLUSIONS

Exposure to jet fuel is ubiquitous throughout the U.S. military, but a detailed picture of JP-8 exposure and adverse human health effects associated with this exposure are not well understood. Although JP-8 was selected as a military fuel because it contains lower quantities of aromatics, the greatest health concern of JP-8 is due to its carcinogenic potential from benzene and naphthalene (IARC, 1982; NTP, 2000). The National Research Council concluded that inhalation exposure of JP-8 vapor does not pose a carcinogenic risk, but dermal exposure may pose a risk for skin cancer and sensitization (NRC, 2003). Consequently, differentiating between the absorption, distribution, metabolism, and elimination of JP-8 compounds via inhalation and dermal routes of exposure will be a critical aspect for conducting human risk assessments for JP-8. Additionally, the standardization of techniques used to quantify JP-8 exposure and the implementation of safety practices to minimize JP-8 exposure are required. Human exposure to JP-8 predominantly occurs on military bases, but it is entirely possible that jet-fuel exposure also occurs at airports, fuel storage facilities, and residential areas surrounding military bases.

REFERENCES

Armbrust Aviation Group. (1998). *World Jet Fuel Almanac.* Palm Beach Gardens, FL: Armbrust Aviation Group.

ATSDR. (1998). Toxicological Profile for Jet Fuels (JP-5 and JP-8). Atlanta: USDHHS.

Bell, I.R., Brooks, A.J., Baldwin, C.M., Fernandez, M., Figueredo, A.J., and Witten, M.L. (2005). JP-8 jet fuel exposure and divided attention test performance in 1991 Gulf War veterans. *Aviat. Space Environ. Med.,* 76(12): 1136–1144.

Callister, J.D., King, R.E., and Retzlaff, P.D. 1996. Cognitive assessment of USAF pilot training candidates. *Aviat. Space Environ. Med.,* 67: 1124–1129.

Carlton, G.N., and Smith, L.B. (2000). Exposures to jet fuel and benzene during aircraft fuel tank repair in the U.S. Air Force. *Appl. Occup. Environ. Hyg.,* 15(6): 485–491.

Chao, Y.-C.E., and Nylander-French, L.A. (2004). Determination of keratin protein in a tape-stripped skin sample from jet fuel exposed skin. *Ann. Occup. Hyg.,* 48(1): 65–73.

Chao, Y.-C.E., Gibson, R.L., and Nylander-French, L.A. (2005). Dermal exposure to jet fuel (JP-8) in US Air Force personnel. *Ann. Occup. Hyg.,* 49(7): 639–645.

Chao, Y.-C.E., Kupper, L.L., Serdar, B., Egeghy, P.P., Rappaport, S.M., and Nylander-French, L.A. (2006). Dermal exposure to jet fuel JP-8 significantly contributes to the production of urinary naphthols in fuel-cell maintenance workers. *Environ. Health Perspect.,* 114(2): 182–185.

Defense Fuel Supply Center. (1997). *1997 Fact Book.* DFSC, Ft. Belvoir, VA.

Egeghy, P.P., Hauf-Cabalo, L., Gibson, R., and Rappaport, S.M. (2003). Benzene and naphthalene in air and breath as indicators of exposure to jet fuel. *Occup. Environ. Med.,* 60: 959–976.

Henz, K. (1998). Survey of Jet Fuels Procured by the Defense Energy Support Center 1990–1996. Defense Logistics Agency, Ft. Belvoir, VA.

Horst, R.L., and Kay, G.G. (1991). COGSCREEN: Personal computer-based tests of cognitive function for FAA medical recertification. *Sixth Biennial Int. Symp. Aviat. Psychol.,* Columbus, OH.

IARC. (1982). Some industrial chemicals and dyestuffs. *IARC Monogr. Eval. Carcinog. Risk Hum.,* 29: 95–148.

Kay, G.G. (1992). COGSCREEN: Development and application of a computer-based cognitive screening test. Symposium: Assessment of Stressor-Induced Changes In Cognitive Performance, *100th Annual Convention, American Psychological Association,* Washington, D.C.

Kay, G.G., Berman, B., Mockoviak, S.H., Morris, C.E., Reeves, D., Starbuck, V., Sukenik, E., and Harris, A.G. (1997). Initial and steady-state effects of diphenhydramine and loratadine on sedation, cognition, mood and psychomotor performance. *Arch. Int. Med.,* 157: 2350–2356.

Kim, D., Andersen, M.E., and Nylander-French, L.A. (2006a). Dermal absorption and penetration of jet fuel components in humans. *Toxicol. Lett.,* 165: 11–21.

Kim, D., Andersen, M.E., and Nylander-French, L.A. (2006b). A dermatotoxicokinetic model of human exposures to jet fuel. *Toxicol. Sci.,* 93(1): 22–33.

Kim, D., Andersen, M.E., Chao, Y.-C. E., Egeghy, P.P., Rappaport, S.M., and Nylander-French, L.A. (2007). PBTK modeling demonstrates contribution of dermal and inhalation exposure components to end-exhaled breath concentrations of naphthalene. *Environ. Health Perspect.,* 115(6): 894–901.

Knave, B., Persson, H.E., Goldberg, J.M., and Westerholm, P. (1976). Long-term exposure to jet fuel: an investigation on occupationally exposed workers with special reference to the nervous system. *Scand. J. Work Environ. Health,* 2: 152–164.

Knave, B., Olson, B.A., Elofsson, S., et al. (1978). Long-term exposure to jet fuel. II. A cross-sectional epidemiologic investigation on occupationally exposed industrial workers with special reference to the nervous system. *Scand. J. Work Environ. Health,* 4: 19–45.

Knave, B., Mindus, P., and Struwe G. (1979). Neurasthenic symptoms in workers occupationally exposed to jet fuel. *Acta Psychiatr. Scand.,* 60: 39–49.

Loffler, H., Dreher, F., and Maibach, H.I. (2004). Stratum corneum adhesive tape stripping: Influence of anatomical site, application pressure, duration and removal. *Br. J. Dermatol.,* 151: 746–752.

Mapou, R.L., Kay, G.G., Rundell, J.R., and Temoshook, L. (1993). Measuring performance decrements in aviation personnel infected with the human immunodeficiency virus. *Aviat., Space, and Environ. Med.*, 64: 158–164.

Mattorano, D.A., Kupper, L.L., and Nylander-French, L.A. (2004). Estimating dermal exposure to jet fuel (naphthalene) using adhesive tape strip samples. *Ann. Occup. Hyg.,* 48(2): 139–146.

McDougal, J.N., and Rogers, J.V. (2004). Local and systemic toxicity of JP-8 from cutaneous exposures. *Toxicol. Lett.,* 149: 301–308.

NRC (National Research Council). (2003). Toxicologic assessment of jet-propulsion fuel 8. National Research Council of the National Academies. The National Academies Press, Washington, D.C.

NTP. (2000). NTP Technical Report on the Toxicology and Carcinogenesis Studies of Naphthalene (CAS No. 91-20-3) in F344/N Rats (Inhalation Studies). NTP TR 500, NIH Publication No. 01-4434, U.S. Department of Health and Human Services.

Pleil, J.D., Smith, L.B., and Zelnick, S.D. (2000). Personal exposure to JP-8 jet fuel vapors and exhaust at air force bases. *Environ. Health Perspect.,* 108: 183–192.

Porter, H.O. (1990). Aviators intoxicated by inhalation of JP-5 fuel vapors. *Aviat. Space Environ. Med.,* 61: 654–656.

Reddy, M.B., Stinchcomb, A.L., Guy, R.H., and Bunge, A.L. (2002). Determining dermal absorption parameters in vivo from tape strip data. *Pharm. Res.,* 19: 292–298.

Reutman, S.R., LeMasters, G.K., Knecht, E.A., Shukla, R., Lockey, J.E., Burroughs, G.E., and Kesner, J.S. (2002). Evidence of reproductive endocrine effects in women with occupational fuel and solvent exposures. *Environ. Health Perspect.,* 110(8): 805–811.

Rhodes, A.G., LeMasters, G.K., Lockey, J.E., Smith, J.W., Yiin, J.H., Egeghy, P., and Gibson, R. (2003). The effects of jet fuel on immune cells of fuel system maintenance workers. *J. Occup. Environ. Med.*, 45(1): 79–86.

Risby, T.H. Tu, R.H., and Gibson, R. 2000. U.S. Air Force JP-8 Study. Unpublished data.

Ritchie, G.D., Still, K.R., Alexander, W.K., et al. (2001a). A review of the neurotoxicity risk of selected hydrocarbon fuels. *J. Toxicol. Environ. Health, B*, 4: 223–312.

Ritchie, G., Bekkedal, M.Y.V., Bobb, A.J., and Still, K.R. (2001b). Biological and Health Effects of JP-8 Exposure. Naval Health Research Center Detachment (Toxicology). Report No. TOXDET 01-01.

Ritchie, G., Still, K., Rossi III, J., Bekkedal, M., Bobb, A., and Arfsten, D. (2003). Biological and health effects of exposure to kerosene-based jet fuels and performance additives. *J. Toxicol. Environ. Health B Crit. Rev.*, 6: 357–451.

Serdar, B., Egeghy, P.P., Waidyanatha, S., Gibson, R., and Rappaport, S.M. (2003). Urinary biomarkers of exposure to jet fuel (JP-8). *Environ. Health Perspect.*, 111(14): 1760–1764.

Serdar, B., Egeghy, P.P., Gibson, R., and Rappaport, S.M. (2004). Dose-dependent production of urinary naphthols among workers exposed to jet fuel (JP-8). *Am. J. Indus. Med.,* 46: 234–244.

Smith, L.B., Bhattacharya, A., LeMasters, G., et al. (1997). Effect of chronic low-level exposure to jet fuel on postural balance of US Air Force personnel. *J. Occup. Environ. Med.,* 39: 623–632.

Subcommittee on Jet-Propulsion Fuel 8. (2003). Toxicologic Assessment of Jet-Propulsion Fuel 8. The National Academies Press, Washington, D.C.

Surber, C., Schward, F.P., and Smith, E.W. (1999). Tape stripping technique. In Bronough, H., and Maibach, H.I. (Eds.), *Percutaneous Absorption-Drug-Cosmetics-Mechanisms-Methodology.* Marcel Dekker, New York, pp. 395–409.

Struwe, G., Knave, B., and Mindus, P. (1983). Neuropsychiatric symptoms in workers occupationally exposed to jet fuel — A combined epidemiological and casuistic study. *Acta Psychiatr. Scand.,* 303(Suppl.): 55–67.

Taylor, J.L., O'Hara, R., Mumenthaler, M.S., and Yesavage, J.A. (2000). Relationship of CogScreen-AE to flight simulator performance and pilot age. *Aviat. Space Environ. Med.,* 71:373–380.

Tu, R.H., Mitchell, C.S., Kay, G.G., and Risby, T.H. (2004). Human exposure to the jet fuel, JP-8. *Aviat Space Environ Med*, 75: 49–59.

Zeiger, E., and Smith, L. (1998). The First International Conference on the Environmental Health and Safety of Jet Fuel. *Environ. Health Perspect.*, 106: 763–764.

Index

H

I

J

L

M

N